STEEL METALLURGY FOR THE NON-METALLURGIST

IRON - CARBON/CEMENTITE PHASE DIAGRAM

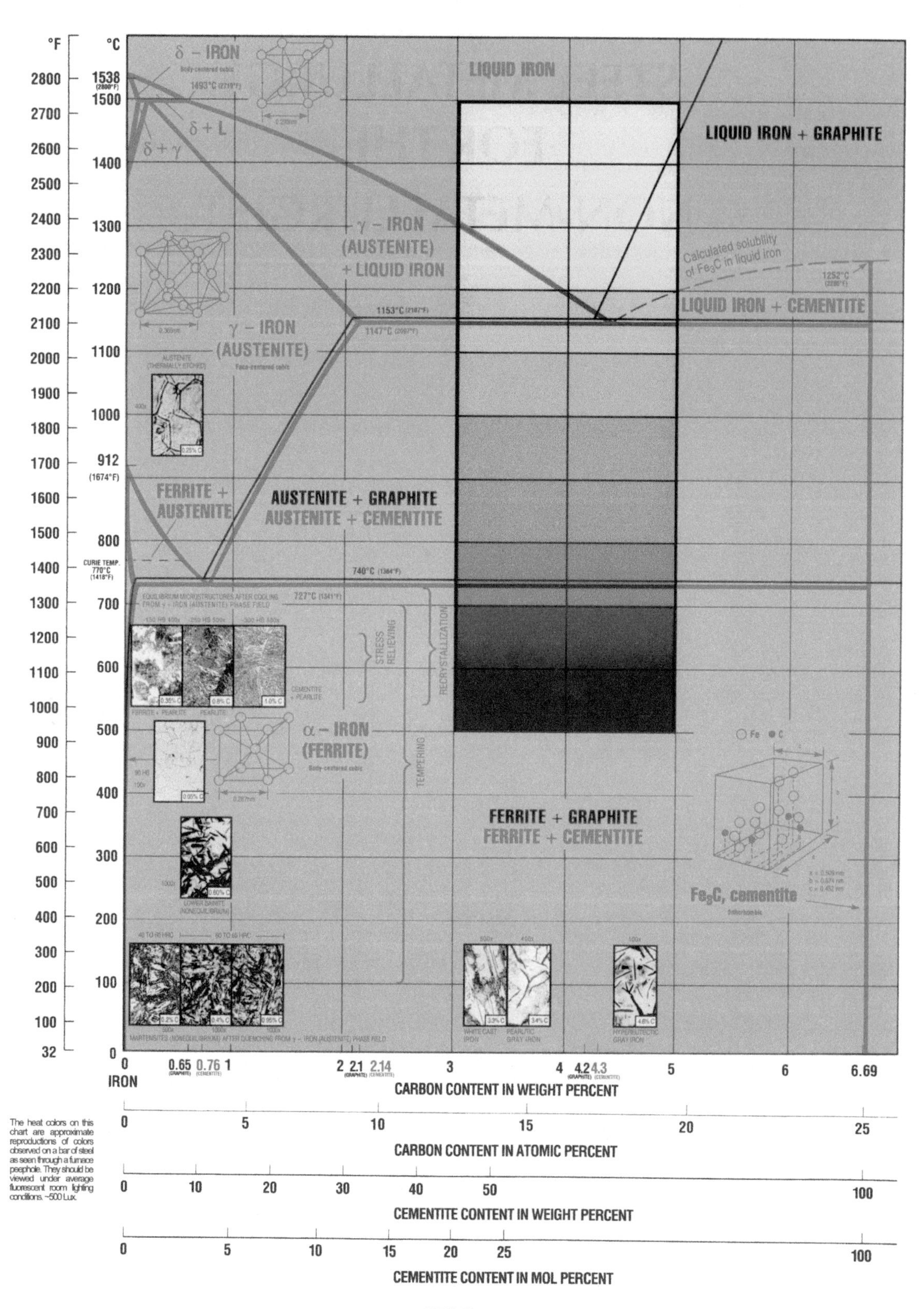

The heat colors on this chart are approximate reproductions of colors observed on a bar of steel as seen through a furnace peephole. They should be viewed under average fluorescent room lighting conditions. ~500 Lux.

Compiled by G.E. Weisch, Case Western Reserve University

Phase diagram from *Binary Alloy Phase Diagrams*, Second Edition edited by T.B. Massalski, H. Okamoto, P.R. Subramanian, L. Kacprzak

Materials Park, Ohio 44073-0002
Call 440-338-5151 Fax: 440-338-4634

Microstructures courtesy of G. Vander Voort, Carpenter Technology Corporation, and A. Benscoter, Lehigh University

The microconstituents formed in plain carbon steels on rapid quenching austenite into isothermal baths at the temperatures shown.

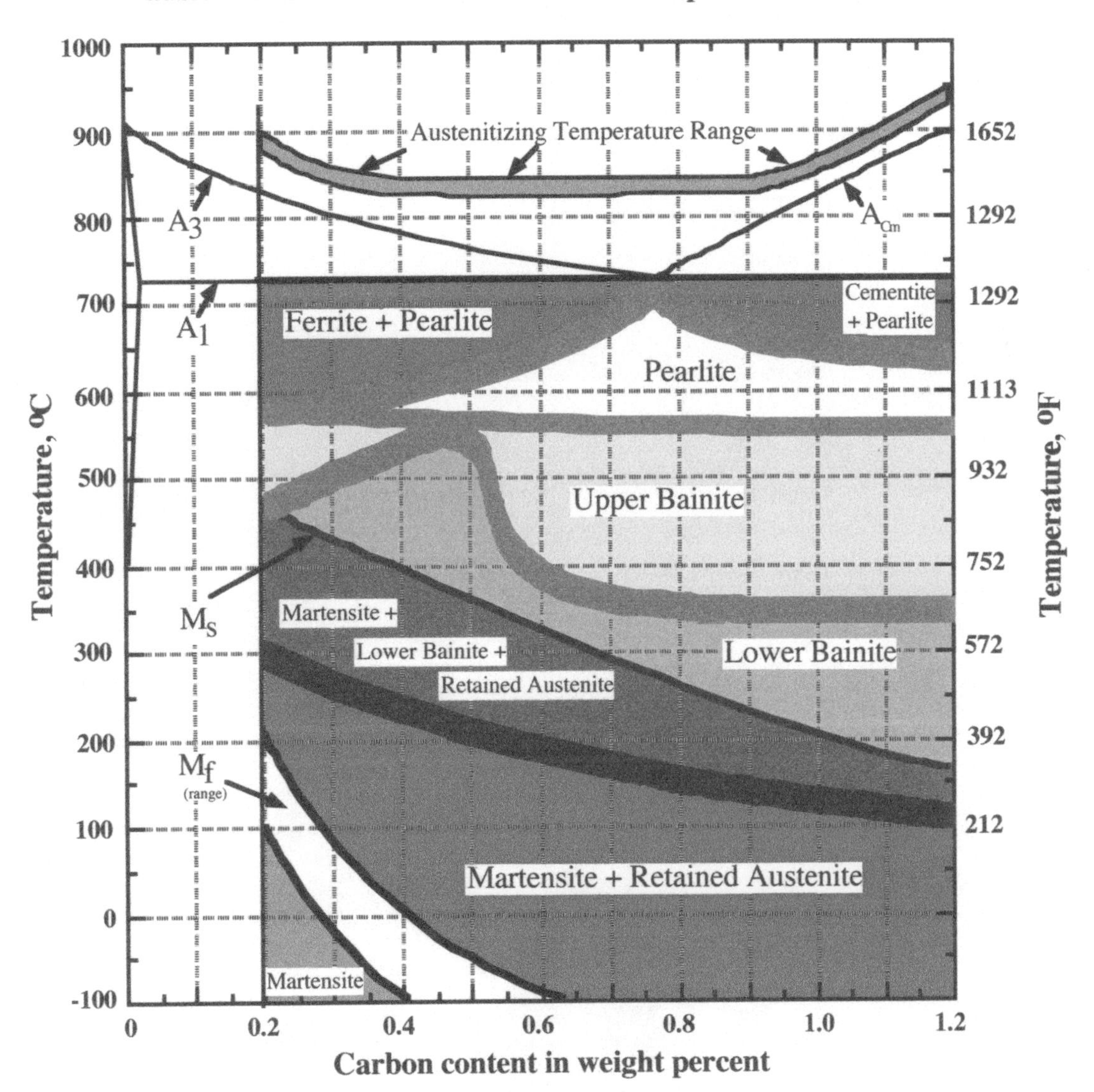

STEEL METALLURGY FOR THE NON-METALLURGIST

JOHN D. VERHOEVEN

ASM International®
Materials Park, Ohio 44073-0002
asminternational.org

First printing, November 2007
Second printing, April 2008
Digital printing, October 2009
Fourth printing, November 2010
Fifth printing, October 2018
Sixth printing, December 2019

Comments, criticisms, and suggestions are invited, and should be forwarded to ASM International.

Prepared under the direction of the ASM International Technical Book Committee (2006–2007), James C. Foley, Chair.

ASM International staff who worked on this project include Scott Henry, Senior Manager of Product and Service Development; Steve Lampman, Technical Editor; Eileen De Guire, Associate Editor; Ann Briton, Editorial Assistant; Bonnie Sanders, Manager of Production; Madrid Tramble, Senior Production Coordinator; Diane Grubbs, Production Coordinator; and Kathryn Muldoon, Production Assistant

Library of Congress Control Number: 2007932445

ISBN-13: 978-0-87170-858-8 (print)
ISBN-10: 0-87170-858-2 (print)
ISBN: 978-1-61503-056-9 (pdf)
ISBN: 978-1-62708-264-8 (electronic)

SAN: 204-7586

ASM International®
Materials Park, OH 44073-0002
www.asminternational.org

Printed in the United States of America

Contents

Preface

This book is an attempt to explain the metallurgical aspects of steel and its heat treatment to non-metallurgists, starting, from simple concepts taught in high-school-level chemistry classes and then building to more complex concepts involved in heat treatment of nearly all types of steel as well as cast iron. It was inspired by the author having worked with practicing bladesmiths for the past 15 to 20 years.

Most of the chapters in the book contain a summary at the end. These summaries provide a short review of the contents of each chapter. It may be useful to read these summaries before and perhaps after reading the chapter contents.

The Materials Information Society, ASM International, has published a book, *Heat Treater's Guide: Practices and Procedures for Irons and Steels*, 2nd ed., 1995, that contains a wealth of information on available steels and is extremely useful to those who work and heat treat steel. A major goal of this book is to provide the necessary background that will permit a metal worker, not trained in metallurgy, to understand how to use the information in the ASM book, as well as other Handbooks published by ASM International.

I would like to acknowledge the help of two bladesmiths who have contributed to this book in several ways: Alfred Pendray and Howard Clark. Both men have helped me understand the level of work being done by U.S. bladesmiths, and they have also contributed to some of the experiments used in this book. They are also responsible for the production of this book, because of their encouragement to write it. In addition, I would like to acknowledge many useful discussions with fellow metallurgist William Dauksch, retired vice president of Nucor Steel, and my colleague, Prof. Brian Gleeson, who made many useful suggestions on the stainless steel chapter.

I am particularly indebted to Iowa State University and their Materials Science and Engineering Department for providing me with the opportunity to teach metallurgical engineering students about steel for over two decades, as well as to the Ames Laboratory, DoE, that provided access to optical and electron microscopes and supported most of my research activity. Many of the pictures and all of the methods of presentation in this book result from my experience teaching both laboratory and lecture courses to students and doing research at Iowa State University and its Ames Laboratory.

Ames, Iowa, February 2007

About the Author

Dr. John Verhoeven is a Distinguished Emeritus Professor in the Engineering College at Iowa State University. He earned a B.S. in chemical engineering in 1957 and his M.S. and Ph.D. in metallurgical engineering in 1959 and 1963, all from the University of Michigan. His professional career was spent at Iowa State University teaching metallurgy in the Department of Materials Science and Engineering and doing research at the Ames Laboratory of the U.S. Department of Energy.

Dr Verhoeven's research was in the primary area of physical metallurgy. He has over 200 research publications in refereed journals and owns eighteen patents. He received three Advanced Sustained Research Awards from the DoE, in 1981, 1987 and 1988. He is a fellow of the ASM.

Dr Verhoeven's teaching led to the publication of a textbook, *The Fundamentals of Physical Metallurgy*, John Wiley, 1975, which was used widely in the late 1970s and 1980s. At Iowa State, he was awarded the Outstanding Teaching Award in the Engineering College in 1976 and in the Metallurgical Engineering Program in 1980, and the Iowa Legislature outstanding teaching award in the Engineering College at ISU for sustained outstanding teaching, November 1991. He was appointed an Anson Marston Distinguished Professor of Engineering in 1985.

CHAPTER 1

Pure Iron

STEELS ARE OVER 95% Fe, so a good starting point for understanding steel is to study the nature of solid iron. Consider the following experiment. A bar of pure iron (e.g., 25 mm, or 1 in., in diameter) is sectioned to form a thin disk in the shape of a quarter. A face of the disk is now polished on polishing wheels, starting first with a coarse-grit polish and proceeding in steps with ever finer grits until the face has the appearance of a shiny mirror. The shiny disk is now immersed for approximately 20 to 30 s in a mixture of 2 to 5% nitric acid (HNO_3) with methyl alcohol, a mixture often called nital—"nit" for the acid and "al" for the alcohol. This process is known as etching, which causes the shiny surface to become a dull color. If the sample is now viewed in an optical microscope at a magnification of 100×, it is found to have the appearance shown on the right of Fig. 1.1. The individual regions, such as those numbered 1 to 5, are called iron grains, and the boundaries between them, such as that between grains 4 and 5 highlighted with an arrow, are called grain boundaries. The average size of the grains is quite small. At the 100× magnification of this picture, a length of 200 microns (0.008 in.) is shown by the arrow so labeled. The average grain diameter for this sample has been measured to be 125 microns (0.005 in.), where 1 micron = 0.001 mm. Although a small number, this grain size is much larger than most commercial irons. (It is common to use the term μm for micron, and 25 μm = 0.001 in. = 1 mil. The thickness of aluminum foil and the diameter of a hair are both approximately 50 μm.)

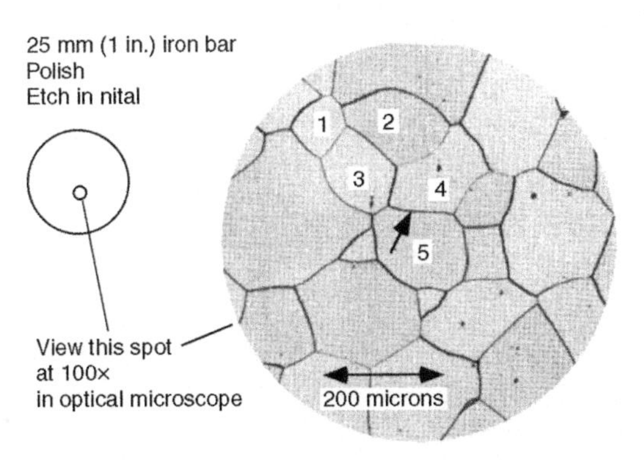

Fig. 1.1 Optical microscope image of the surface of a polished and etched iron bar showing grains and grain boundaries. Original magnification: 100×. Source: Ref 1.1

The basic building blocks of solids such as salt and ice are molecules, which are units made up of two or more atoms, for example, sodium + chlorine in table salt (NaCl) and hydrogen + oxygen in ice (H_2O). In metals, however, the basic building blocks are the individual atoms of the metal, that is, iron (Fe) atoms in a bar of iron or copper (Cu) atoms in copper wire. Each one of the grains in Fig. 1.1 is what is called a crystal. In a crystal made up of atoms, all of the atoms are uniformly arranged on layers. As shown in Fig. 1.2, if lines are drawn connecting the centers of the atoms, a three-dimensional array of little cubes stacked together to fill space is generated. In iron at room temperature, the cubes have an atom at each of the eight corners and one atom right in the middle of the cube. This crystal structure is called a body-centered cubic (bcc) structure, and the geometric arrangement of atoms is often called a bcc lattice. Notice that the crystal lattice can be envisioned as three sets of intersecting planes of atoms, with each plane set parallel to one face of the cube. Iron with a bcc structure is called ferrite. Another name for ferrite is alpha iron, or α-iron, where α is the Greek symbol for the first letter in the Roman alphabet, "a."

The nature of a grain boundary is illustrated at the bottom center of Fig. 1.2. The boundary is a planar interface, generally curved, along which two grains of different orientation intersect.

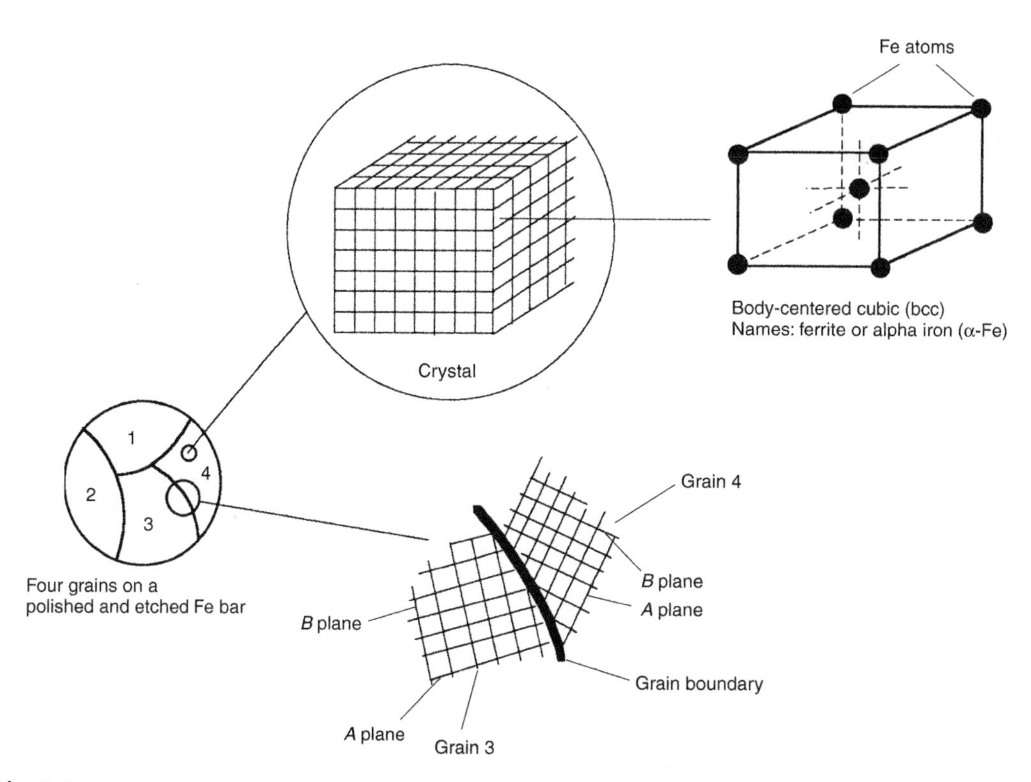

Fig. 1.2 Crystal structure of the grains and the nature of the grain boundaries

The *A* planes in grain 4 make a much steeper angle with the horizontal than do the *A* planes of grain 3. If grain 4 were rotated clockwise to cause its *A* planes to line up with the *A* planes of grain 3, then the grain boundary would go away, and the two grains would become one larger grain. An interesting question is why these grain boundaries show up on the etched surface? When a metal is etched in an acid, atoms are chemically removed from the surface. The rate of removal of iron atoms by nital depends on the orientation of the crystal that is facing the acid. Because each grain presents a different orientation, each grain etches down at a different rate. For example, the planes that form the faces of the body-centered cubes etch far slower than other crystal planes. Hence, after a period of etching, small steps develop at the grain boundaries, as shown in Fig. 1.3. For example, at the boundary of a fast-etching grain, one would see a step up to the surrounding grains. The step generally causes light to be scattered away from the eye, and the boundaries appear as dark lines.

If iron is heated to 912 °C (1670 °F), a somewhat magical effect occurs; the crystal structure changes spontaneously from bcc to a new structure called face-centered cubic (fcc). This structure is shown in Fig. 1.4, where it is seen that, as the name suggests, the atoms lie on the corners of a cube as well as one atom at each of the six faces of the cube. Like the low-temperature bcc structure, this structure has two names, either austenite or gamma iron (γ-iron), where γ is the Greek symbol for the third letter of the Roman alphabet, "c."

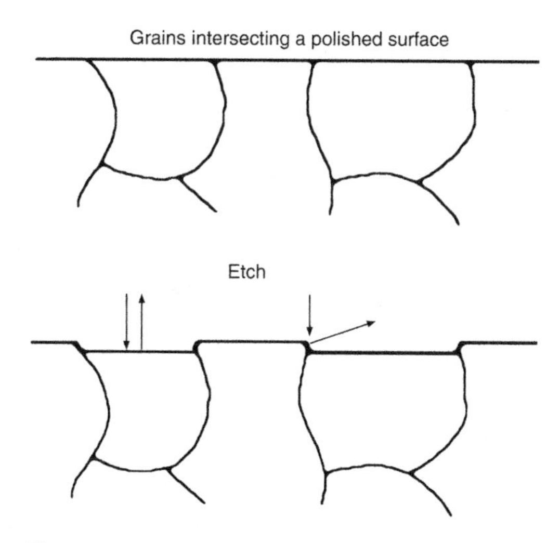

Fig. 1.3 Revealing steel microstructure by etching with nital

HISTORICAL NOTE

The first three letters of the Greek alphabet are alpha, beta, and gamma (α, β, γ), but there is no structure of iron called beta iron. When the structure of iron was being discovered in the late 19th century, the magnetic transition in iron that occurs at 770 °C (1420 °F) caused scientists to theorize a structure of iron they called beta iron, which was later shown not to exist.

When the ferritic iron is heated to 912 °C (1670 °F), the old set of ferrite grains changes (transforms) into a new set of austenite grains. Imagine that the ferrite grain structure shown in Fig. 1.2 has just reached the transformation temperature. One would first see the formation of a new set of very small austenite grains forming on the old ferrite grain boundaries, and then the growth of these grains until all the old ferrite grains were gone. Two important effects occur when ferrite changes to austenite: (1) Just as it takes heat energy to transform ice to water, it takes heat energy to change the ferrite grains into austenite grains. Therefore, on heating, the iron temperature will remain close to 912 °C (1670 °F) until all the ferrite grains are transformed. (2) The ferrite-to-austenite (α-to-γ) transformation is accompanied by a volume change. The density of austenite is 2% higher than ferrite, which means that the volume per atom of iron is less in austenite.

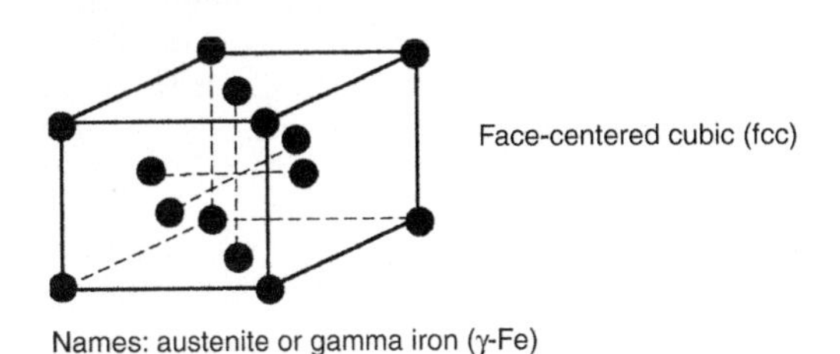

Fig. 1.4 Crystal structure of iron that forms at high temperatures

It is helpful to represent the aforementioned ideas geometrically on a diagram where temperature is plotted on a vertical scale and identifies the changes that occur at significant temperatures, as shown in Fig. 1.5. The following are two experiments that illustrate these changes. Experiment 1: Heat a bar of iron above 770 °C (1420 °F). As it cools, place a magnet near it. When the temperature reaches 770 °C (1420 °F), the hot sample will begin to pull toward the magnet. As the diagram of Fig. 1.5 shows, bcc iron (α-iron) is only magnetic below 770 °C (1420 °F), and fcc iron (γ-iron) is never magnetic. Experiment 2: Obtain a piece of black (nongalvanized) iron picture wire and string it horizontally between two electrical posts spaced 0.9 m (3 ft) apart. Hang a weight from the center of the wire. Pass an electrical current through the wire, heating it above 912 °C (1670 °F), which will be past a red color to

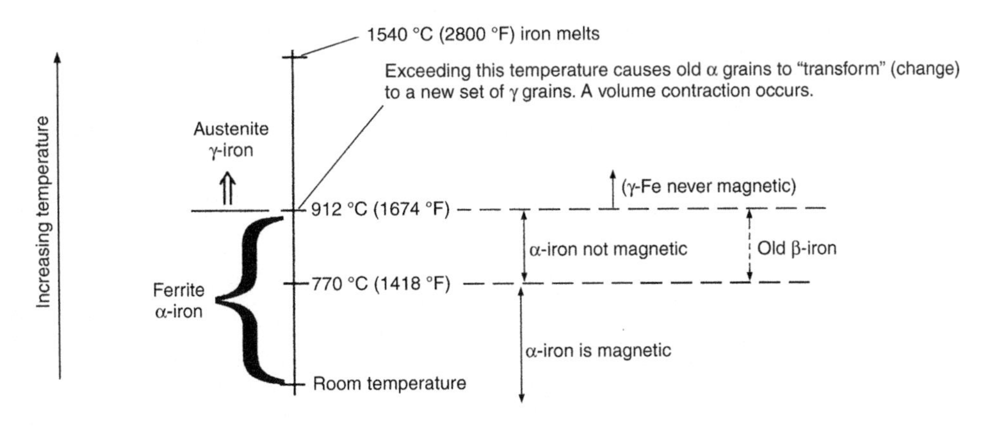

Fig. 1.5 Changes occurring in iron with increasing temperature

an orange-yellow color. It will be necessary to increase the voltage slowly using a high-current-capacity variable power source, such as a variac. As the wire is heated, it expands and the weight drops. Now, turn off the power and watch the wire cool in a darkened room. The following two effects occur at the 912 °C (1670 °F) transformation temperature: (1) As the wire cools, the accompanying volume contraction will raise the center weight, but this rising will be temporarily reversed when the wire expands as the less dense ferrite forms. (2) Heat liberated by the transformation will cause a visible pulse of color temperature increase to be seen in a darkened room. Both of these effects can be observed in reverse order on heating, but they are less dramatic due to the difficulty of heating rapidly.

To understand why heat is given off when austenite changes to ferrite on cooling, think about the water-ice transformation. Clearly, it is necessary to cool water to make it transform to ice (freeze). This means heat is removed from the liquid at the freezing temperature. The same effect occurs when metals freeze; heat is removed from the metal. So, when a hot metal cools to its freezing point, heat is given off from the freezing liquid. The transformation from liquid to solid is a *phase transformation* between the liquid phase and the solid phase. Phase transformations that occur on cooling liberate heat. When austenite transforms to ferrite on cooling, there is a solid-solid (rather than a solid-liquid) phase transformation, and heat is liberated. On heating, the reverse occurs; heat is absorbed when ferrite goes to austenite.

Summary of the Major Ideas in Chapter 1

1. A piece of iron consists of millions of small crystals, all packed together.
2. Each crystal is called a *grain*. A typical grain diameter is 30 to 50 μm (0.0012 to 0.002 in.), where 25 microns = 25 μm = 0.001 in. = 1 mil.
3. The boundaries between the crystals are called *grain boundaries*.
4. Below 912 °C (1670 °F), iron is called either *ferrite* or *alpha* (α) iron. The iron atoms in ferrite are arranged on a body-centered cubic (bcc) geometry. It is common to call the arrangement of atoms a body-centered cubic (bcc) *lattice*.
5. Above 912 °C (1670 °F), iron is called either *austenite* or *gamma* (γ) iron. The iron atoms in austenite are arranged on a face-centered cubic (fcc) lattice.
6. Heating ferrite to 912 °C (1670 °F) causes tiny grains of austenite to form on the ferrite grain boundaries. Continued heating causes these new austenite grains to grow, converting all the old ferrite grains into a new set of smaller austenite grains. On cooling below 912 °C (1670 °F), the same type of change occurs, but in reverse order, where ferrite grains replace austenite grains.

REFERENCE

1.1. *Atlas of Microstructures of Industrial Alloys,* Vol 7, *Metals Handbook,* 8th ed., American Society for Metals, 1972, p 4

CHAPTER 2

Solutions and Phase Diagrams

IN ORDER TO UNDERSTAND how the strength of steels is controlled, it is extremely useful to have an elementary understanding of two topics: solutions and phase diagrams.

Solutions

The idea of a solution can be explained by the following simple example. Put a teaspoon of either sugar or salt into a glass of water. Initially, most of the solid white sugar or salt will float on the water or sink to the bottom of the glass. However, after adequate stirring with the spoon, all of the sugar or salt will disappear, and only clear water will remain. The sugar or salt has *dissolved* into the water, and the final liquid is said to be a *solution* of sugar or salt in water.

What has happened to the sugar or salt? Consider the salt, because its molecule (sodium chloride [NaCl] for table salt or calcium chloride [$CaCl_2$] for the salt used on icy roads) is more simple than sugar. When the salt goes into solution, the individual molecules break apart into their component atoms, and these atoms in turn are pulled into the liquid water and are trapped in the water as charged atoms (called ions) between the water molecules (H_2O) that make up the liquid (water). For table salt, the NaCl molecules decompose following the reaction $NaCl \Rightarrow Na^+ + Cl^-$, where the symbol Na^+ refers to a positively charged sodium atom, called an ion, and the symbol Cl^- refers to a negatively charged chlorine atom. The solid salt cannot be seen because the chemical bonds that held the atoms together in the solid have been broken, causing the former solid structure to disappear as its component atoms became incorporated into the liquid water between the water molecules.

In general, when a solution is formed by dissolving something into a liquid, the freezing temperature of the liquid will decrease. This, of course, is why calcium chloride ($CaCl_2$) salt is used on streets and sidewalks during the winter months. The salt dissolves into the rain water and drops its freezing temperature, so that ice will not form until the temperature has been lowered below the pure water freezing temperature of 0 °C (32 °F). This effect can be represented graphically, as shown in Fig. 2.1. Here, temperature is plotted on the vertical axis and the amount of salt dissolved into the water on the horizontal axis. The amount of salt dissolved in the water is given as a weight percent of the total liquid solution. Two terms are often used to describe the amount that has been dissolved. A 10 wt% sodium chloride value can be called the *concentration* of salt in the solution or the *composition* of the solution. Notice in Fig. 2.1 that after a certain maximum amount of salt has dissolved in the solution, the freezing temperature suddenly begins to rise quite rapidly. This maximum composition is called the *eutectic*

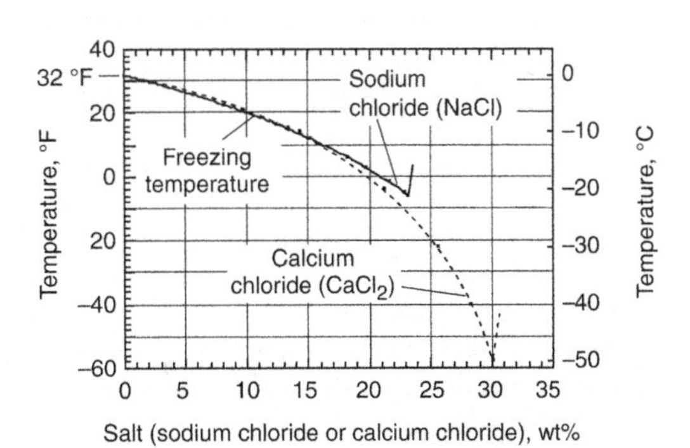

Fig. 2.1 Lowering of the freezing temperature of water with increasing salt (sodium chloride, calcium chloride) content versus the weight percent of two different salts dissolved in the water

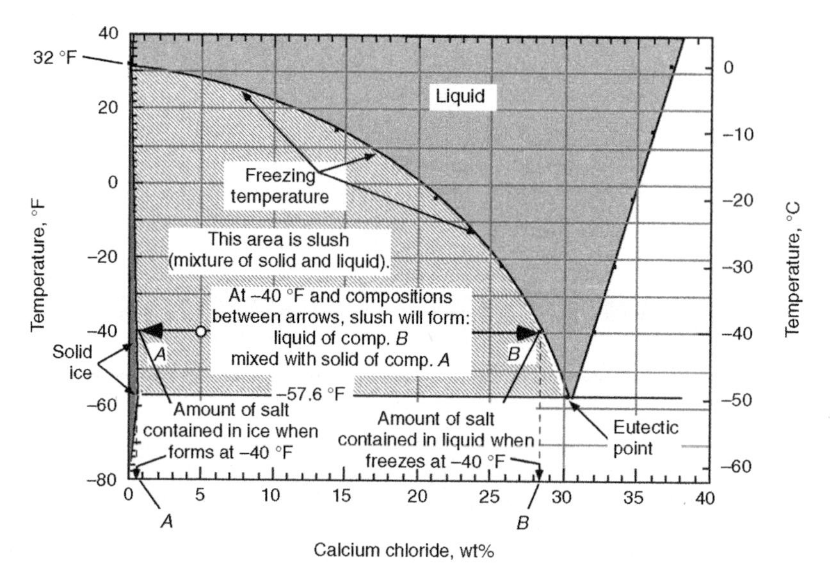

Fig. 2.2 A portion of the water-calcium chloride phase diagram

composition, which is discussed more in Chapter 3, "Steel and the Iron-Carbon Phase Diagram." The curves show why it is preferable to use calcium chloride over sodium chloride to prevent ice formation on sidewalks.

Phase Diagrams

The geometric arrangement of the molecules in water (called the molecular structure) is the same, on average, from point to point in the water. The water is called a liquid *phase*. Similarly, the molecular structure in ice is the same from point to point in the ice, and the ice is called a solid phase. However, the molecular structures of ice and water are very different from each other, one being a liquid and the other a solid. Therefore, liquid water and ice are two different phases.

As shown in Fig. 2.1, the freezing point of water is suppressed as salt is dissolved into the water, forming a solution. The molecular structure of the salt solution is essentially the same as that of pure water, because the salt ions fit in between the H_2O molecules without disturbing their geometric arrangement relative to each other. So, pure water and the salt solution generated by dissolving salt in the water are the same phase.

A phase diagram is a temperature versus composition map that locates on the map the temperature-composition coordinates where the various phases can exist. Figure 2.2 presents part of the water-calcium chloride phase diagram. Just as in Fig. 2.1, the diagram has the temperature plotted vertically and the composition (in weight percent salt) plotted horizontally. The freezing temperature line for the liquid salt solution is the same as in Fig. 2.1, only now the sharp rise in freezing temperature beyond the eutectic point is extended to the maximum temperature of the diagram. The shaded region labeled "Liquid" above the freezing temperature line maps out all possible temperature-composition coordinates that are found to be liquid. Notice that at the extreme left of the diagram there is a thin shaded region labeled "Solid ice." This solid ice has the same molecular structure as pure water ice, but now it contains a very small amount of salt dissolved in it. It is the same phase as the pure water ice. Hence, this thin region is a map of where solid ice occurs on the diagram.

Now, consider a salt solution containing 5 wt% calcium chloride that is cooled to −40 °F (small circle in Fig. 2.2). This point on the phase diagram map does not lie in either the liquid or the solid (ice) regions. This solution at –40 °F (−40 °C) cannot be all solid or all liquid. Figure 2.2 shows a horizontal line at −40 °F that terminates with arrow heads labeled *A* and *B*. The diagram requires this 5% solution at −40 °F to be a mixture of solid ice having composition *A* (approximately 0.7% salt*) and liquid water having composition *B* (approximately 28.3% salt).

*The real value is much less than 0.7%. The value was increased so it would be clear on the diagram.

It may be thought of as a slush, a water-ice mixture. A solution corresponding to any point in the shaded slush region of the phase diagram is going to be a slush, consisting of a mixture of water with ice. Thus, each region on the phase diagram predicts what phases are present, in this case, solid plus liquid. In addition, in the two-phase region (the shaded slush region), the diagram also predicts the composition of each of the two phases when a temperature is chosen. However, in the one-phase solid or one-phase liquid region, it does not predict the composition of the phase when a temperature is chosen. It only requires that all solid or all liquid be present. In these cases, the compositions are the overall average composition that one starts with, a number not predicted by the phase diagram.

Summary of the Major Ideas in Chapter 2

1. A liquid solution occurs when a substance dissolves in a liquid, such as salt into water. A solid solution is similar, such as when salt dissolves in ice.
2. The substance that dissolves into solution (salt in salt solutions, carbon in steel) loses its identity and is hidden from view as its atoms become incorporated into the solution.
3. Solutions have the same molecular structure or atomic structure from point to point within themselves. Each solution is called a phase.
4. A phase diagram is a map having coordinates of temperature on the vertical axis and composition or concentration on the horizontal axis. The phase diagram map identifies those temperature-composition coordinates where a certain phase will exist.
5. A phase diagram also maps out temperature-composition coordinates where only phase mixtures can exist. For example, in the Fig. 2.2 phase diagram, the shaded slush region locates where slush mixtures of solid ice and liquid water exist.

CHAPTER 3

Steel and the Iron-Carbon Phase Diagram

STEEL IS MADE by dissolving carbon into iron. Pure iron melts at an extremely high temperature, 1540 °C (2800 °F), and at such temperatures, carbon readily dissolves into the molten iron, generating a liquid solution. When the liquid solution solidifies, it generates a solid solution, in which the carbon (C) atoms are dissolved into the solid iron. The individual carbon atoms lie in the holes between the iron atoms of the crystalline grains of austenite (high temperatures) or ferrite (low temperatures). If the amount of carbon dissolved in the molten iron is kept below 2.1 weight percent, the result is steel, but if it is above this value, one has cast iron. Although liquid iron can dissolve carbon at levels well above 2.1% C by weight (2.1 wt%), solid iron cannot. This leads to a different solid structure for cast irons (iron with total %C greater 2.1 wt%), which is discussed in more detail in Chapter 16, "Cast Irons."

In addition to carbon, all modern steels contain the element manganese (Mn) and low levels of the impurity atoms of sulfur (S) and phosphorus (P). Hence, steels can be thought of as alloys of three or more elements, given as Fe + C + *X*, where Fe and C are the chemical element symbols for iron and carbon, and *X* can be thought of as third-element additions and impurities. In the United States, most steels are classified by a code developed by the American Iron and Steel Institute (AISI). It is customary to partition steel compositions into two categories: plain carbon steels and alloy steels. In plain carbon steels, *X* consists only of manganese, sulfur, and phosphorus, whereas in alloy steels, one or more additional alloying elements are added. Table 3.1 lists the compositions of two common plain carbon steels, and it can be seen that the amount of carbon (in weight percent) is related to the last two numbers of the code. The manganese level is not related to the code and must be looked up in a table. The first two numbers of the code, "10," identify the steel as a plain carbon steel. These two numbers are changed for alloy steels, and the table lists one example for a chromium (Cr) alloy steel. A reference book, such as Ref 3.1, will provide the full composition. The alloy steels are discussed in Chapters 6, 13, and 14.

Solid solutions are similar to the liquid salt solution discussed in Chapter 2, "Solutions and Phase Diagrams"; that is, after the substance is dissolved, its presence is no longer evident to an observer as it had been previous to the dissolving. This may be illustrated for steel as shown in Fig. 3.1. On the left is shown a starting condition of a round iron containing only three grains. The iron is surrounded by a thin layer of graphite and heated to 1000 °C (1830 °F). After a period of several hours, the graphite layer disappears. The carbon atoms of the graphite have migrated into the solid iron by a process called diffusion, which is discussed in Chapter 7, "Diffusion—A Mechanism for Atom Migration within a Metal." All of the carbon atoms fit into the holes that exist between the iron atoms of the face-centered cubic (fcc) austenite present at this

Table 3.1 Composition (weight percent) of some steels

Steel type	American Iron and Steel Institute code (AISI No.)	%C	%Mn	% other	%S	%P
Plain carbon	1018	0.18	0.75	...	0.05 (max)	0.04 (max)
	1095	0.95	0.40	...	0.05 (max)	0.04 (max)
Alloy	5160	0.60	0.82	0.8 Cr	0.05 (max)	0.04 (max)

temperature. This solid solution of carbon in iron is steel.

Figure 1.4 locates the center of the iron atoms in the fcc lattice. If each of the little solid dots of Fig. 1.4 are allowed to expand until they touch each other, the result is the model of fcc iron shown in Fig. 3.2. The expanded iron atoms touch each other along the face diagonals of the cube. The small open circles locate the center of the void spaces between the iron atoms. If these small circles are expanded until they touch the iron atoms, their maximum diameter will equal 41.4% of the iron atom diameter. This means that atoms smaller than approximately 42% of the iron atom diameter should fit into the holes between the iron atoms. Carbon atoms are small, but the diameter of carbon atoms is estimated to be 56% of the diameter of iron atoms in austenite. Hence, when carbon dissolves in iron, it pushes the iron atoms apart a small amount. The more carbon dissolved, the further the iron atoms are pushed apart. Thus, there is a limit to how much carbon can dissolve in iron.

In pure iron, the difference in ferrite and austenite is a difference in their atomic structures. As illustrated in Fig. 1.2 and 1.4, the iron atoms are arranged with a body-centered cubic (bcc) crystal structure in ferrite and an fcc crystal structure in austenite. In both ferrite grains and austenite grains, this atomic structure does not change within the grain. Thus, similar to the ice and water example in Chapter 2, both ferrite and austenite are individual *phases*.

When carbon is added to austenite to form a solid solution, as illustrated in Fig. 3.1, the solid solution has the same fcc crystal structure as in pure iron. As discussed with reference to Fig. 3.2, the carbon from the graphite just fits in between the iron atoms. The crystal structure remains fcc, the only change being that the iron atoms are pushed very slightly farther apart. Both pure austenite and austenite with carbon dissolved in it are the same phase. Thus, austenite with carbon dissolved in it and ferrite with carbon dissolved in it are two different phases, both of which are steel.

Low-Carbon Steels (Hypoeutectoid Steels)

The iron-carbon phase diagram provides a temperature-composition map of where the two phases, austenite and ferrite, will occur. It also indicates where mixtures of these two phases can be expected, just like the slush region on the ice-salt phase diagram. A portion of the iron-carbon phase diagram is shown in Fig. 3.3, and it is seen that there is a strong similarity to the salt diagram of Fig. 2.2. In pure iron, austenite transforms to ferrite on cooling to 912 °C (1674 °F). This transition temperature is traditionally called the A_3 temperature, and the diagram shows that, just as adding salt to water lowers the freezing point of water, adding carbon to iron lowers the A_3 temperature. Whereas the maximum lowering

HISTORICAL NOTE

The iron age dates to approximately 1000 B.C., when our ancestors first learned to reduce the plentiful iron oxide ores found on earth into elemental iron. The iron was made in furnaces that were heated by charcoal fires that were not able to get hot enough to melt the iron. They produced an iron called bloomery iron, which is similar to modern wrought iron. Even though charcoal was used in these furnaces, very little carbon became dissolved in the iron. So, to make steel, carbon had to be added to the bloomery iron. (Even to this day, steel cannot be economically made directly from the iron ore. The modern two-step process first makes high-carbon pig iron, and the second step reduces the carbon composition to the steel range.) Our ancestors did not know the nature of the element carbon until approximately 1780 to 1790 A.D., and steel and cast iron played a key role in its discovery. Until shortly before that time, the production of steel was primarily the result of blacksmiths heating bloomery iron in charcoal fires. This process is tricky because a charcoal fire can just as easily remove carbon as add it (Chapter 7, "Diffusion—A Mechanism for Atom Migration within a Metal"). Therefore, steel was made on a hit-and-miss basis from the start of the iron age, and the quality of such steel varied widely. The successful blacksmiths guarded their methods carefully.

occurs at what is called the eutectic point in water-salt, a similar maximum lowering occurs in iron-carbon, but here it is called the *eutectoid* point and also the *pearlite* point. The eutectoid point represents the temperature and composition on the phase diagram where a eutectoid reaction occurs, that is, a reaction where one solid transforms into two solids. The eutectoid point in the iron-carbon system has a composition of 0.77% C, and steels with compositions less than this value are called *hypoeutectoid* steels. The eutectoid temperature is traditionally called the A_1 temperature.

Steels that are 100% austenite must have temperature-composition coordinates within the upper central dark area of Fig. 3.3. Steels that are ferrite must have temperature-composition coordinates in the skinny dark region at the left side of Fig. 3.3. The maximum amount of carbon that will dissolve into ferritic iron is only 0.02%, which occurs at the eutectoid temperature of 727 °C (1340 °F). This means that ferrite is essentially pure iron, because it is always

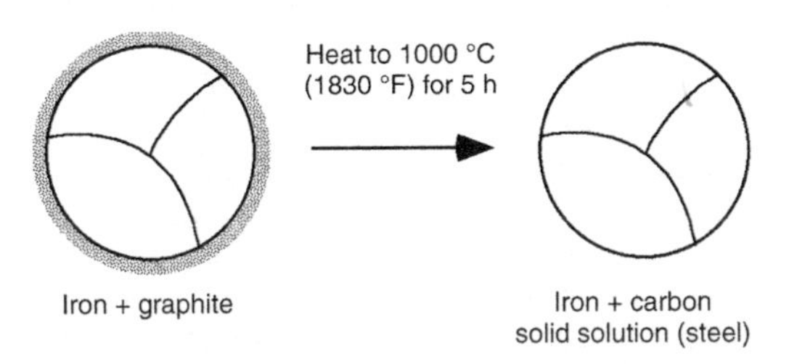

Fig. 3.1 Converting a rod of iron containing only three grains to steel by dissolving carbon in it

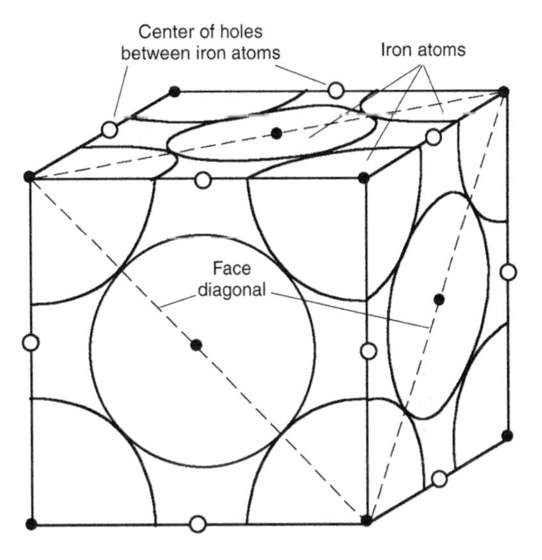

Fig. 3.2 Location of iron atoms in face-centered cubic austenite. Small circles locate the centers of holes between the iron atoms

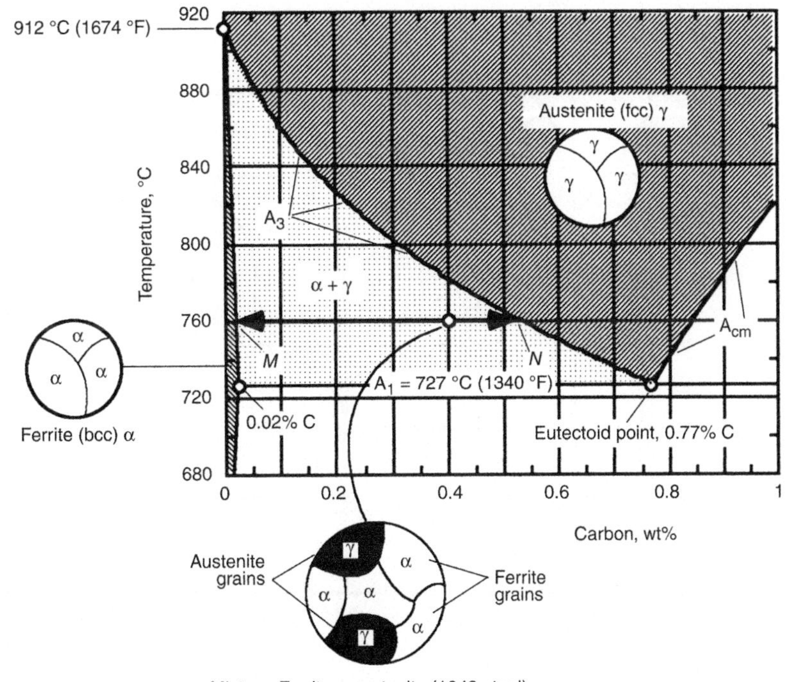

Fig. 3.3 Portion of iron-carbon phase diagram for hypoeutectoid steel alloys (%C less than 0.77)

99.98% or purer with respect to carbon. Notice, however, that austenite may dissolve much more carbon than ferrite. At the eutectoid temperature, austenite dissolves 0.77% C, which is roughly 38 times more carbon than ferrite will hold at this temperature. Austenite holds more carbon than ferrite, because the holes between iron atoms are larger in the fcc structure than the bcc structure.

Remember that in the salt-water phase diagram, the shaded slush region mapped the temperature-composition points where one obtains slush, a mixture of water and ice. Similarly, the central shaded region of Fig. 3.3 labeled $\alpha + \gamma$ maps temperature-composition points where steel consists of a mixture of ferrite and austenite. Suppose a hot stage microscope is used to look at a polished steel having a composition of 0.4% C after it had been heated to 760 °C (1400 °F). Because this temperature-composition point lies in the central shaded region labeled $\alpha + \gamma$, the steel must be a mixture of ferrite and austenite, an example of which is shown at the bottom of Fig. 3.3. The phase diagram also gives information about the composition of the two phases. The austenite grains must have the composition given as *N* and the ferrite the composition given as *M* in Fig. 3.3.

As a further illustration of the usefulness of the phase diagram, consider the following simple experiment. A steel of composition 0.4% C is first heated to 850 °C (1560 °F) and held for approximately 10 min. After this short hold, all the grains in this steel would consist of pure fcc austenite grains with a composition of 0.4% C. To make the illustration simple, imagine looking at the steel in a hot stage microscope and seeing a region of only three grains, as shown in Fig. 3.4. Now, imagine the hot stage temperature is reduced to 760 °C (1400 °F), and the sample slowly cools to this temperature. What happens to the microstructure? According to the phase diagram, the steel, after cooling, must become two phases, a mixture of austenite and ferrite. Experiments show that the ferrite that forms in the pure austenite as it cools virtually always forms on the austenite grain boundaries. This is illustrated in the lower microstructure of Fig. 3.4, where the ferrite has formed as a number of α grains along the prior-austenite grain boundaries. Comparing the microstructure at the right side of Fig. 3.4 to that at the bottom of Fig. 3.3 reveals similarities and differences. They are similar in that both contain a mixture of ferrite and austenite grains with the same volume fraction of each, but they are different in that the distribution of the ferrite grains is quite different. The microstructure of Fig. 3.3 will generally be formed on heating the steel from room temperature to 760 °C (1400 °F). Thus, this example illustrates one of the fascinating aspects of steels: the microstructure is controlled by heat

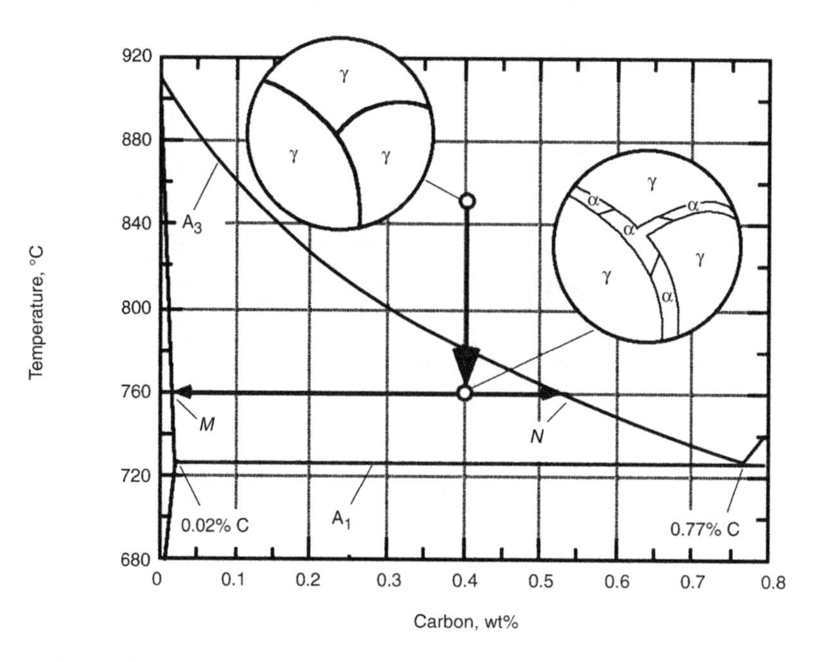

Fig. 3.4 Portion of iron-carbon phase diagram and change in microstructure on cooling a 1040 steel from 850 to 760 °C (1560 to 1400 °F)

treatment. As is shown later, the mechanical properties of steels are controlled by microstructure. The *microstructure* of a steel generally refers to the specific shape, size, distribution, and phase types of the grains in the steel. Remember, the phases present in Fig. 3.3 and 3.4 are only present at the high temperature of 760 °C (1400 °F), not at room temperature. Room-temperature phases are discussed later.

High-Carbon Steels (Hypereutectoid Steels)

As the %C dissolved in austenite increases, the iron atoms are pushed further apart. This stretches the chemical bonds that hold the iron atoms together, generating a form of energy called strain energy. There is a limit to how much strain energy the austenite can stand. The amount of carbon dissolved in austenite when this limit is reached is called the solubility limit. Question: The solubility limit in austenite at 820 °C (1508 °F) is 1 wt% C. If an alloy containing 1.5 wt% C is heated to 820 °C (1508 °F), only 1% of this 1.5% C will be dissolved in the austenite. What happens to the remaining 0.5% C? This excess carbon becomes incorporated into a new phase called cementite. The new phase, cementite, has one major difference from austenite or ferrite. It is a chemical compound that exists at only one composition. Chemical compounds are generally represented by their elemental formulas, such as NaCl for sodium chloride table salt. The composition for NaCl is 50 atomic percent sodium, which corresponds to 39.3 weight percent sodium. The chemical element formula for cementite is Fe_3C. For each atom of carbon in the compound, there are three atoms of iron, giving an atomic composition of 25 at.% C. The corresponding weight percent carbon in cementite is 6.7 wt% C. Other than this limitation to one composition, cementite has several similarities to austenite and ferrite. It is a crystal having its atoms arranged in regularly repeating geometrical arrays. The crystal structure is a little more complex than either the bcc of ferrite or the fcc of austenite, but it is well known. Also, it is a separate phase and is present as discrete grains. So, all the excess 0.5% C in this example at 820 °C (1508 °F) will be present as separate cementite grains mixed in with the austenite grains; that is, the microstructure will be a two-phase mixture of austenite and cementite.

The iron-carbon phase diagram extended to higher-carbon compositions where cementite becomes important is shown in Fig. 3.5. As before, the region on the temperature-composition map corresponding to austenite is shown as the central dark region. Because cementite exists at only one composition, it is shown on the phase

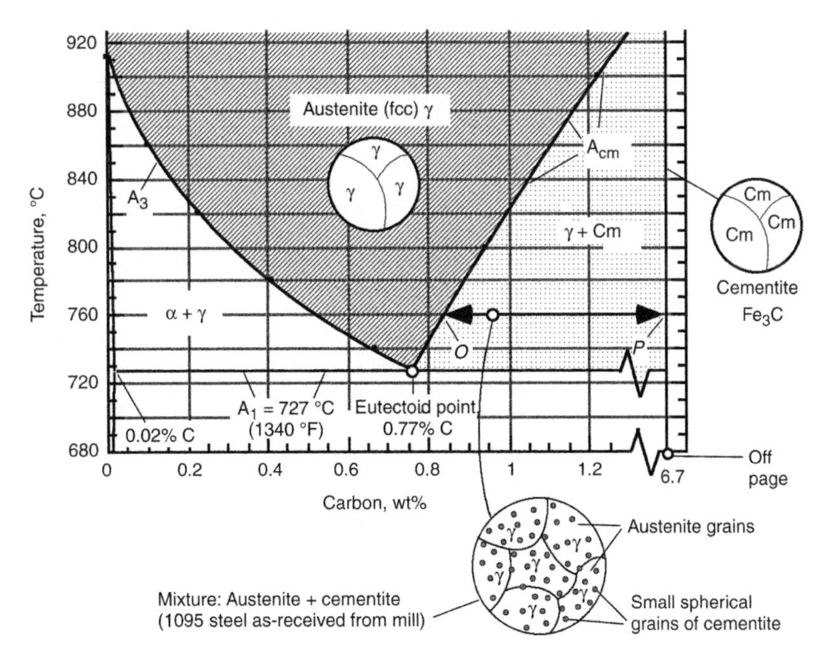

Fig. 3.5 Extension of the iron-carbon phase diagram to hypereutectoid steel alloys (%C greater than 0.77)

diagram as a vertical line located at its composition, 6.7% C. Notice that the composition axis at the bottom of the diagram has a break in it just beyond 1.2%, and the value of 6.7% is located next to the break. If the break were not inserted, the 6.7% composition would appear roughly one foot to the right. To envision the true diagram, imagine extending the right portion approximately one foot, which expands the shaded two-phase region labeled γ + cm into a much larger area.

The line on the diagram labeled A_{cm} defines the solubility limit for carbon in austenite. Notice that at 820 °C (1508 °F), this line gives a point at 1 wt% C, which is the maximum amount of carbon that can be dissolved in austenite at 820 °C (1508 °F). Alloys having %C compositions to the right of the A_{cm} line are in the shaded two-phase region labeled γ + cm and must consist of a mixture of austenite and cementite grains. For example, consider a 1095 steel (0.95 wt% C) received from a steel mill. If this steel is heated to 760 °C (1400 °F), the temperature-composition point will be at the open circle in Fig. 3.5 with the horizontal arrowed line passing through it. Because the temperature-composition point lies in the shaded two-phase region labeled γ + cm, this steel must consist of a mixture of austenite having composition *O* (0.85% C) and cementite of composition *P* (6.7% C). The diagram does not describe what the microstructure will look like; however, experiments show that the microstructure will be as illustrated at the bottom of Fig. 3.5. All of the cementite appears as small, spherically shaped grains distributed fairly randomly over the austenite grains, which have much larger sizes and the typical curved grain boundaries.

To further illustrate the use of the phase diagram in understanding how microstructure changes during heat treatment, consider an experiment where the as-received 1095 steel is heated to 850 °C (1560 °F) and held for 20 min or so. As shown on the phase diagram of Fig. 3.6, this temperature-composition point corresponds to the single-phase austenite region. Assuming the structure could be observed in a hot stage microscope, the small region under observation at high magnification may consist of just three grains, which would have an appearance similar to that shown in Fig. 3.6. If the temperature of the hot stage is lowered to 760 °C (1400 °F), then this temperature-composition point lies in the two-phase austenite + cementite region, which means that cementite grains must form on cooling. Just as is the case for the hypoeutectoid steel of Fig. 3.3, where ferrite forms on the austenite grain boundaries during cooling, here cementite forms on the austenite grain boundaries during cooling. As shown in Fig. 3.6,

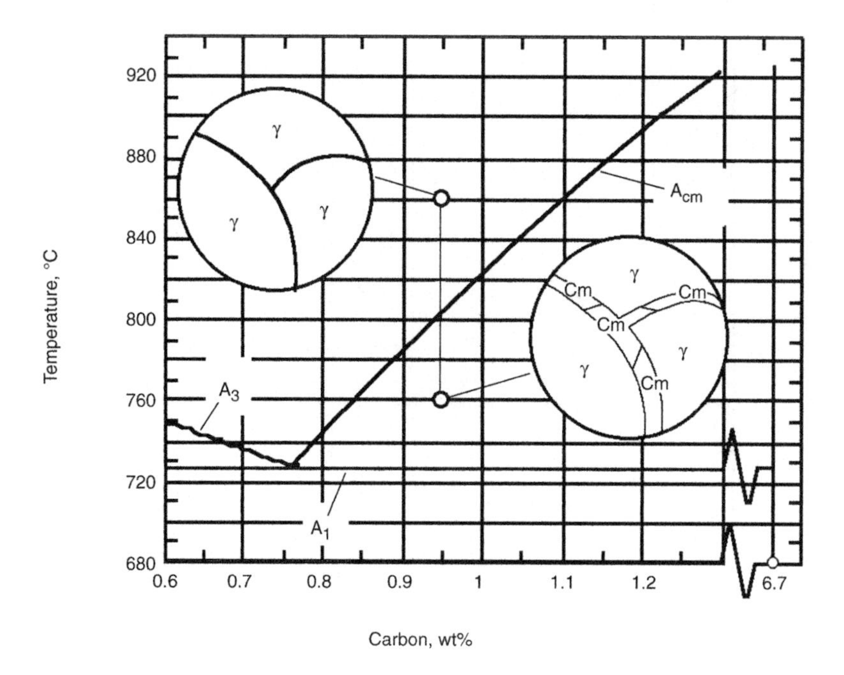

Fig. 3.6 Portion of iron-carbon phase diagram and change in microstructure on cooling a 1095 steel from 860 to 760 °C (1580 to 1400 °F)

the final microstructure has all of the prior-austenite grain boundaries from the 850 °C (1560 °F) structure filled with thin, plate-shaped grains of cementite. Notice the dramatic difference in this microstructure from that shown in Fig. 3.5, where the 1095 steel was heated directly from room temperature to 760 °C (1400 °F). Both microstructures contain the same volume fraction of cementite and austenite, but the distribution of the cementite is quite different. Unlike austenite and ferrite, cementite is very brittle. Consequently, the 1095 structure of Fig. 3.6 with its interconnected cementite plates is not as tough as the Fig. 3.5 structure with its small, isolated cementite grains. Again, this is an example of how heat treatment can change microstructure, which, in turn, changes mechanical properties.

Eutectoid Steel (Pearlite)

In the previous two sections, steels having compositions on either side of the eutectoid composition of 0.77 wt% C were considered. Steels having the composition of 0.77% C (a 1077 steel) generate a unique microstructure called pearlite. Figure 3.7 presents the iron-carbon phase diagram on which the area below the A_1 line of 727 °C (1340 °F) is shaded dark. This entire area is a two-phase region. Any steel cooled slowly into the temperature-composition coordinates of this area must consist of a mixture of the two phases, ferrite and cementite (α + Cm). The microstructures of steels in this two-phase region vary widely, and pearlite is just one of many microstructures that can occur.

To understand the pearlite microstructure, consider a 1077 steel that is heated in a hot stage microscope to 800 °C (1470 °F). As shown in Fig. 3.7, the steel will consist of all austenite grains after just a minute or two at 800 °C. After cooling to a temperature below A_1 and holding for 5 to 10 min or so, the austenite grains will be completely replaced by a new set of pearlite grains, as shown in the figure. Contrary to all the grain structures discussed up to this point, the pearlite grains are not a single phase. Rather, they consist of a mixture of two phases, ferrite and cementite (α + Cm), having a unique microstructure. To observe the true details of the microstructure, a small region within a pearlite grain must be observed under very high magnification, as shown in the lower right of Fig. 3.7. The structure consists of alternating plates of ferrite and cementite. The ferrite plates are much fatter than the cementite plates, occupying 90% of the volume compared to only 10% for the cementite. At the pearlite grain boundaries, there is an abrupt change in the orientation

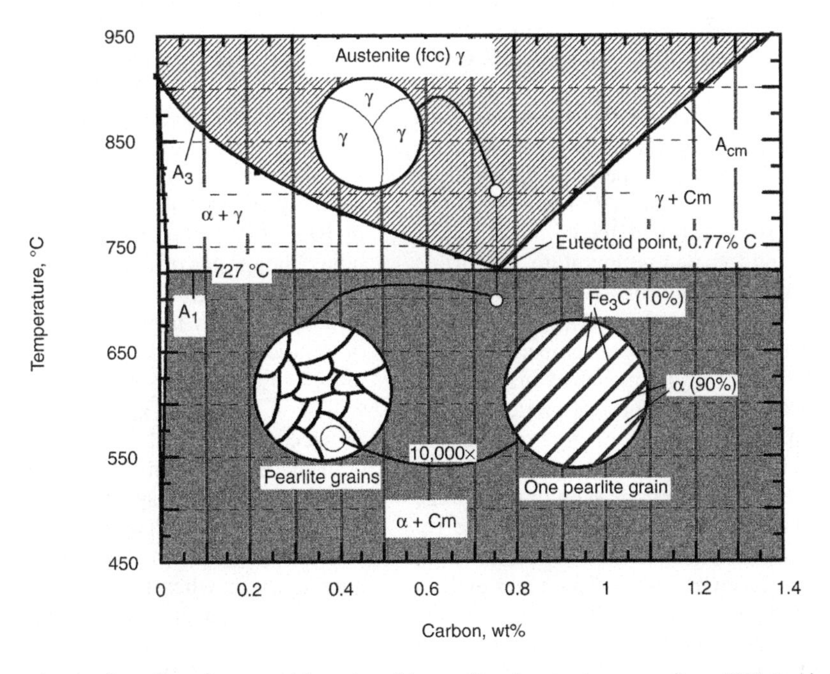

Fig. 3.7 Portion of iron-carbon phase diagram and formation of the pearlite microstructure on cooling a 1077 steel below the A_1 temperature of 727 °C (1340 °F)

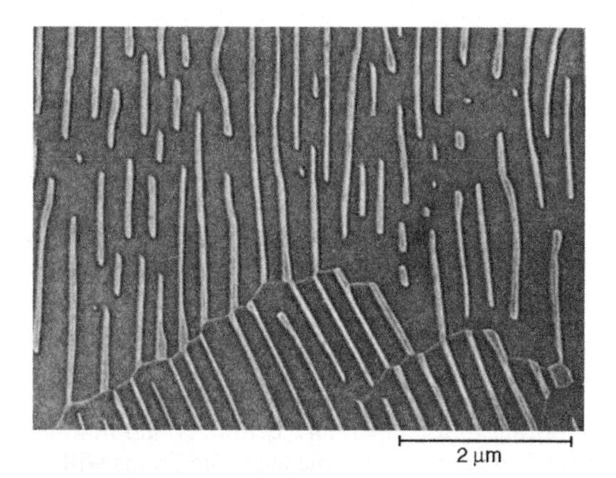

Fig. 3.8 Electron microscope image of pearlite after polishing and etching in nital. Original magnification: 11,000 ×. Source: Ref 3.2

of the plates, as is shown for a real sample in Fig. 3.8, which is a transmission electron microscope picture at a magnification of 11,000 times. In this picture, the cementite plates are the light phase and the ferrite plates are the dark phase. (Note that the cementite plates are only 0.1 μm, or 4×10^{-6} in., thick, too thin to be resolved on an optical microscope. Although cementite is brittle, pearlite is not, due largely to the fine size of the cementite plates.)

If the 1077 steel is now cooled from the 700 °C (1300 °F) temperature shown in Fig. 3.7 to room temperature, the microstructure will not change significantly, no matter how slow or fast the sample is cooled. The ferrite will remain a nearly pure bcc iron with less than 0.02% C dissolved in it, and the cementite will remain at 6.7% C with an unchanged crystal structure. According to the phase diagram, when austenite is cooled below the A_1 temperature and held for a short period, the austenite will be completely replaced by some form of ferrite + cementite, and on cooling to room temperature, no further changes in the resulting ferrite + cementite microstructure will occur. Austenite is never seen in plain carbon steels at room temperature. (An exception to this rule is quenched high-carbon steels, which contain mixtures of martensite and retained austenite.) This means that microstructures containing austenite, such as those shown in Fig. 3.2 to 3.5, can only be seen in a hot stage microscope, because the austenite will be replaced by other structures on cooling. At high cooling rates (such as obtained in water and oil quenching), these include the martensite and bainite structures discussed subsequently. At air cooling rates and slower (such as furnace cooling), the austenite will transform into some form of ferrite + cementite.

The A_1, A_{e_1}, A_{c_1}, and A_{r_1} Nomenclature

The iron-carbon phase diagram shown in the previous figures is called an *equilibrium* phase diagram. This means that the transformation temperatures given by the A lines were determined at extremely slow cooling or heating rates, where equilibrium conditions are obtained. The A_3, A_{cm}, and A_1 lines that appear on equilibrium phase diagrams are often labeled the A_{e_3}, $A_{e_{cm}}$, and A_{e_1} lines, respectively, where the "e" indicates equilibrium conditions. In this book, the "e" will generally not be used. The absence of the "e" implies equilibrium conditions.

Consider again the experiment illustrated in Fig. 3.7. The diagram predicts that when the austenite grains cool to 727 °C (1340 °F), pearlite will start to form from the austenite. This only occurs if the austenite is cooled extremely slowly. At even modest cooling rates such as 3 °C/min (5 °F/min), the transformation temperature is lowered by approximately 20 °C (36 °F). This means that the iron-carbon diagram can only be used as a rough guide for estimating transformation temperatures. Not only is the transformation from austenite to pearlite on cooling shifted down in temperature, the reverse transformation from pearlite to austenite on heating is shifted up in temperature. A simple experiment was run to demonstrate the magnitude of this shift on cooling and heating and how it can be measured. A hole was drilled in a small piece of 1018 steel and a thermocouple secured in the hole. The output of the thermocouple was measured with a digital voltmeter every 2 s and sent to a computer, where it was converted to a file of temperature versus time. The sample was placed in a small resistance furnace and heated to 870 °C (1600 °F) and then removed from the furnace. The output of the thermocouple for both the heating and cooling cycle is shown in Fig. 3.9 and 3.10, respectively.

As explained in Chapter 1, "Pure Iron," when austenite changes to ferrite on cooling, heat is liberated, and when ferrite changes to austenite on heating, heat is absorbed. Consider first the heating curve of Fig. 3.9. As is discussed in Chapters 5 and 7, the A_3 and the A_1 lines of the

> **HISTORICAL NOTE**
>
> The same French scientist, Floris Osmond, who is responsible for the name of martensite is also responsible for the use of the letters "r" and "c" for the shift in the A lines on cooling and heating. At the end of the 19th century, he was the first scientist to use thermocouples to measure the effect of heating and cooling rates. The letter "r" is from the French word for cooling, *refroidissement,* and the letter "c" is from the French word for heating, *chauffage.*

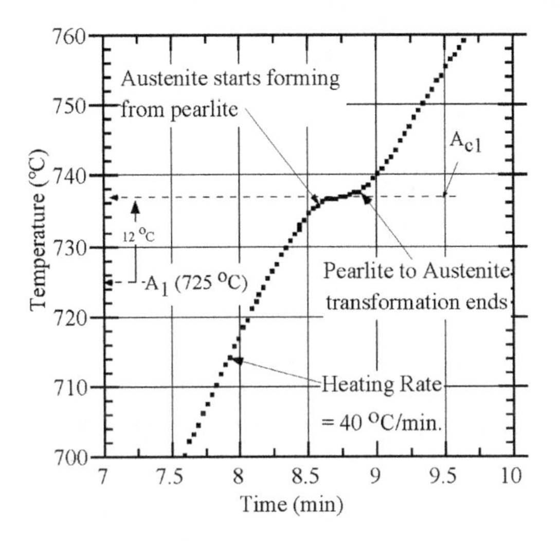

Fig. 3.9 Heating (*chauffage*) curve for a 1018 steel

pure iron-carbon diagram are shifted down in temperature by the 0.75% Mn and 0.2% Si present in 1018 steel to the values of A_1 = 725 °C (1337 °F) and A_3 = 824 °C (1515 °F). The heating data of Fig. 3.9 show that at approximately 737 °C (1359 °F), the rate of temperature rise abruptly decreases. This is due to the heat absorbed by the sample as the pearlite part of the steel transforms to austenite. It means that the A_1 line has shifted up by 12 °C (25 °F), from 725 to 737 °C (1337 to 1359 °F). It is customary to label the actual temperature of the transformation on heating as A_{c_1}, as shown in the figure. The amount of the upward shift depends on the heating rate. If the heating rate were increased above the value of 40 °C/min (72 °F/min), the value of A_{c_1} would increase.

On cooling, an opposite effect occurs: the heat liberated when austenite transforms to ferrite or pearlite slows down the rate of decrease in the sample temperature. Figure 3.10 illustrates this effect for both the austenite ⇒ ferrite transformation below the A_3 temperature and the austenite ⇒ pearlite transformation below the A_1 temperature. The former transformation begins to occur at 762 °C (1404 °F), which is 62 °C (110 °F) below the A_3 temperature of 824 °C (1515 °F), and the latter occurs at 652 °C (1206 °F), which is 73 °C (130 °F) below the A_1 temperature of 725 °C (1337 °F). As shown in Fig. 3.10, it is customary to label the actual transformation temperatures that occur on cooling as the A_{r_3} and A_{r_1} temperatures. Notice that the cooling rate for the data of Fig. 3.10 is three times larger than the heating rate for the data of Fig. 3.10. This larger rate accounts for the increased shift in the A_1 temperature from −73 °C (−130 °F) on cooling to only + 12 °C (22 °F) on heating. (The temperature range of Fig. 3.9 did not extend high enough to measure the increased A_3 temperature, which is called the A_{c_3} temperature.)

Figure 3.11 presents a graphical summary of the aforementioned ideas, showing the shift up of the A lines on heating, with labels having subscript "c" added, and the shift down on cooling, with the labels have the subscript "r" added. These shifts in transformation temperatures can become important in manufacturing operations involving rapid heating and cooling. An example is the shift up in transformation temperature with heating in processes such as flame and induction hardening.

Summary of the Major Ideas in Chapter 3

1. In the United States, steels are specified with the American Iron and Steel Institute (AISI) code. As shown in Table 3.1, the code tells the carbon composition and whether the steel is an alloy steel or a plain

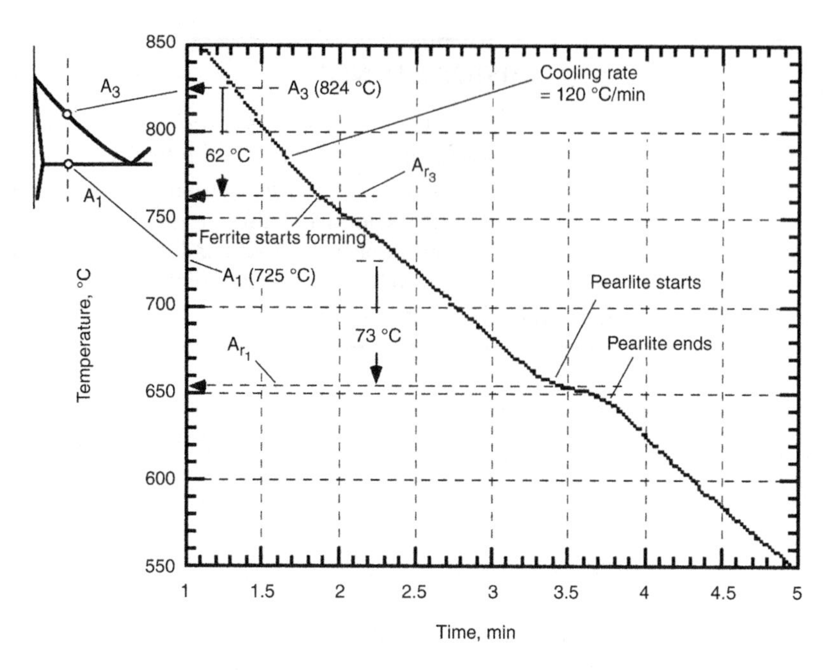

Fig. 3.10 Cooling (*refroidissement*) curve for a 1018 steel

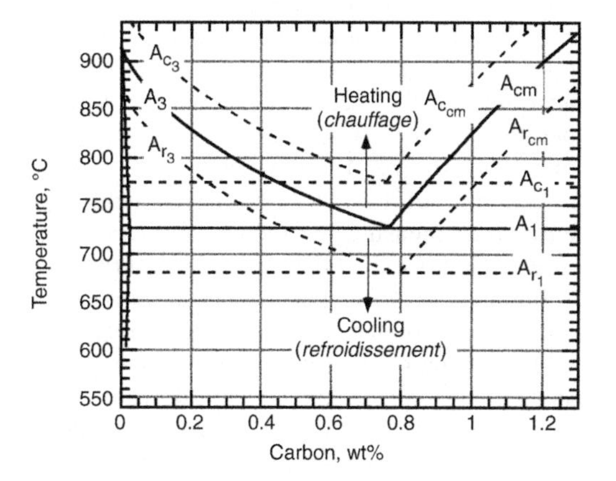

Fig. 3.11 Nomenclature for the A lines shifted up on heating and down on cooling

carbon steel. A reference, such as Ref 3.1, should be consulted to determine the full composition.

2. Steel is made by adding carbon to iron, and the main structures are ferrite and austenite, the same as those of pure iron discussed in Chapter 1. At low-carbon compositions, the carbon is dissolved into the iron, forming solid solutions of ferrite (body-centered cubic, bcc) at low temperatures and austenite (face-centered cubic, fcc) at high temperatures. At high-carbon compositions, a compound of carbon with iron, cementite, or Fe_3C, forms in the steel. These three structures are called phases, and the iron-carbon phase diagram maps the temperature-composition values where the three different phases are stable.
3. The iron-carbon phase diagram is similar to the salt-water phase diagram of Chapter 2. As carbon is added, the austenite-to-ferrite transformation drops from the 912 °C (1674 °F) value of pure iron and attains a minimum value at 727 °C (1340 °F). Beyond the minimum, the transformation temperature increases, and cementite begins to form. The minimum occurs at a composition of 0.77% C, which is called the eutectoid point.
4. It is common to classify steels with carbon compositions less than the eutectoid value as hypoeutectoid steels and those with more than 0.77% C as hypereutectoid steels.
5. As a hypoeutectoid steel cools from the austenite region, grains of ferrite will begin to form on the old austenite grain boundaries. At any specific temperature below that at which ferrite starts to form and above the eutectoid temperature, the phase diagram allows the calculation of the volume fraction of ferrite formed and the composition of both the ferrite formed and the remaining austenite.
6. The microstructure of a steel refers to the size, shape, and distribution of the phases that make up the steel. The microstructure is

controlled by the history of the steel, that is, how it was heated and cooled and any prior mechanical deformation.

7. As a hypereutectoid steel cools from the austenite region, grains of cementite will begin to form on the old austenite grain boundaries. Cementite is a compound of iron and carbon having the chemical formula Fe_3C (6.7 wt% C or 25 at.% C). Like austenite and ferrite, it is a phase in steel composed of small grains, but it is different in that it has only one chemical composition and is quite brittle. At any specific temperature below which cementite starts to form and above the eutectoid temperature, the phase diagram allows the calculation of the volume fraction of cementite formed and the composition of the remaining austenite.
8. When steels having the eutectoid composition of 0.77 wt% C are cooled below the eutectoid temperature of 727 °C (1340 °F), they transform into the pearlite structure. Pearlite consists of two-phase grains composed of alternating thin plates of cementite and fat plates of ferrite. The plate thickness depends on cooling rate, and in air-cooled samples, the cementite plates are so thin they cannot be seen in an optical microscope (thickness less than 0.2 μm, or 8×10^{-6} in.).
9. The three important lines on the iron-carbon phase diagram are labeled as: A_3, the austenite ⇔ ferrite transformation line; A_{cm}, the austenite ⇔ cementite transformation line; and A_1, the austenite ⇔ pearlite transformation line. Only when heating or cooling is very slow do these transformations occur at the A_{cm}, A_3, and A_1 temperatures given by the equilibrium phase diagram. On cooling (*refroidissement*), the transformation lines shift down in temperature, and the subscript "r" is added, as A_{r_3}, $A_{r_{cm}}$, and A_{r_1}. On heating (*chauffage*), the transformation lines shift up in temperature, and the subscript "c" is added, as A_{c_3}, $A_{c_{cm}}$, and A_{c_1}. The amount of the shift up or down scales with the heating or cooling rates and is quite large even at modest rates (Fig. 3.9, 3.10).

REFERENCES

3.1. *Practices and Procedures for Iron and Steels, Heat Treater's Guide*, 2nd ed., ASM International, 1995

3.2. *Atlas of Microstructures of Industrial Alloys*, Vol 7, *Metals Handbook*, 8th ed., American Society for Metals, 1972, p 48

CHAPTER 4

The Various Microstructures of Room-Temperature Steel

THE OPTICAL MICROSCOPE is the principal tool used to characterize the internal grain structure of steels. Traditionally, the structure revealed by the microscope is called the microstructure. The mechanical properties of a given steel are strongly influenced by the microstructure of the steel, and this chapter reviews the common steel microstructures and how they are achieved by the heat treatment of the steel. Chapter 3, "Steel and the Iron-Carbon Phase Diagram," discusses steel microstructures that occur at high temperatures, where it is generally not possible to observe such structures clearly, even in a hot stage microscope. This chapter discusses only the room-temperature microstructures that are observed in optical and electron microscopes.

Optical Microscope Images of Steel Grains

As discussed in Chapter 1, "Pure Iron," the grain structure of iron and steel samples is revealed by polishing the surface to a mirror quality, etching in an acid solution, and then examining in an optical microscope. The etch removes atoms from the polished surface, and the rate of removal depends on both the crystal orientation of the grain and the type of grain, for example, ferrite versus austenite. The gray level observed in the microscope for a given grain depends on the degree of smoothness of the grain surface after the etching. As shown in Fig. 4.1, grains that are single phase, such as ferrite, austenite, and cementite, will have their atoms removed uniformly from point to point, and so, the original polished surface will remain smooth within a given grain after etching.

An optical microscope uses reflected light to generate an image. A beam of light is directed down onto the surface (open arrowheads in Fig. 4.1), and the image is generated either on film or in the eye by light reflected back up along the same direction. Figure 4.1 uses a symbol for the eye located directly above the sample surface, showing the light path generating the image. The reflected light is shown by the dark arrows. For a smooth surface, a very large fraction of the incoming light is reflected back up the image path, thereby forming a bright (white) appearance. Hence, the single-phase grains of ferrite, austenite, and cementite all appear at the white end of the gray-level range in an optical image. Because they all appear white, it is often not possible to distinguish between them by their appearance (gray levels) without further information.

Now consider pearlite. The common etches used for steels, nital (nitric acid in alcohol) and picral (picric acid in alcohol), etch the ferrite plates of pearlite much faster than the cementite plates. Thus, after etching, the cementite plates protrude from the ferrite plates. The cementite plates are very fine, and they scatter the incoming light away from the image path (Fig. 4.2), thereby generating a low gray-level (dark) image. So, pearlite will generally appear gray to black in an optical microscope, but not always. If the spacing of the cementite plates is large enough, the cementite plates will appear as dark lines with white ferrite plates between them. The optical microscope can only resolve distances to approximately 0.2 μm (8 × 10^{-6} in., or 0.008 mils) at the highest useful magnifications of approximately 1000×. Therefore, when the spacing of pearlite is less than 0.2 μm

Type of grain	Surface after etching	Appearance in the optical microscope
Ferrite	Smooth	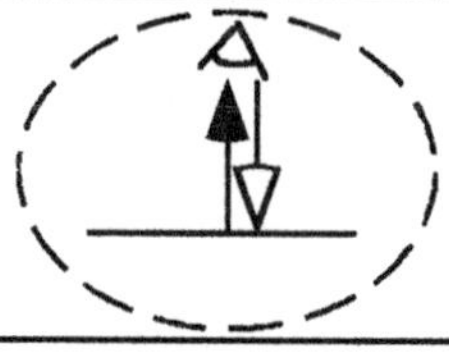 White
Austenite	Smooth	White
Cemenite	Smooth	White
Pearlite	Cm plates protrude	 Dark
Martensite (fresh)	Smooth	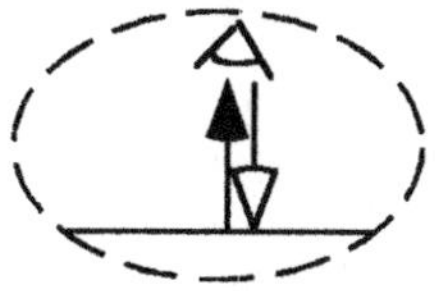 Mostly white
Bainite and tempered martensite	Cm particles protrude	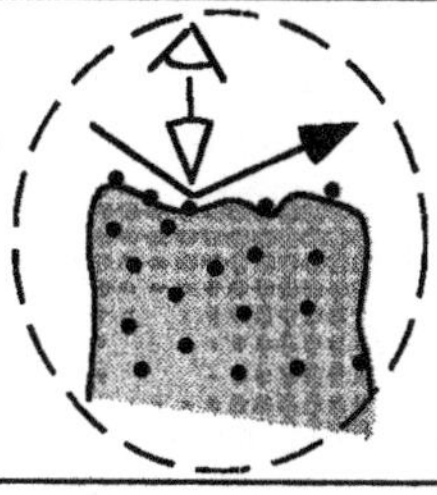 Dark

Fig. 4.1 Appearance of various types of steel grains in an optical microscope. Cm, cementite

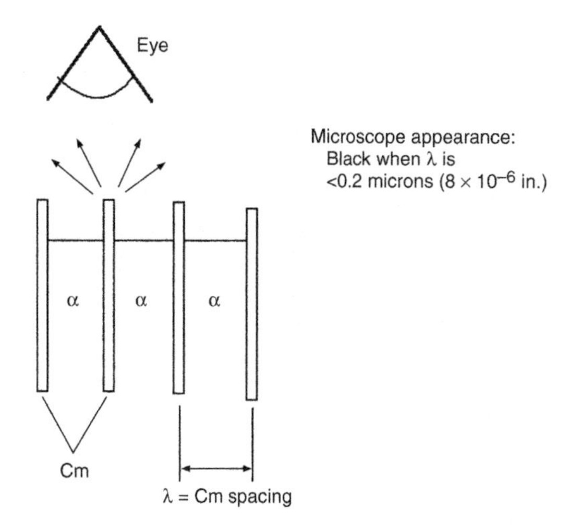

Fig. 4.2 Light scattering at an etched pearlite surface showing surface morphology of the ferrite (α) and cementite (Cm) phases

(8 × 10^{-6} in., or 0.008 mils), the optical microscope image shows pearlite grains with a mottled dark gray level, as shown in Fig. 4.3. The spacing of the cementite plates in pearlite depends on how fast the sample was cooled from the austenite range past the A_1 temperature; faster cooling gives finer spacings. Perhaps the most common way to do such cooling is simply to remove the sample from the furnace and allow it to cool in the air. The cooling rate then depends on how big the sample is. Even for fairly large samples, air cooling produces cementite spacings smaller than 0.2 μm (8 × 10^{-6} in., or 0.008 mils). So, pearlite in samples that were air cooled virtually always appears with dark gray levels, similar to that seen in Fig. 4.3. The picral etch used in Fig. 4.3 dissolves ferrite more uniformly as its crystal orientation changes from one grain to the next than does the nital etch.

Therefore, picral gives a more uniform gray level than nital and is the preferred etch for pearlite.

Room-Temperature Microstructures of Hypo- and Hypereutectoid Steels

Reconsider the experiment shown in Fig. 3.4, where a 0.4% C austenite is cooled from 850 to 760 °C (1560 to 1400 °F). It was shown that the structure formed after such cooling will consist of thin sheets of ferrite grains laying on the prior-austenite grain boundaries that originally formed at 850 °C (1560 °F). The question to consider now is what happens to the remaining austenite grains if the sample is cooled to room temperature.

Fig. 4.3 Optical micrograph of pearlite. Picral etch. Original magnification: 500 ×

To analyze this problem, consider a three-step cooling process from 760 °C (1400 °F). The sample is first cooled to 740 °C (1360 °F), second cooled to 727 °C (1340 °F, the A_1 temperature), and third cooled to room temperature. The previous analysis (Fig. 3.4) showed that the austenite present at 760 °C (1400 °F) has a carbon composition of *N*, which is also shown in Fig. 4.4. After the first cooling from 760 to 740 °C (1400 to 1360 °F), the phase diagram requires that the carbon composition of the austenite increase to the value of *N′*, shown in Fig. 4.4. After the second cooling to 727 °C (1340 °F), the carbon composition of the austenite grains must further increase to exactly the eutectoid composition (also known as the pearlite composition) of 0.77% C. At this point, the microstructure of the steel will appear as shown on the left of Fig. 4.5. The only apparent difference between this structure and that shown in the lower right of Fig. 3.4 is that the thin ferrite grains are a little thicker, but now, the carbon composition of the austenite grains lies at the eutectoid point, 0.77% C. The eutectoid point is the point on the phase diagram where a eutectoid reaction occurs, that is, a reaction where one solid

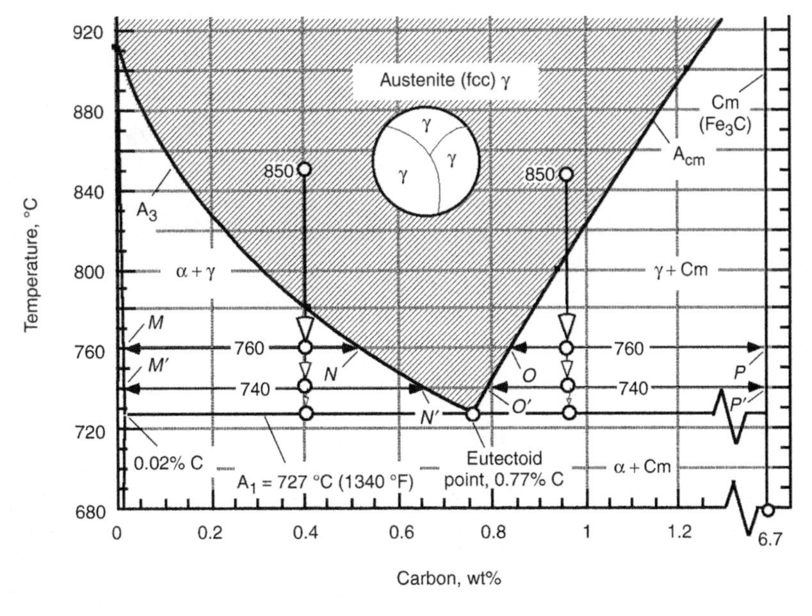

Fig. 4.4 Phase diagram analysis of the decomposition of 0.4 and 0.95% C austenite on cooling to the pearlite temperature, 727 °C (1340 °F)

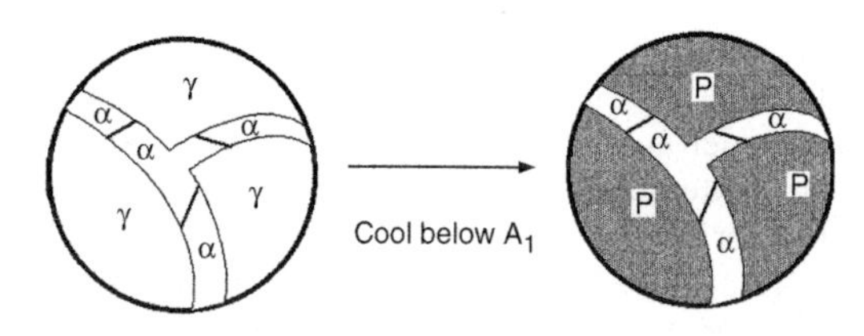

Fig. 4.5 Austenite grains transform to pearlite on cooling below 727 °C (1340 °F)

transforms into two solids. So, in this example, on the final cooling to room temperature, these austenite grains will transform into pearlite grains (ferrite + cementite), giving a microstructure similar to that shown on the right of Fig. 4.5.

In the optical microscope, the old austenite grains appear dark because they have transformed to pearlite, and the ferrite grains will appear white and outline the pearlite grains. Figure 4.6 presents an actual micrograph of such a structure for a 1045 steel. Because the ferrite forms prior to the pearlite and because the pearlite forms at the 727 °C (1340 °F) eutectoid temperature, the ferrite is often called *proeutectoid* ferrite. (It is possible to understand why the %C in austenite must increase on cooling by recalling that ferrite can dissolve almost no carbon. On cooling in the two-phase region, more ferrite must form, so when a small element of austenite transforms into ferrite, virtually all the carbon atoms in that volume element must be ejected into any remaining austenite, thereby increasing its %C. This increase cannot exceed 0.77% C, because at this composition, austenite decomposes to pearlite on further cooling, and pearlite has an average composition of 0.77% C.)

Now, consider a steel whose %C is greater than 0.77, that is, a hypereutectoid steel. A common plain carbon steel is 1095 steel (also known as either drill rod or W1 tool steel). After heating to 850 °C (1560 °F), its temperature-composition coordinate is located in Fig. 4.4. It is now cooled following the same three-step process that was considered previously for the 0.40% C steel. The action that occurs for the 1095 steel is very similar to that of the 1040 steel, except that the new phase formed on the austenite grain boundaries is cementite and not ferrite. At 760 °C (1400 °F), thin plate-shaped grains of cementite form on the prior-austenite grain boundaries, and the austenite grains must have the composition *O* shown in Fig. 4.4, approximately 0.85% C. At 850 °C (1560 °F), the austenite grains had the original composition of 0.95% C. This composition is reduced to 0.85% C after the cementite forms, because the cementite, at 6.7% C, must suck up carbon atoms as it forms. On cooling to 740 °C (1360 °F), more cementite is formed, and the austenite composition drops to point *O'*. On further cooling to the A_1 temperature of 727 °C (1340 °F), the austenite composition will have dropped to the eutectoid composition of 0.77% C. At this point, the structure would appear in a hot stage microscope as shown on the left of Fig. 4.7, which is similar to Fig. 4.5 but with the prior-austenite grain boundaries decorated with cementite grains rather than ferrite grains. Cooling below 727 °C (1340 °F) transforms the austenite grains to the pearlite structure, as shown on the right of Fig. 4.7, and this structure will not change on cooling. So, it would appear the same in a hot stage microscope at temperatures just below 727 °C (1340 °F) as it does at room temperature where it is viewed easily. Figure 4.8 presents an optical micrograph of a 1095 steel that was given the heat treatment shown in Fig. 4.4, first held at 850 °C (1560 °F) and then air cooled.

Notice the similarities between the hypoeutectoid steel of Fig. 4.6 and the hypereutectoid steel

Fig. 4.6 A 1045 steel air cooled from 850 °C (1560 °F) to room temperature, with dark austenite grains surrounded by white pearlite grains. Nital etch. Original magnification: 100 ×

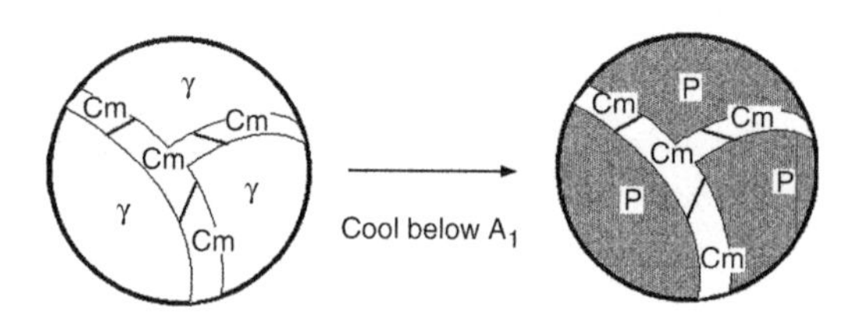

Fig. 4.7 Austenite grains transform to pearlite on cooling below 727 °C (1340 °F)

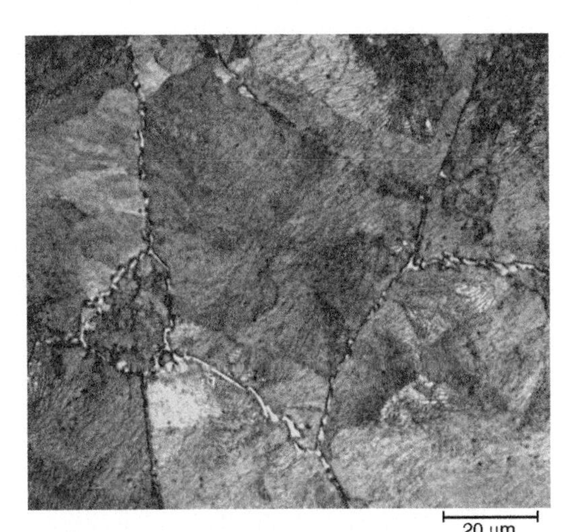

Fig. 4.8 Hypereutectoid 1095 steel air cooled from 850 °C (1560 °F) to room temperature, with dark austenite grains surrounded by thin cementite grains. Nital etch. Original magnification: 500 ×

of Fig. 4.8. In each case, the set of prior-austenite grain boundaries are filled with a proeutectoid phase, ferrite in the low-carbon steel and cementite in the high-carbon steel. The phase diagram provides information about the volume fraction of the proeutectoid phase that will be present between the pearlite grains. Suppose the steel were a 1075 steel. The 0.75% C composition is very close to the pearlite composition of 0.77, and so, this steel must be nearly 100% pearlite. Similarly, suppose one had a 1005 steel. Because the 0.05% C is so close to the pure ferrite composition of 0.02% C, this steel must be mostly ferrite with only a small percent pearlite. Hence, the volume fraction of pearlite depends on where the overall composition lies between pure ferrite, 0.02% C, and pure pearlite, 0.77% C, along the A_1 temperature line. Let X equal the overall composition of a hypoeutectoid steel. Then, the volume fraction* of ferrite in the steel is:

$$\text{Fraction ferrite} = (0.77 - X)/(0.77 - 0.02) \quad \text{(Eq 4.1)}$$

For the 1045 steel example, this gives a value of 0.43 or 43%. For hypereutectoid steels, the same type of rule applies, with the volume fraction cementite measured by the relative position of the overall composition point along the A_1 line between pure pearlite at 0.77% C and pure cementite at 6.7% C. The formula now becomes:

$$\text{Fraction cementite} = (X - 0.77)/(6.7 - 0.77) \quad \text{(Eq 4.2)}$$

For 1095 steel, Eq 4.2 gives a value of 0.03 or 3% cementite.

The room-temperature microstructures of steel vary widely, and those shown in Fig. 4.6 and 4.8 should be regarded as just two examples of many different possible structures. The following generalizations hold, depending on whether the steel is cooled slowly (for example, air cooling or furnace cooling) or more rapidly (for example, water quenched or oil quenched).

Microstructures with Slow Cooling. If the steel has been cooled at rates of air cooling or slower, its structure will be some mixture of ferrite and cementite. The cementite will usually be present as a component of the pearlite grains, but not always. By proper heat treatment, it is possible to form the cementite as isolated grains, such as shown for the microstructure at the bottom of Fig. 3.5. This structure is called spheroidized because the cementite is present as small spherical grains. A pearlitic steel is much stronger and more difficult to machine than a spheroidized steel. That is why steel mills generally supply hypereutectoid steels in the spheroidized condition. The microstructure of such steels consists of spherical cementite particles in a matrix of ferrite grains. (For further discussion, see the section "Spheroidized Microstructures" in this chapter.)

The micrographs of Fig. 4.6 and 4.8 show the proeutectoid phases localized along prior-austenite grain boundaries. As the steel composition moves further away from the eutectoid composition, this morphology becomes less common. For example, Fig. 4.9 shows the microstructure of a 1018 steel that has been furnace cooled from the austenite region (900 °C, or 1650 °F). Notice that the dark pearlite occupies a much smaller volume fraction than in the 1045 steel of Fig. 4.6. The white ferrite grains dominate the structure and show no obvious alignment along the prior-austenite grains where they first formed. The ferrite grains are simply too big to show traces of those old boundaries. Also, notice that the ferrite and pearlite appear to lie along alternating bands. This steel is called a *banded* steel. Virtually all hypoeutectoid steels show this pearlite/ferrite banding if the steel has been heavily deformed, followed by slow cooling from the austenite range. All wrought steels

*Strictly speaking, this formula gives weight fraction ferrite. However, weight fraction is extremely close to volume fraction, because the densities of cementite and ferrite are very nearly the same.

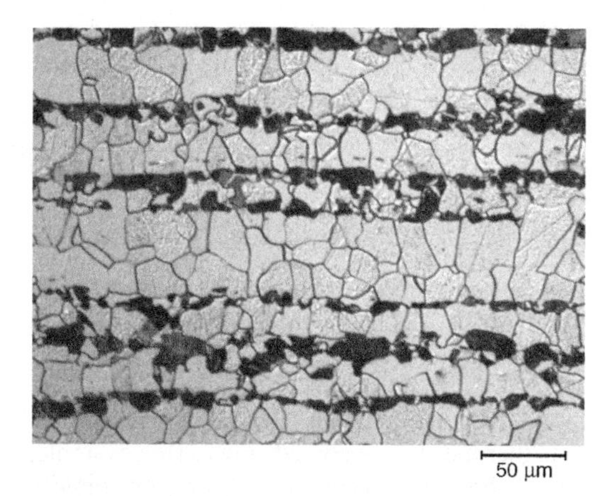

Fig. 4.9 A wrought 1018 steel slow cooled from 900 °C (1650 °F) to room temperature. The structure is dominated by white ferrite grains with only a small volume fraction of dark pearlite grains. Deformation direction is horizontal. Nital etch. Original magnification: 240×

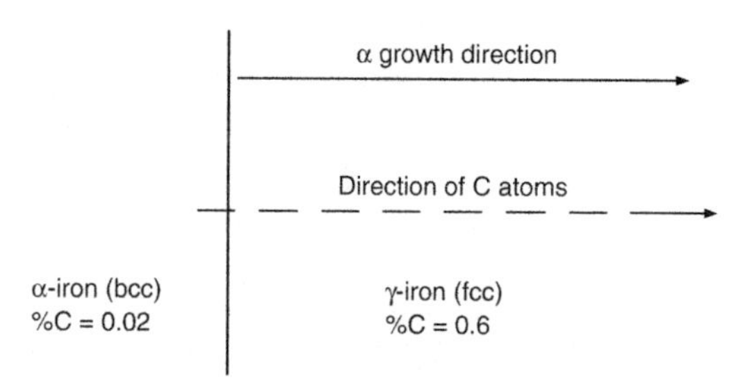

Fig. 4.10 Sketch of advancing interface as austenite transforms to ferrite

are heavily deformed by some method, usually a mixture of hot and cold rolling. If such steels are slow cooled (usually just a little slower than air cooling) and if they are sectioned parallel to the deformation direction, they will virtually always appear banded. Wrought steels having round cross sections will not show banding when sectioned at right angles to the deformation direction (the axis of the round), but they will show banding when sectioned along their axis.

Microstructures with Rapid Cooling. If the steel is cooled rapidly from the austenite region, the type of phases, their relative amounts, or their compositions can no longer be estimated from the phase diagram, as was done in the previous section. At the slower end of these faster cooling rates, mixtures of pearlite + ferrite for low-carbon steels and pearlite + cementite for high-carbon steels still result, but the amount of pearlite depends on the cooling rate. At increasing cooling rates, one of two new types of structures begins to form: bainite or martensite. These structures are the subject of the next section.

Microstructure of Quenched Steel

Perhaps the most fascinating aspect of steel is that it may be strengthened to amazingly high levels by quenching (rapid cooling). The strength levels are higher than the strongest commercial alloys of aluminum, copper, and titanium by factors of roughly 4.7, 2.2, and 2.1, respectively. Steels are generally quenched by immersing the hot metal into liquid coolants, such as water, oil, or liquid salts. Increased strengths do not occur unless the hot steel contains the austenite phase. The very rapid cooling prevents the austenite from transforming into the thermodynamically preferred ferrite + cementite structure. A new structure, called martensite*, is formed instead, and this martensite phase is responsible for the very high strength levels.

Martensite. As explained in Chapter 1, "Pure Iron," austenite has a face-centered cubic (fcc) crystal structure, and ferrite has a body-centered cubic (bcc) crystal structure. The steel phase diagram shows that the fcc structure will dissolve much more carbon than the bcc structure. At the A_1 temperature, the %C that can dissolve in fcc iron is higher than in bcc iron by the ratio of 0.77/0.02 = 38.5. As previously discussed, carbon atoms are much smaller than iron atoms, and the dissolved carbon atoms lie in the interstices (holes) between the larger iron atoms. The fcc structure dissolves more carbon atoms because some of the holes in this structure are larger than any of the holes in the bcc structure. Figure 4.10 shows austenitic iron in a 1060 steel (0.6% C) transforming to ferritic iron as the interface (vertical line) moves to the right. After the interface has moved, for example, 25 mm (1 in.), the %C in that 25 mm (1 in.) region must drop from 0.6 to 0.02% C. At slow cooling rates, the carbon is able to move ahead of the interface into the austenite iron along the direction of the dashed arrow by the diffusion process discussed in Chapter 7, "Diffusion—A Mechanism for Atom Migration within a Metal."

*The names ferrite, austenite, pearlite, eutectoid, and martensite were suggested by two men, an American, Henry Marion Howe, and a Frenchman, Floris Osmond, in the period from 1890 to 1903. In the evolution of science, the names suggested by researchers often fall by the wayside. An example of this is Howe's suggestion that martensite be called hardenite. To this author, it seems unfortunate that Osmond's preference for martensite was eventually adopted, because the term *hardenite* so aptly describes the outstanding property of the martensite phase.

However, if this transformation is forced to occur very rapidly by quenching, there is not enough time for the carbon atoms to rearrange themselves, and some or all of them get trapped in the ferrite, causing its composition to rise well above 0.02%, which makes its crystal structure become distorted from the bcc form. The resulting distorted crystal structure is martensite. Figure 4.11 compares the unit cell of bcc ferrite to that of the distorted unit cell of martensite. The unit cell of the martensite crystal is similar to the bcc unit cell in that it has an atom at its center and one atom at each of the eight corners. However, the unit cell is no longer a cube. One of its edges, called the *c*-axis in Fig. 4.11, is longer than the other two, called the *a*-axes. This structure is called body-centered tetragonal (bct).

The *a* and *c* lengths of the unit cell of martensite can be measured using x-ray diffraction techniques. As shown in Fig. 4.12, it is found that as the %C dissolved in the martensite increases, the *c*-axis becomes proportionately larger than the *a*-axis. The increased carbon content in the martensite is obtained by quenching austenites of higher %C levels. The results of Fig. 4.12 show that as the %C goes up, the resulting distortion from the cubic structure (*c* becomes progressively bigger than *a*) increases due to the trapped carbon in the bct martensite structure. The strength and hardness of the martensite is found to increase dramatically as the %C increases, as shown by the hardness data of Fig. 4.13. One way to rationalize this increased hardness is to think of the chemical bonds holding the iron atoms together as springs. As the %C increases, the springs will be extended by larger amounts, thereby making it more difficult to extend them further, that is, making the structure harder.

Two Types of Martensite. The appearance of martensite in the optical microscope changes as the %C changes. The right of Fig. 4.14 illustrates schematically that quenched pure austenite with compositions above or below the eutectoid value of 0.77% C can produce structures that are all martensite. As shown on the left of this figure, it is customary to partition the form of martensite into two types, depending on %C. For %C from 0 to 0.6, the martensite is called *lath martensite*, and for %C above 1, it is called *plate martensite*. At the intermediate %C levels from 0.6 to 1.0, there is a mixture of these two types of martensite. The appearance of martensite in the optical microscope is a little disappointing. Lath martensite has a somewhat fuzzy

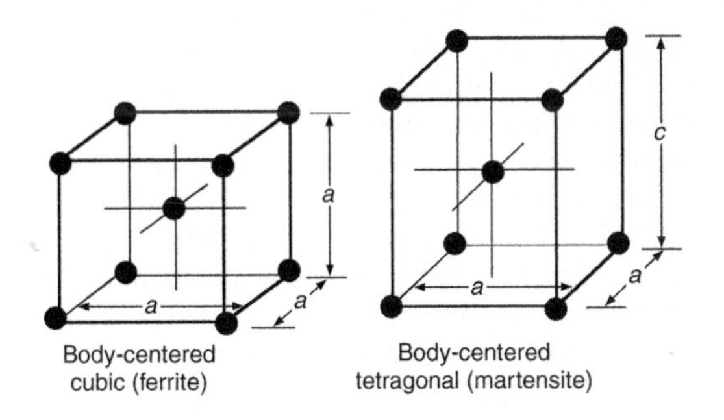

Fig. 4.11 Comparison of the crystal structures of ferrite and martensite

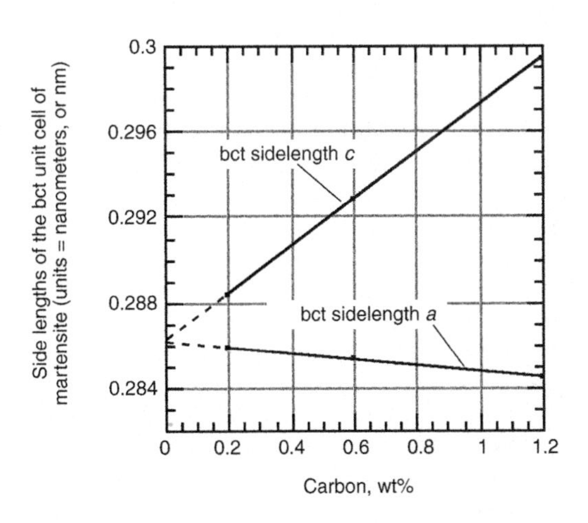

Fig. 4.12 Actual *a* and *c* side lengths of the body-centered tetragonal (bct) unit cell of martensite (1 nm = 1000 μm = 4 × 10^{-8} in.)

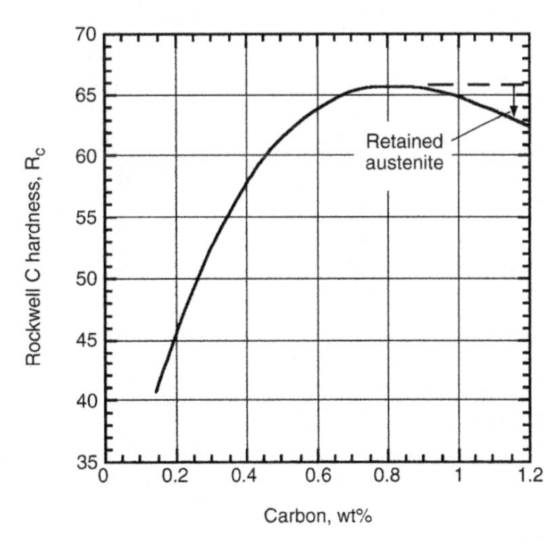

Fig. 4.13 Rockwell hardness of fresh martensite as a function of carbon content

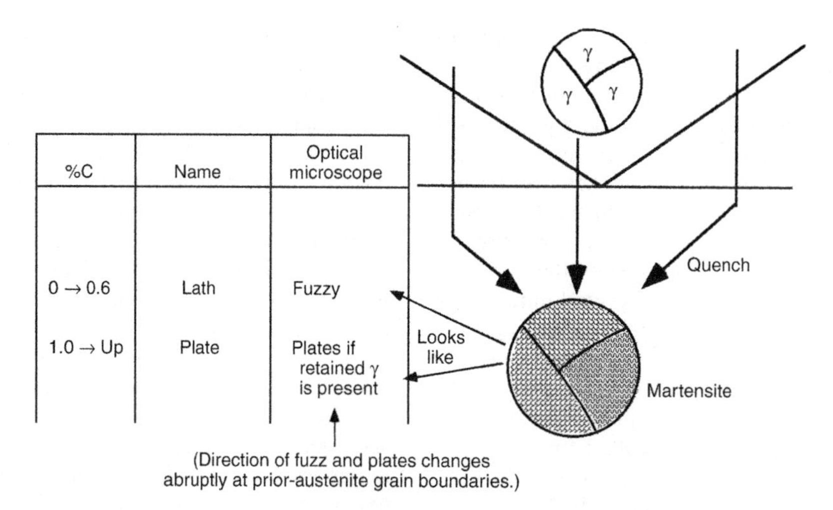

Fig. 4.14 Martensite forms on quenching austenite with varying wt%C. The carbon content determines whether the morphology will be lath or plate

and nondistinct appearance, as shown in Fig. 4.15(a) for a 1018 steel. In plate martensite, it is possible to see the individual plates, but only if the %C becomes larger than approximately 1% C, as is illustrated for a 1.4% C martensite shown in Fig. 4.15(b). The reason the plates are visible is because this structure is not 100% martensite. The white regions surrounding the plates are grains of austenite that did not transform on quenching. It is customary to call these grains retained austenite. This austenite is soft compared to the martensite, and it is due to this austenite that the hardness is found to decrease at compositions above approximately 0.9% C in Fig. 4.13. If a plate martensite contains little to no retained austenite, the individual plates do not stand out (Fig. 4.15c), although they can be imaged clearly in a type of microscope called a transmission electron microscope. Comparison of Fig. 4.15(a) and (c) illustrates that it is quite difficult (except for steel metallographers) to distinguish between a pure lath martensite and a mixed lath + plate martensite based only on the appearance of the micrographs.

When martensite is removed from the quench bath, it is called fresh martensite. The hardness data of Fig. 4.13 refers only to fresh martensite. A big problem with fresh martensite is that if the carbon content is greater than approximately 0.2 to 0.3% C, the steel will be very brittle. This brittleness can be removed at the expense of a loss of some hardness if the quenched steel is heated slightly, a process known as *tempering*. Therefore, quenched steels are almost always tempered to improve the toughness of the steel, and the resulting martensite is called *tempered martensite*. The increased temperature of tempering allows the carbon atoms trapped in the bct structure to move a little. This atom motion does two things: (1) It allows the bct structure to change to the bcc structure. (2) It allows formation of very small particles of carbides (often having compositions of $Fe_{2.5}C$ and called epsilon carbide), too small to see in the optical microscope. Although too small to see individually, the carbides cause the etching process to roughen the surface, so that tempered martensite appears dark in the optical microscope. As illustrated in Fig. 4.1, fresh martensite etches with a fairly smooth surface to give a mostly white appearance, in contrast to the dark appearance of a tempered martensite. The tempering temperature needed to darken the martensite is fairly low, on the order of 150 °C (300 °F). The martensites of Fig. 4.15 were lightly tempered.

The M_s and M_f Temperatures. There are two very important temperatures that must be understood when doing heat treatment processing that produces martensite: the *martensite start temperature, M_s*, and the *martensite finish temperature, M_f*. When austenite transforms to another phase on cooling, it is found that the new phase virtually always forms first, or nucleates, on the old austenite grain boundaries. For example, if a 1077 steel is converted to all austenite at 850 °C (1560 °F) and then air cooled to 650 °C (1200 °F), small grains of pearlite will nucleate on the old austenite grain boundaries shortly after the temperature falls below 727 °C (1340 °F). When the temperature reaches 650 °C (1200 °F), these

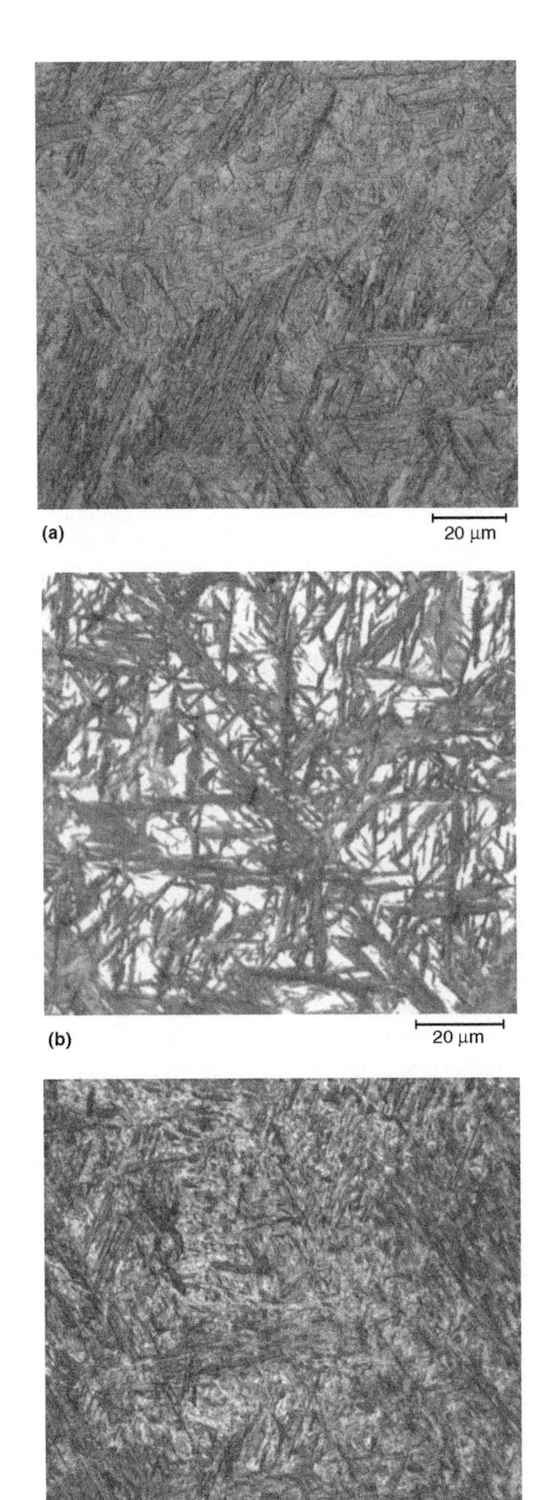

Fig. 4.15 Optical micrographs of (a) lath 0.18 %C, (b) plate 1.4 %C, and (c) mixed structures of martensite. 0.9 %C. Nital etch

grains will have grown, but there will still be some austenite trapped between them. If the sample is held at 650 °C (1200 °F), the pearlite grains will continue to grow until all of the austenite is consumed. Now, consider an alternate heat treatment where the sample is quenched from 850 °C (1560 °F) into a hot liquid at 650 °C (1200 °F). This treatment will cause the temperature to drop much more rapidly to 650 °C (1200 °F). The austenite decomposition sequence is the same; small pearlite grains nucleate on the old austenite grain boundaries and grow into the remaining austenite until it is gone. The faster cooling will produce three differences: (1) the pearlite grains will grow faster, (2) the cementite spacing of the pearlite will be much finer, and (3) the pearlite grains will be smaller because more of them will nucleate on the old austenite grain boundaries.

Now, suppose that the 1077 steel is quenched into a bath at a much lower temperature, for example, a water bath at room temperature. At such a low quench temperature, martensite can be expected to form and harden the steel. There are similarities and differences between the formation of martensite at room temperature and pearlite at 650 °C (1200 °F). Like the pearlite, the martensite will start to form (nucleate) along the old austenite grain boundaries, but, unlike the pearlite, the martensite grows into the austenite at tremendously faster rates. Whereas the pearlite grows into the austenite at rates of approximately 50 μm/s (0.002 in./s), at 650 °C (1200 °F) (and slower at higher temperatures), the martensite grows at near the speed of sound (4510 m/s, or 14,800 ft/s) into austenite, at whatever temperature it forms. Also, unlike the pearlite, which will completely replace the austenite by simply holding the sample long enough at the reduced temperature, the martensite will not transform all of the austenite unless the quench temperature is below the temperature called the martensite finish temperature, M_f. Furthermore, martensite will not start to form at all unless the quench temperature lies below the martensite start temperature, M_s. If the quench temperature lies between M_s and M_f, then only a fraction of the austenite will transform to martensite, and the remaining austenite just sits there and is called retained austenite. Figure 4.16 presents a graph that illustrates these ideas. Notice that an M_{50} temperature is defined as the temperature where 50% martensite has formed. So, in a steel quenched to the M_{50} temperature, 50% of the austenite would transform to martensite in a matter of milliseconds after the temperature

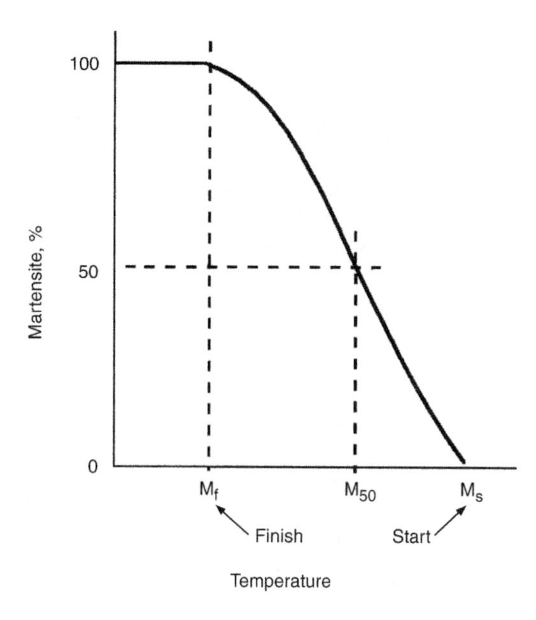

Fig. 4.16 Quench temperature controls the amount of martensite

reaches the M_{50} temperature. However, the remaining 50% of the austenite surrounding the martensite will remain as retained austenite so long as the temperature is not changed.

Notice that the steel phase diagram of Fig. 3.7 predicts that all austenite should disappear when the steel temperature goes below the A_1 temperature. Furthermore, it requires that the phases ferrite + cementite should be present below the A_1 temperature. However, with quenched steels, both of these rules are violated, because austenite is present and there is no ferrite or cementite present. This illustrates that the phase diagram correctly maps the phases present at temperature-composition coordinates only if the steel is cooled at slow to moderate rates. At higher cooling rates, a new phase, martensite, not predicted by the phase diagram, is found. Such phases are said to be metastable, that is, not stable. If martensite is heated, as in the tempering process, it converts to more stable phases. The stable phases are given on the phase diagram, so heating martensite structures up to temperatures below A_1 converts both the martensite and the retained austenite into ferrite + cementite.

Martensite and Retained Austenite. The M_s temperature of steels depends very strongly on the %C in the austenite. This fact can be illustrated with two different types of graphs, shown in Fig. 4.17 and 4.18. In Fig. 4.17, the M_s and M_f temperatures are plotted for increasing %C in plain carbon steels. The M_f temperature is not well established, because experiments show a wide scatter in measured values. Notice the horizontal line labeled T_{room} (room temperature) in Fig. 4.17. Most quenching is done to room temperature, so this line allows the estimation of what %C values are required to produce retained austenite in quenched plain carbon steels. According to Fig. 4.17, a room-temperature quench will begin to lie above the M_f temperature, at %C values exceeding approximately 0.3 to 0.4% C. It is possible to measure the percent retained austenite in quenched steels fairly accurately by using x-ray diffraction techniques. Figure 4.18 shows the results of the measurement of the volume percent retained austenite in quenched plain carbon steels as the carbon content increases. Similar to the M_f temperature, there is a wide scatter in the data. For example, in a 1.4% C alloy, the percent retained austenite can be anywhere from 28 to 45%. This figure can be used to note several items of interest: (1) Fully lath martensite steels (%C less than 0.6) will not have significant amounts of retained austenite in them. (2) Fully plate martensite steels (%C greater than 1) will have significant and rapidly increasing amounts of retained austenite in them as %C increases. Notice that the example of 1077 steel quenched to room temperature is expected to have a mixed lath/plate structure and somewhere between 6 to 10% retained austenite. It is generally difficult to see the retained austenite between the martensite plates in an optical microscope until the percentage becomes larger than approximately 10%.

In the previous discussion, the assumption has been that the quenching process would drop the temperature of the austenite below the M_s temperature so rapidly that no ferrite, pearlite, or cementite would form prior to reaching the M_s. Consider a 1095 steel and note that its composition is fairly close to the pearlite composition. Figure 4.18 predicts that, if quenched rapidly enough from the austenite region to room temperature to avoid pearlite formation, this steel will transform into martensite with approximately 13% retained austenite. Now, suppose that the steel is quenched a little too slowly to avoid formation of pearlite prior to reaching the M_s temperature. In this case, it is possible that the austenite may begin to transform to pearlite before any martensite forms. If this happens, the pearlite will begin to form along the old austenite grain boundaries, but when the temperature goes below M_s, the surrounding austenite transforms to the martensite + retained

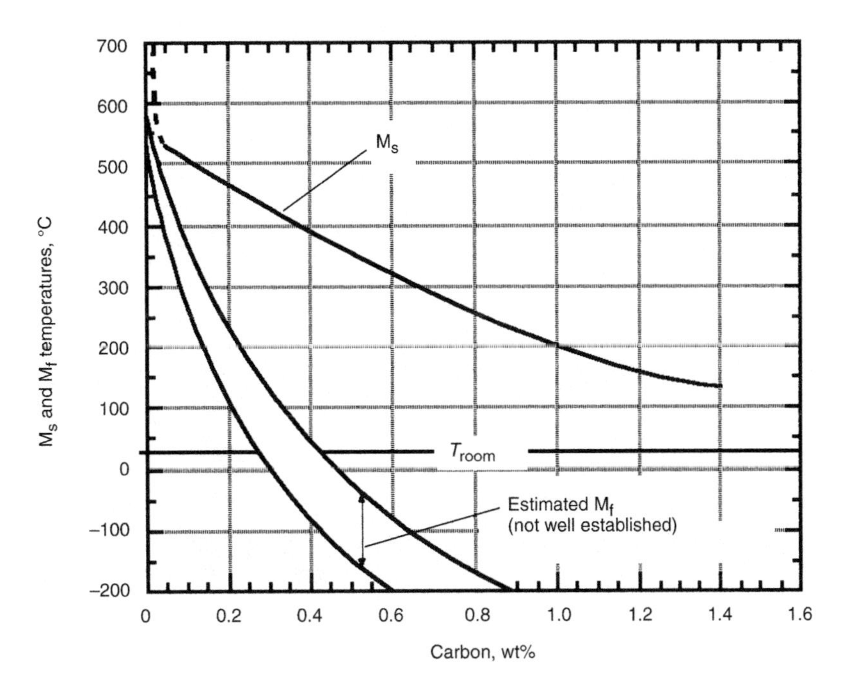

Fig. 4.17 Both martensite transformation temperatures, M_s and M_f, fall rapidly as wt%C in austenite increases

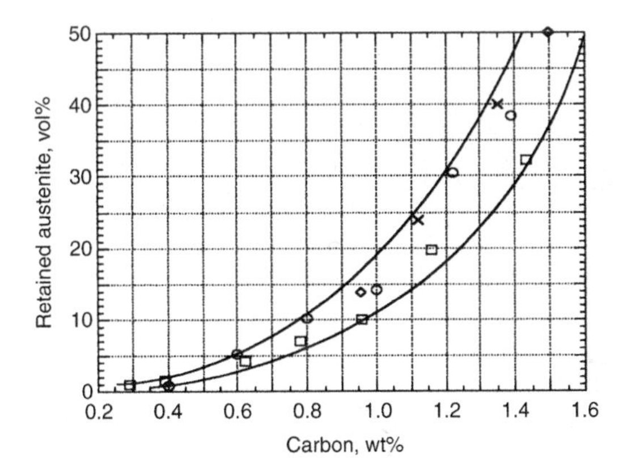

Fig. 4.18 Volume percent retained austenite versus carbon content in plain carbon steels quenched to room temperature. Data points from several references establish the width of the scatter band

austenite in a matter of milliseconds. This rapid formation of martensite traps the first formed pearlite along the old austenite grain boundaries. The result is a structure such as that shown in Fig. 4.19, with the dark pearlite surrounded by the white martensite + retained austenite. (Note: The sample was not adequately tempered to darken the martensite, so the fresh martensite + retained austenite region appears white.) Thin white lines were superimposed on this micrograph to show the probable location of the prior-austenite grain boundaries. Notice that the pearlite grains grow out from the austenite grain boundaries in little round-shaped structures. These are called pearlite nodules, and they are characterized by spherical-shaped growth fronts.

Bainite. If this steel is cooled from the austenite temperature a little faster, it is possible to observe the formation of a new steel microstructure, which is called bainite. Like pearlite, bainite forms from austenite, and it also forms first along the old austenite grain boundaries. Consequently, at fast cooling rates, there will be a competition along the old austenite grain boundaries, with pearlite forming in some places and bainite forming in other places, as illustrated in Fig. 4.20. The bainite is the lighter constituent growing out from the prior-austenite grain boundaries. Again, a thin white line was traced along the old austenite grain boundaries on the micrograph. Similar to pearlite, bainite is a mixture of ferrite + cementite. However, the internal structure of bainite is very different from pearlite. Whereas the cementite is present in pearlite as plates distributed between plates of ferrite, in bainite it is present as filaments and/or small particles dispersed in a ferrite matrix. Another difference in the two structures is illustrated in Fig. 4.20. The growth front of the pearlite is spherical but that of the bainite appears to be needlelike. Remember that Fig. 4.20 is a two-dimensional

Fig. 4.19 Pearlite nodules (dark areas) formed on prior-austenite grain boundaries, indicated by white lines. Slow-quenched 1095 steel. Nital etch. Original magnification: 600 ×

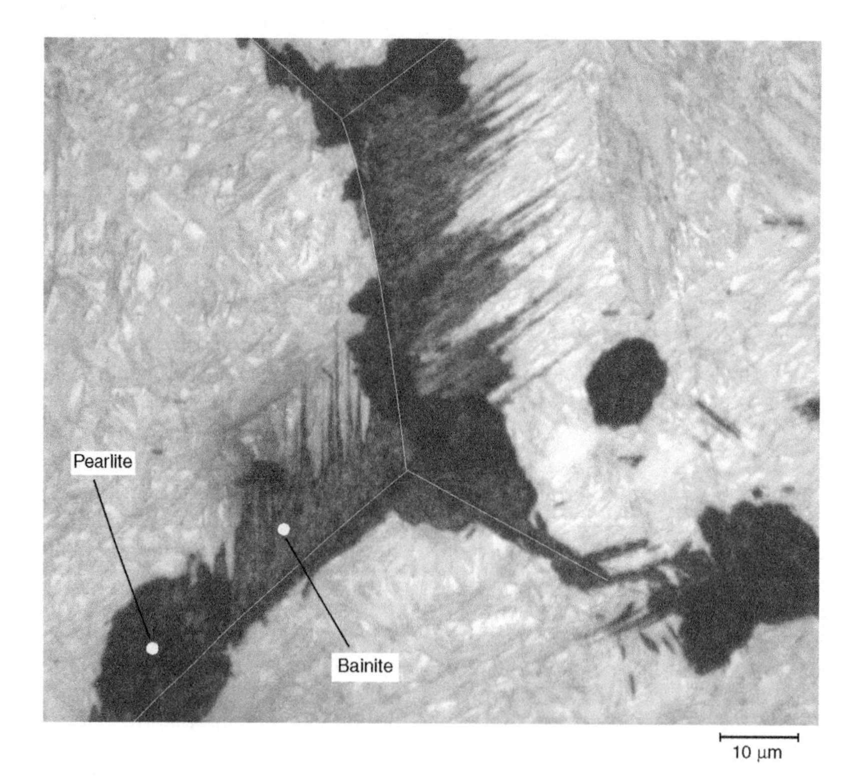

Fig. 4.20 Mixed pearlite and bainite structures formed on prior-austenite grain boundaries, indicated by white lines. Faster-quenched 1095 steel. Mixed nital-picral etch. Original magnification: 1000 ×

cut through the structure. If the bainite growth front were really needle shaped, the needle shape would be apparent only if the cut were to lie parallel to the needle axes, which is unlikely. However, if the growth front were plate shaped, it would look like needles in the two-dimensional view for many different orientations of the cut. The growth front of bainite is plate shaped here, as in most steel bainites.

The optical micrograph of Fig. 4.20 was taken at the highest useful magnification of 1000×. At higher magnifications, obtained by enlargement, no new structural variation is apparent, because these magnifications are beyond the 0.2 μm (200 nm, or 8×10^{-6} in.) resolution limit of the optical microscope. The plate and particle structures of the cementite in the pearlite and bainite regions of Fig. 4.20 are not seen because both features are too small to be resolved in the optical microscope. As mentioned before, the cementite particles are put into relief by etching and they scatter the light, making the image dark in the optical microscope. The pearlite appears darker than the bainite because the light scattering is more effective from the arrays of fine plates than from the small particle of the bainite.

There is a type of electron microscope that became available in the late 1960s, called a scanning electron microscope (SEM), that produces images of surfaces with resolution much higher than the optical microscope, down to 2 nanometers (nm), where 2 nm = 0.002 μm (8×10^{-8} in.). Figure 4.21 is an SEM micrograph of the pearlite at the point located in Fig. 4.20, and Fig. 4.22 is an SEM micrograph of the bainite at the point located in Fig. 4.20. In each micrograph, the cementite is the dark phase. The difference in the two microstructures is clearly illustrated in these micrographs. As the cooling rate is increased, the spacing of the cementite plates in pearlite decreases. The plate spacing in Fig. 4.21 is 0.05 μm (50 nm, or 2×10^{-6} in.), which is the smallest spacing ever formed by fresh, undeformed pearlite. As illustrated in Fig. 4.22, the cementite in bainite is generally in the form of ribbon-shaped filaments and/or ribbon-shaped particles. Both the ribbon axes of the bainite and the plane of the plates of pearlite are aligned close to the direction of growth of these structures.

In the late 1920s, Edgar Bain initiated the study of quenched steels by a method called isothermal transformation. Isothermal means constant temperature. This method generally requires the use of thin pieces of steel. First, the thin steel is heated to transform it to austenite, and then it is quenched into a liquid quench bath, usually molten salt, held at the desired isothermal temperature. Because the steel is thin, it will cool to the quench bath temperature throughout its volume before any of the austenite has decomposed. The steel is now held in the molten bath until the austenite is completely transformed, and it is then cooled to room temperature and examined in a microscope. With this method, it is possible to map out what structures can form

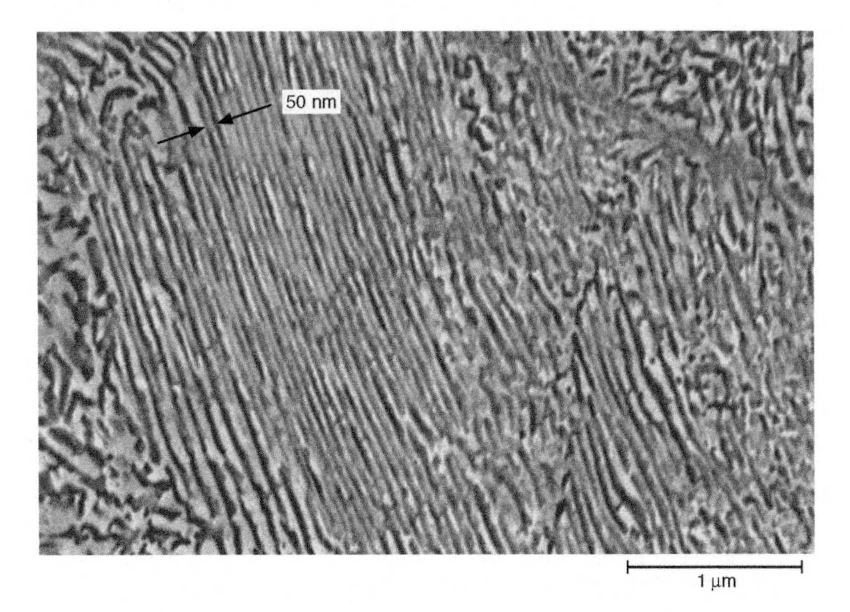

Fig. 4.21 SEM micrograph of pearlite at location shown in Fig. 4.20. Original magnification: 21,000 ×

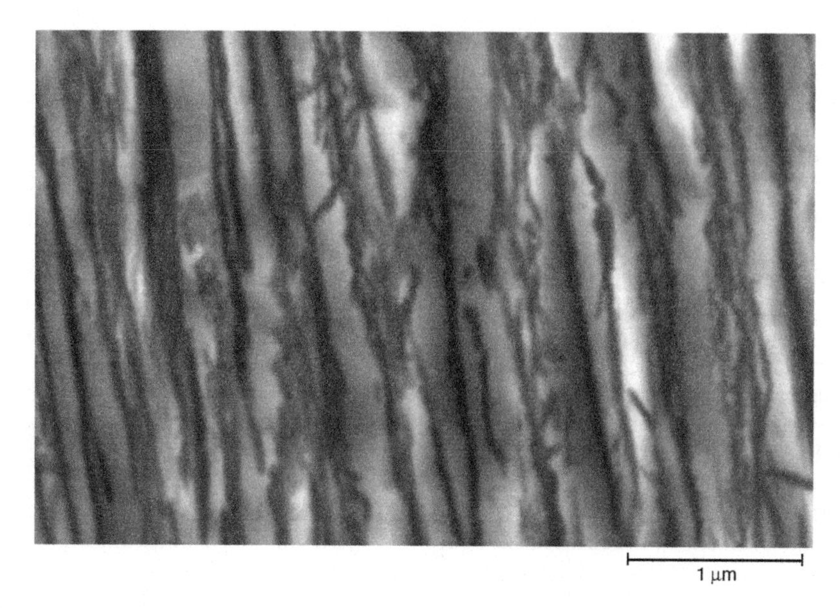

Fig. 4.22 SEM micrograph of bainite at location shown in Fig. 4.20. Original magnification: 20,000 ×

from austenite at all the temperature-composition coordinates on the phase diagram below the A_1 temperature. The results are given in Fig. 4.23.

Using this technique, Bain and his coworkers discovered that there is a temperature range below which pearlite will not form and above which martensite will not form. In this range, they discovered that a new structure forms, which, after approximately 15 years, came to be called bainite. As shown in Fig. 4.23, bainite has two main forms, which are termed upper bainite and lower bainite, in reference to the temperature range in which they form. The bainite of Fig. 4.22 is upper bainite. The main difference between these two structures is that the carbide particles and filaments are much finer and more closely spaced in lower bainite. Pearlite never has any retained austenite in its structure. Bainite, however, may have significant amounts of retained austenite, particularly in certain alloy steels. When the element silicon is added to steels at levels above the usual impurity level, it is found to have a significant effect on bainite. Silicon is present in higher levels in the spring steel AISI 9260 and in all cast irons. It causes the bainites in these metals to contain significant amounts of retained austenite, reduced amounts of carbides, and it changes the type of carbide in the lower bainite from Fe_3C (cementite) to $Fe_{2.5}C$ (epsilon carbide). Epsilon carbide does not appear on the iron-carbon equilibrium phase diagram because it is not a stable phase. It is a metastable phase and can only be formed in quenched steels. It will be encountered again when tempering is discussed.

Figure 4.23 can be used to illustrate an important aspect of pearlitic steels that is often not well understood. Just because a microstructure has the pearlite form does not mean that the composition of the steel has the pearlite composition of 0.77% C. Consider a 1095 steel. The diagram shows that if this steel is quenched into the temperature range of approximately 550 to 650 °C (1020 to 1200 °F) and held until all austenite has decomposed, it will become a 100% pearlitic structure. The difference between the structure of this steel and a 1077 pearlite is that the cementite plates of the 1095 pearlitic steel are slightly fatter than those of the 1077 pearlite. Similarly, it is possible to make a 1060 steel with a fully pearlitic structure by quenching to the 550 to 650 °C (1020 to 1200 °F) range, and here the difference is that the cementite plates are thinner than in the 1077 pearlite. In both cases, the change in the plate widths are too small to be observed in the optical microscope or even in the SEM.

Figure 4.23 does not tell what structures will form on continuous cooling. Continuous cooling means the sample is cooled from the austenite temperature, not by quenching, but by removal from the furnace and cooling at slower rates, such as air cooling, or simply a slow quench, such as perhaps oil quenching. Air cooling is often called *normalizing*. As the name implies, continuous cooling means that the steel

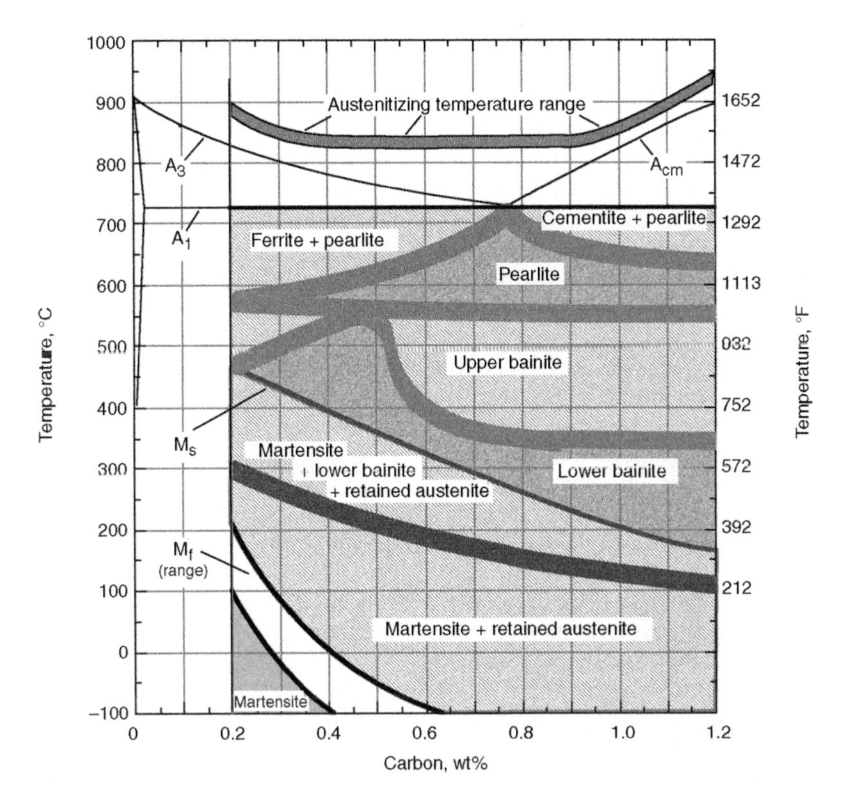

Fig. 4.23 Austenite decomposition products for plain carbon steels during isothermal transformation (quenching and holding) at various temperatures below A_1

temperature falls in a continuous manner rather than abruptly plummeting, as happens in quenching. Experiments show that on fast air cooling, a 1095 steel will become fully pearlitic but a 1060 steel will not. The 1060 steel will have ferrite in the prior-austenite grain boundaries surrounded by pearlite, similar to that of Fig. 4.6. As cooling rates are increased above that of fast air cooling, bainite will begin to appear, and a great variety of microstructures are possible involving mixtures of ferrite, pearlite, bainite, and eventually martensite.

Spheroidized Microstructures

The cementite phase present in most steels is in the form of fine plates of pearlite for hypoeutectoid steels and in this form plus thicker plates and globules formed on austenite grain boundaries in hypereutectoid steels. These cementite shapes result because they are the forms that cementite assumes when austenite transforms to pearlite at the A_{r_1} temperature and when it nucleates on austenite grain boundaries in hypereutectoid steels at the $A_{r_{cm}}$ temperature. It is possible, however, to heat treat steels and change the shape of the cementite into a spherical form, sometimes called spheroidite. Cementite is very hard but very brittle, and by changing its form to isolated spheres in a matrix of, for example, ferrite or tempered martensite, the mechanical properties of the steel come closer to matching that of the matrix phase: soft and machinable with a ferrite matrix or strong and less brittle with a martensite matrix.

There are two heat treating processes that are effective for producing a spheroidized cementite structure. In the first process, the steel is simply heated into the fully austenite region and then quenched to martensite. For both hypo- and hyper-eutectoid compositions, the steel is then heated to a temperature just below the A_1 temperature for approximately 1 h, and on cooling, a microstructure of fine spheres of cementite in a ferrite matrix emerges. The room-temperature martensite is not a stable phase (it does not appear on the phase diagram), and at elevated temperatures, it decomposes into fine carbides (cementite) in a ferrite matrix. For hypereutectoid steels, spheres of cementite can be produced in a martensite or bainite matrix by heating the quenched steel to temperatures between A_1 and A_{cm} for approximately 1 h, where the martensite will decompose into fine carbides in an austenite matrix. The martensite or bainite matrix is then formed from the austenite by quenching at the appropriate rate. The volume fraction of spherical cementite in the steel depends on the temperature of the 1 h hold, ranging from a maximum amount for temperatures just above A_1 to near zero for temperatures just below A_{cm}.

The second method is most effective with hypereutectoid steels. It requires no quenching and is therefore used in industry to produce the spheroidized structures of cementite + ferrite in which these steels are supplied from the steel mill. In this method, a trick is used that prevents pearlite from forming on cooling austenite below the A_1 temperature. The trick is to produce small particles of cementite in the austenite and then cool it below A_1 at a slow rate (usually a little slower than air cooling). This treatment causes the austenite to transform directly into a spheroidized cementite + ferrite structure rather than into the plate-shaped pearlite structure. (The transformation is sometimes called the divorced eutectoid transformation, or DET.) Small particles of cementite are naturally formed in the austenite of a hypereutectoid steel when it is heated above A_1 and below A_{cm}. Experiments show that such is the case as long as the starting steel does not contain large particles of cementite (for example, if it consists of pearlite, bainite, martensite, or is already spheroidized as ferrite + cementite). In addition, the steel cannot be heated above approximately 800 to 840 °C (1470 to 1540 °F) during austenitization to have the DET dominate the pearlite transformation. The method will also work with hypoeutectoid steels but requires more control. The reason is that when a hypoeutectoid steel is heated above A_1, the cementite plates of its pearlite are transformed completely into austenite in progressively shorter time frames as the temperature is raised. Because the method requires particles of cementite to be present in the cooling austenite, hypoeutectoid steels cannot be held for very long above the A_1 temperature before cooling. An effective way to achieve spheroidization with hypoeutectoid steels is to cycle the steel approximately 50 °C (90 °F) above and below A_1 a few times.

Figure 4.24 presents an example of a 52100 bearing steel produced by the second method described in the previous paragraph. The steel was originally in a pearlitic condition with no cementite other than in the pearlite. It was then held at 795 °C (1460 °F) for 30 min and furnace cooled at 370 °C/h (670 °F/h) to 680 °C

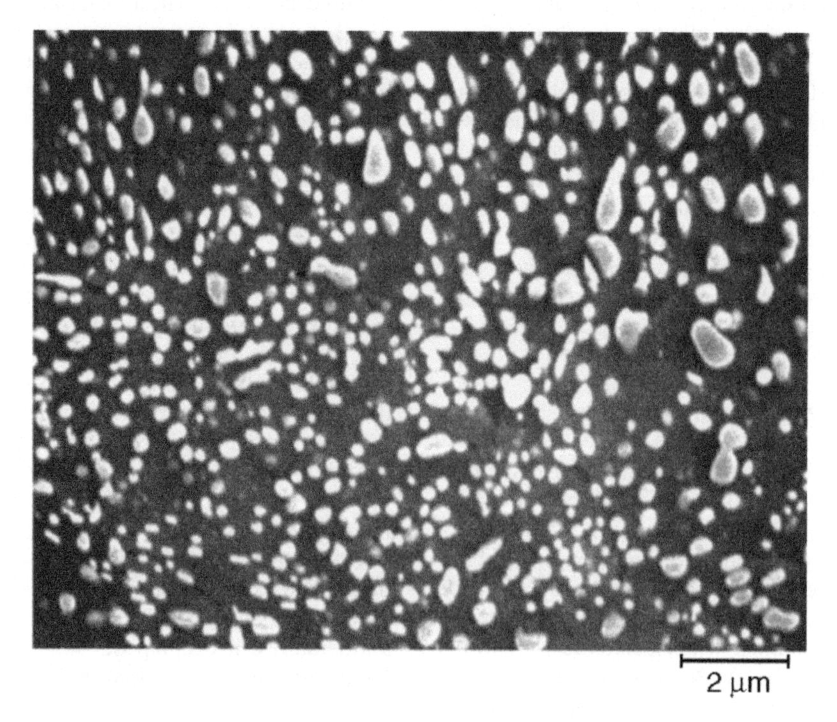

Fig. 4.24 SEM micrograph of a 52100 steel showing fine spherical cementite in a ferrite matrix. Original magnification: 6000 ×

(1260 °F), where the DET had been completed, and then air cooled to room temperature. The slow cooling to 680 °C (1260 °F) caused the austenite to transform to ferrite + cementite by the DET rather than by the pearlite transformation, thereby forming a completely spheroidized structure of very fine cementite particles in a ferrite matrix.

Examples of spheroidized steels include most hypereutectoid steels, such as bearing steels and tool steels, which are supplied from the mill with a spheroidized structure of cementite in a ferrite matrix to facilitate machining. Also, bearing steels and tool steels are often used in practice for applications in which the hard cementite particles enhance wear resistance. Often, these steels are heat treated to produce a spheroidized cementite structure in a matrix controlled by the final cooling, which may consist of fine pearlite, bainite, or tempered martensite.

Summary of the Major Ideas in Chapter 4

1. The term *microstructure* refers to the distribution of the grains in the steel, and its description includes such things as grain size and shape and the identification of the microconstituents present. Microstructure is determined mainly with the optical microscope by examination of polished and etched surfaces. All of the important microconstituents of steel (the phases ferrite, austenite, cementite, and fresh martensite, plus the phase mixtures pearlite, bainite, and tempered martensite) have characteristic appearances on a polished and etched surface, which allows evaluation of the microstructure in the optical microscope. Mechanical properties of steels are strongly influenced by their microstructure.
2. The room-temperature microstructure of slow-cooled (nonquenched) steels depends strongly on the carbon content of the steel. Eutectoid steels (0.77% C) are usually all pearlite (Fig. 4.3). Hypoeutectoid steels (%C less than 0.77) are often arrays of pearlite grains with ferrite in their boundaries (Fig. 4.6). Hypereutectoid steels (%C greater 0.77) are often arrays of pearlite grains with cementite in their boundaries (Fig. 4.8). At the low end of the hypoeutectoid steels (%C less than approximately 0.2), ferrite grains become the dominant component, with the pearlite distributed in various forms. Wrought steels often show a banded microstructure, with alternating bands of ferrite grains and pearlite grains aligned in the direction of the prior mechanical deformation (Fig. 4.9).
3. The cementite in slow-cooled steels generally is contained as thin plates in the pearlite component. However, it is possible to change the pearlitic cementite into arrays of small, isolated spheres in a matrix of ferrite grains using special heat treatments described in the section "Spheroidized Microstructures" in this chapter. Such structures are referred to as spheroidized steels (Fig. 4.24) and are often the form in which high-carbon steels are delivered from the mill, because they are much easier to machine than when the cementite is in the pearlitic form.
4. When austenite is rapidly cooled by quenching, two additional microconstituents may appear: martensite and bainite.
5. Martensite is a nonequilibrium phase (it does not appear on the phase diagram). Martensite wants to be body-centered cubic ferrite, but the carbon in the austenite distorts the crystal structure to body-centered tetragonal. The more carbon in the austenite, the more the distortion and the harder (stronger) the martensite (Fig. 4.13).
6. Martensite has two appearances in the optical microscope: lath and plate. Lath forms from 0 to 0.6% C and plate at 1% C and above, with a mixed structure in between (Fig. 4.15).
7. The martensite formed in the quench bath is called fresh martensite. If the %C is greater than approximately 0.3, it will be too brittle for most uses. Therefore, most martensites are tempered by heating to low temperatures. The tempering causes very small carbides to form in the martensite, which reduces strength but enhances ductility. It also makes tempered martensite etch dark in the optical microscope.
8. To form martensite, austenite must be quenched to temperatures below the M_s temperature. As the quench temperature drops below M_s, progressively more martensite forms until the M_f temperature is reached, where 100% martensite forms. Between M_s and M_f, a structure of martensite plus retained austenite is found (Fig. 4.15 b, 4.16).

9. Both M_s and M_f drop as %C is increased, and samples quenched to room temperature can contain appreciable amounts of retained austenite in higher carbon steels. The data of Fig. 4.17 and 4.18 show that significant amounts of retained austenite begin to form in plain carbon steels quenched to room temperature for carbon contents above approximately 0.4% C.
10. At quench rates a little slower than that required to form martensite, microconstituents called bainite form. Bainite is similar to pearlite in that it consists of ferrite plus carbides. The carbide component consists of aligned filaments or particles, as opposed to the aligned plates in pearlite.
11. Bainite occurs in two forms, upper bainite and lower bainite, depending on the temperature where it forms. The carbides are finer in lower bainite. Strengths approaching martensite, with toughness often better than tempered martensite having the same hardness, can be obtained.
12. A unique property of martensite is that it forms much more rapidly than any of the other austenite products—ferrite, cementite, pearlite, or bainite. Martensite grows at approximately half the speed of sound, so when the austenite temperature is below M_s, martensite forms in milliseconds.
13. When a steel is cooled from the austenite region, the phase diagram tells which austenite products want to form first: ferrite in hypoeutectoid steels, pearlite in eutectoid steels, and cementite in hypereutectoid steels. These products form first on austenite grain boundaries. To form 100% martensite, a steel must be quenched so fast that these products do not form on the austenite grain boundaries. If one of these products does form on the austenite grain boundaries before the austenite reaches the M_s temperature, it will be trapped there by the martensite, which forms very quickly, resulting in structures similar to those in Fig 4.19 and 4.20.
14. If steels are quenched to a constant temperature in a salt bath so quickly that the austenite has not yet decomposed, then the austenite will transform isothermally, at one temperature. Such isothermal transformation of steels has been studied extensively, and it was found that the austenite product or products that form depends on the quench temperature and the %C in the steel, as summarized in Fig. 4.23.

CHAPTER 5

Mechanical Properties

PROBABLY THE MOST IMPORTANT properties of metals that account for their widespread use are their mechanical properties. These properties include a combination of very high strength with the ability to bend rather than break. Various tests have been developed to characterize the strength and ductility (a measure of bendability) of metals and other materials. The majority of such tests used to characterize the mechanical properties of steel are discussed in this chapter.

The Tensile Test

Consider the simple example of pulling apart a small spring a fair amount and then releasing it. The spring springs back to its original length. Now, repeat the process, but this time, pull on the spring and extend it by a large amount. As the spring becomes overextended relative to its design, the increasing force needed to continue extending the spring suddenly drops, and the spring somewhat "gives." When the spring is released, it does not return to its original length. The spring is now permanently stretched to a longer length and probably ruined, at least for its intended purpose.

Long ago, engineers developed a test to evaluate the strength of metals that is related to this simple experiment. A length of the metal, usually a round cylindrical rod, is pulled apart in a machine that applies a known force, F. The machine has grips attached to the ends of the cylindrical metal rod, and the force is applied parallel to the axis of the rod, as shown schematically at the right in Fig. 5.1. As the force increases, the rod gets longer, and the change in length is represented as Δl, where the symbol Δ means "a change in," and the l refers to the original length of the rod. A force of 100 pounds is applied to two rods of the same material, where one is thin and the other thick. The thin rod will elongate more. To compare their mechanical properties independent of rod diameter, engineers define a term called *stress,* which is simply the force divided by the cross-sectional area of the rod. When the same stress is applied to the thin and thick rods, they elongate the same amount, because the actual force applied to the thick rod is now larger than that applied to the thin rod by an amount proportional to its larger area. Because stress is force per area, it has units of pounds per square inch (psi), or, more commonly, kpsi. The symbol "k" for kilo means 1000 times, so 1 ksi = 1000 psi, or 1000 pounds per square inch. For example, 50 ksi is the same as 50,000 psi.

When a metal is pulled along its axis, the applied force is called a tensile force, and the machine that applies the force is called a tensile test machine. Figure 5.1 presents a typical result obtained from a tensile test of a metal. The applied stress is plotted on the vertical axis. The change in length is plotted on the horizontal axis, but it is customary to plot the fractional change in length, $\Delta l/l$, as shown in Fig. 5.1. The fractional change in length is called the *strain*, and the diagram in Fig. 5.1 is called a stress-strain diagram.

The stress-strain diagram can be divided into regions, as shown in Fig. 5.1. the *elastic* region and the *plastic* region. As the stress on the metal rod increases, the rod elongates, and just like the spring, as long as the stress is not too high, release of the stress returns the rod to its original length. This is called elastic deformation. However, if the applied stress reaches a critical level, called the *yield stress* (YS), the metal "gives," just like in the spring example, and two things happen: (1) the increase in stress needed to produce a given small increase in strain becomes

> **NOTE**
>
> Industry in the United States continues to be a holdout in the use of ksi or psi for the units of stress. The rest of the world has adopted an International System of units, called SI units. In this system, stress is given in units of megapascals, or MPa. One pascal is a stress of one newton per square meter, and a megapascal is one million pascals. To convert ksi to MPa, simply multiply stress in ksi by 7 to obtain an excellent approximation of the stress in MPa, for example, 50 ksi ≈ 350 MPa. (The correct multiplier is 6.895, so 50 ksi = 345 MPa.)

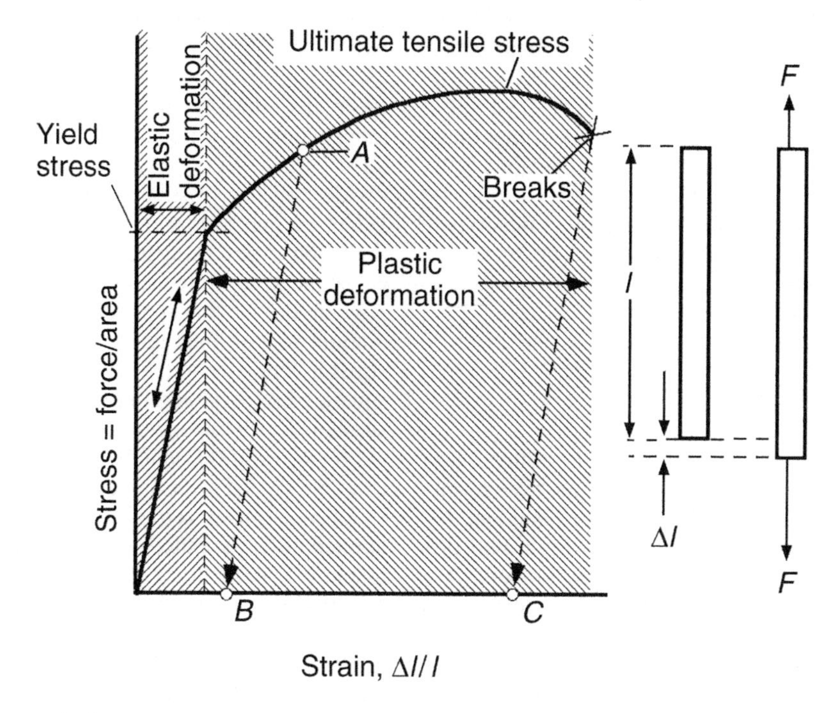

Fig. 5.1 Stress-strain diagram

lower, and (2) on release of the stress, the metal is permanently elongated, as shown by the arrowed line *A-B* in Fig. 5.1. In this case, the rod was stressed to point *A,* and after releasing the stress, the rod had elongated from its original length by a percent given as *B* × 100. As shown in the figure, the increase in stress needed to continue elongating the rod reaches a maximum value in the plastic region and then drops a little before the stress is able to break the rod in two. This maximum stress value is usually called the *ultimate tensile strength* (UTS) or often just the tensile strength (TS).

The stress-strain diagram also provides an additional measure of the mechanical properties of a metal. The ductility is the amount of elongation that occurs after the stress increases beyond the yield stress and before the sample breaks. This elongation is sometimes called permanent elongation, because it remains in the sample after breakage and can be measured easily. The permanent elongation in the sample in Fig. 5.1 after the elastic strain is relaxed is given by point *C*. By simply multiplying the strain at *C* by 100, the *percent elongation* of the metal is obtained. Figure 5.2 presents possible stress-strain diagrams for a ductile metal and a brittle metal. The breakage strain is much larger for the ductile metal, and its percent elongation is much larger. The figure also shows that the diameter of the bar at the fracture surface of a brittle failure remains close to its original value, while that of the ductile failure is reduced. This reduction in diameter by plastic flow near the fracture surface is referred to as necking, and it develops in ductile metals just before fracture. (In addition to percent elongation, ductility is often characterized by percent reduction in area, which is simply the percent by which the original cross-sectional area of the bar is reduced at the fractured neck.) The concept of ductile versus brittle behavior is fairly obvious. Place a metal rod in a vice and beat on the exposed end with a hammer. A brittle metal will break almost immediately, whereas a ductile metal can be bent severely by the hammer blows and may not break after bending 90° or more.

When a bladesmith shapes a knife blade by forging, each blow of the forging hammer causes some permanent deformation of the metal. On each blow, the stress produced in the metal is exceeding the yield stress. As anyone who has hand hammered a metal or bent a metal rod back and forth probably realizes, after the metal has been deformed a little, it becomes more difficult to deform. This effect is called work hardening and can be understood from the stress-strain diagram in Fig. 5.3. The stress-strain diagram for the original (undeformed) metal is shown by the dashed line. If the original rod is deformed to point *A*, it will have strain equivalent to point *B* when the stress is released. If this bar is retested, one finds that its YS has increased from the original value up to point *A*, as shown on the diagram. This increase in yield strength is a measure of the work hardening resulting from the original deformation (work) put into the metal on the first deformation. Notice that although the metal is now stronger in the sense that it has to be stressed more before it will undergo plastic deformation, its UTS is not changed significantly, and it is less ductile, that is, its percent elongation to failure has gone down. It is a general characteristic of metals that processes that increase their yield strength will decrease their ductility. People who have worked metals know this, because heavily work-hardened metals often break easily. A familiar example of this behavior is the commonly used method of breaking a wire by bending it back and forth many times.

The mechanical properties of metals are often characterized by listing values of their yield strength, tensile strength, and percent elongation. The data in Table 5.1 illustrate such a characterization for some plain carbon steels. The units of stress in the table are MPa (ksi). The data on the normalized steels in Table 5.1 illustrate two general characteristics of steels: (1) increasing the %C in steel increases both YS and TS, and (2) increasing %C drops the ductility (percent elongation). Comparing the as-rolled properties to

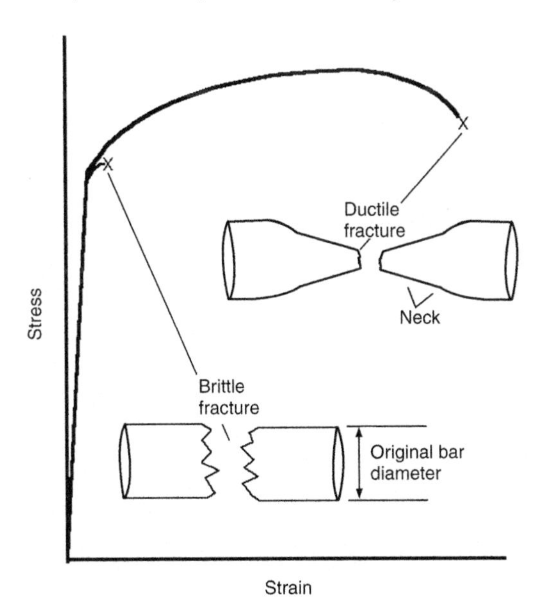

Fig. 5.2 Stress-strain diagram showing tensile stress for a brittle and a ductile metal

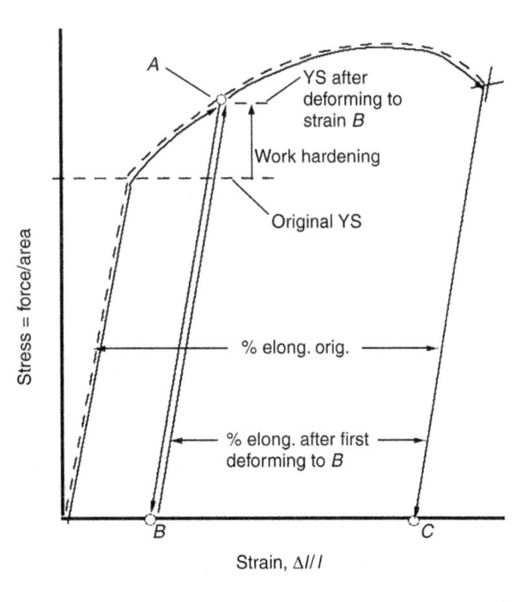

Fig. 5.3 Increase of yield stress (ys) after plastic deformation (work hardening)

the normalized properties illustrates the effects of work hardening: yield strength goes up at higher carbon levels, tensile strength is changed only slightly, and percent elongation drops a little.

The Hardness Test

A problem with the tensile test is that the metal is destroyed by the test. Another test that characterizes the strength of a metal but does not destroy the metal is the hardness test. This test is used widely because it is quick and can be applied to parts that can then be placed into service. Over the years, several different, useful hardness tests have evolved, and the essential features of these tests may be explained by reference to Fig. 5.4. A hard material, called the indenter, is forced into the metal surface with a fixed load (weight). The metal region located under the indenter point is deformed to strains well into the plastic region of Fig. 5.1, so that permanent deformation is generated, causing a crater (called an indent) to be left in the metal surface. The hardness is then defined by some number that is proportional to the size of the indent. In some techniques, the size of the indent is measured from its diameter, and in others, it is measured from the depth of the indent.

Rockwell Hardness. Table 5.2 presents a comparison of various indentation hardness tests that are used for measuring the hardness of steels and other metals. The Rockwell hardness tester is probably the most widely used test in the United States. The table shows the three most common scales for the Rockwell test. Both the "C" and "A" scales use a conical diamond indenter, with the only difference being that the "A" scale uses a lighter load. These testers provide a very quick and easy measurement, with the machine measuring the depth of penetration automatically and providing the hardness value either on a dial or, on newer machines, as a digital readout. The smaller load for the "A" scale reduces the penetration depth and is often used to measure hardness on steel surfaces that have been case hardened with a thin, hard layer on the surface. The reported hardness values are referred to variously as R_C or HRC and similarly, R_A or HRA. The indenter for the "B" scale is a 1.6 mm (0.063 in., or 63 mils) hardened ball, and this scale is sometimes used for soft steels.

Brinell Hardness. A more reliable hardness test for soft-to-medium hard steels is the Brinell test. It uses a larger ball, either hardened steel or tungsten carbide, for indentation, and the diameter of the indent is measured. This test is a two-step test in that the indent must be made first and then the indent diameter measured optically. The size of the indent should be between 3 to 6 mm (0.12 to 0.24 in., or 120 to 240 mils). The Brinell hardness number (referred to variously as HB or BHN) is then determined either from a table or by using an equation on a calculator. An advantage of the Brinell test is that it measures hardness over a much larger area than the Rockwell

Table 5.1 Mechanical properties of some steels

Steel	Treatment	Yield stress MPa	Yield stress ksi	Tensile stress MPa	Tensile stress ksi	Elongation, %
1020	Normalized	345	50	440	64	36
1040	(air cooled)	370	54	595	86	28
1095		505	73	1015	147	9.5
1020	As rolled	330	48	450	65	35
1040		415	60	620	90	25
1095		570	83	965	140	9

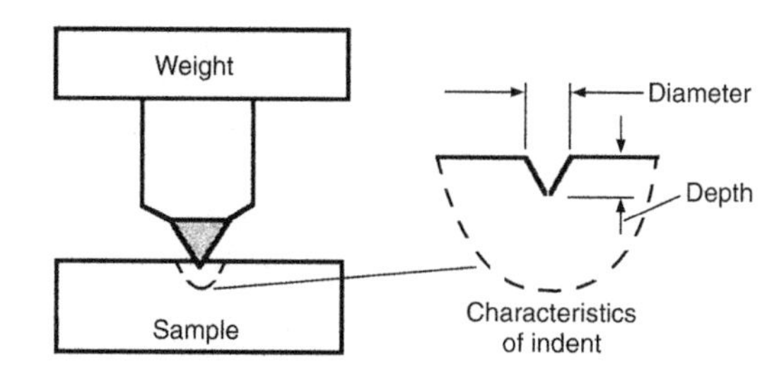

Fig. 5.4 Schematic of essential features of hardness testing

Table 5.2 Comparison of indentation hardness tests

Hardness test	Weight (a) kg	Weight (a) lb	Indenter	Measure	Use
Rockwell C	150	330	120° diamond cone	Depth	Medium to high hardness
Rockwell A	60	130	120° diamond cone	Depth	Hardened surfaces
Rockwell B	100	220	1.6 mm ($\frac{1}{16}$ in.) ball	Depth	Soft
Brinell	500–3000	1100–6615	Large ball, 10 mm (0.4 in.)	Diameter	Soft to medium hardness
Microhardness	1 g–1 kg	0.04 oz–2.2 lb	136° diamond pyramid, diamond pyramid hardness, or Knoop	Diagonal	Microconstituents
Vickers	10–120	22–265	136° diamond pyramid	Diagonal length	Soft to high hardness

(a) g=1 gram; kg=kilogram=1000 grams

test, which is useful for materials with coarse microstructures, such as cast irons. However, this is also a disadvantage for application where the testpiece must be small, such as the tooth on a gear. Another disadvantage is that the test is not useful at the higher hardness ranges of steels. See Ref 5.1 and 5.2 for further discussion of hardness testing.

Table 5.3 presents information (adapted from ASTM standard E 140) listing HRC values measured on steel and the corresponding equivalent hardnesses of various other tests for steel. Remember that these equivalent hardnesses apply to steels only; that is, do not use them for aluminum, copper, and other nonferrous alloys.

Vickers/Diamond Pyramid Hardness. There is another test, developed in England and generally used there in preference to the Rockwell test, called the Vickers hardness test. This test is similar to the Brinell test in that it is a two-step test measuring the diameter of the indent, but with a diamond indenter shaped in the form of a pyramid. The pyramid geometry produces an indent having a square shape, and the size is measured by the length of the diagonals of the square indent. The average diagonal length and the load used are plugged into a formula that calculates the Vickers hardness (HV) number, also called the diamond pyramid hardness (DPH) number. In Table 5.3, the ranges of hardnesses where the various tests are not applicable are shown either with blanks or by using parentheses. Notice that the Vickers test is the only one that applies at all hardness levels.

Microhardness: Diamond Pyramid and Knoop. As shown in the second row from the bottom of Table 5.2, there is a microhardness test available. The test uses very light loads and makes indent diameters small enough to fit into microsized regions. It uses a microscope that allows the indent to be positioned at desired locations on a microstructure, and the size of the indent is measured with the same microscope. There are two different indenters used in microhardness testers: the diamond pyramid indenter of the Vickers test and a special-shaped indenter called a Knoop indenter. Figure 5.5 illustrates the use of the diamond pyramid (Vickers) hardness test on a ferrite-pearlite banded 1045 steel. Indents were placed in a ferrite band and in a pearlite band. Notice that the indent is smaller in the pearlite band, indicating that the pearlite is harder than the ferrite. The weight used in this test had to be adjusted to 50 g to keep the indent small enough to fit inside the bands. The

Table 5.3 Approximate equivalent hardness numbers for several hardness tests on steels

					Ultimate tensile strength	
HRC	HV	HB(a)	HRA	HRB	MPa	ksi
68	940	...	85.6	...	...	...
67	900	...	85.0	...	...	...
66	865	...	84.5	...	...	...
65	832	(739)	83.9	...	...	...
64	800	(722)	83.4	...	...	...
63	772	(705)	82.8	...	...	...
62	746	(688)	82.3	...	...	...
61	720	(670)	81.8	...	...	...
60	697	(654)	81.2	...	...	...
59	674	(634)	80.7	...	...	...
58	653	615	80.1	...	...	...
57	633	595	79.6	...	...	...
56	613	577	79.0	...	...	...
55	595	570	78.5	...	2075	301
54	577	543	78.0	...	2013	292
53	560	525	77.4	...	1951	283
52	544	512	76.8	...	1882	273
51	528	496	76.3	...	1820	264
50	513	481	75.9	...	1758	255
49	498	469	75.2	...	1696	246
48	484	455	74.7	...	1634	237
47	471	443	74.1	...	1579	229
46	458	432	73.6	...	1531	222
45	446	421	73.1	...	1482	215
44	434	409	72.5	...	1434	208
43	423	400	72.0	...	1386	201
42	412	390	71.5	...	1338	194
41	402	381	70.9	...	1296	188
40	392	371	70.4	...	1248	181
39	382	362	69.9	...	1213	176
38	372	353	69.4	...	1179	171
37	363	344	68.9	...	1158	168
36	354	336	68.4	(109.0)	1117	162
35	345	327	67.9	(108.5)	1082	157
34	336	319	67.4	(108.0)	1055	153
33	327	311	66.8	(107.5)	1027	149
32	318	301	66.3	(107.0)	1000	145
31	310	294	65.8	(106.0)	979	142
30	302	286	65.3	(105.5)	951	138
29	294	279	64.7	(104.5)	931	135
28	286	271	64.3	(104.0)	910	132
27	279	264	63.8	(103.0)	883	128
26	272	258	63.3	(102.5)	862	125
25	266	253	62.8	(101.5)	841	122
24	260	247	62.5	(101.0)	827	120
23	254	243	62.0	100.0	807	117
22	248	237	61.5	99.0	786	114
21	243	231	61.0	98.5	772	112
20	238	226	60.5	97.8	758	110
(18)	230	219	...	97.8	731	106
(16)	222	212	...	95.5	703	102
(14)	213	203	...	93.9	676	98
(12)	204	194	...	92.3	648	94
(10)	196	187	...	90.7	621	90
(8)	188	179	...	89.5	600	87
(6)	180	171	...	85.5	579	84
(4)	173	165	...	85.5	552	80
(2)	166	158	...	83.5	531	77
(0)	160	152	...	81.7	517	75

Note: For carbon and alloy steels in the annealed, normalized and quenched and tempered condition and for austenitic steels. Values in parentheses are beyond the normal range.
(a) 3000 kg (6615 lb) load and tungsten carbide ball

microhardness test uses the same equation as the Vickers test to calculate hardness, HV, and the values found were HV = 260 in pearlite and 211 in ferrite. From Table 5.3, these correspond to HRC values of 24 in pearlite and 13.6 in ferrite.

Figure 5.6 presents results from a similar test on a ferrite-pearlite banded 1144 steel, using the Knoop indenter. (Like a 1044 steel, 1144 contains 0.44% C, but it also contains elevated sulfur levels, and, as a result, contains significant

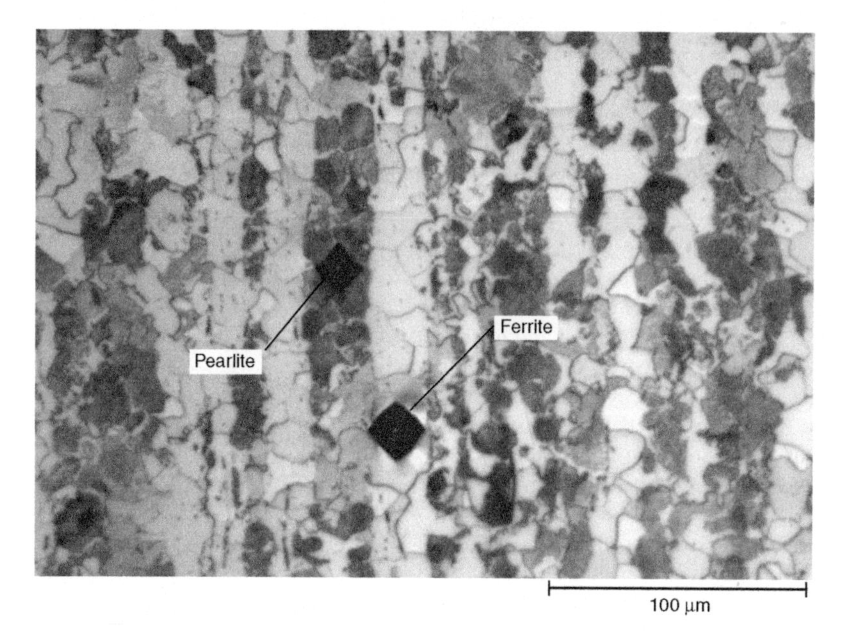

Fig. 5.5 Diamond pyramid (Vickers) hardness indentations in ferrite and pearlite of banded 1045 steel. Original magnification: 230×

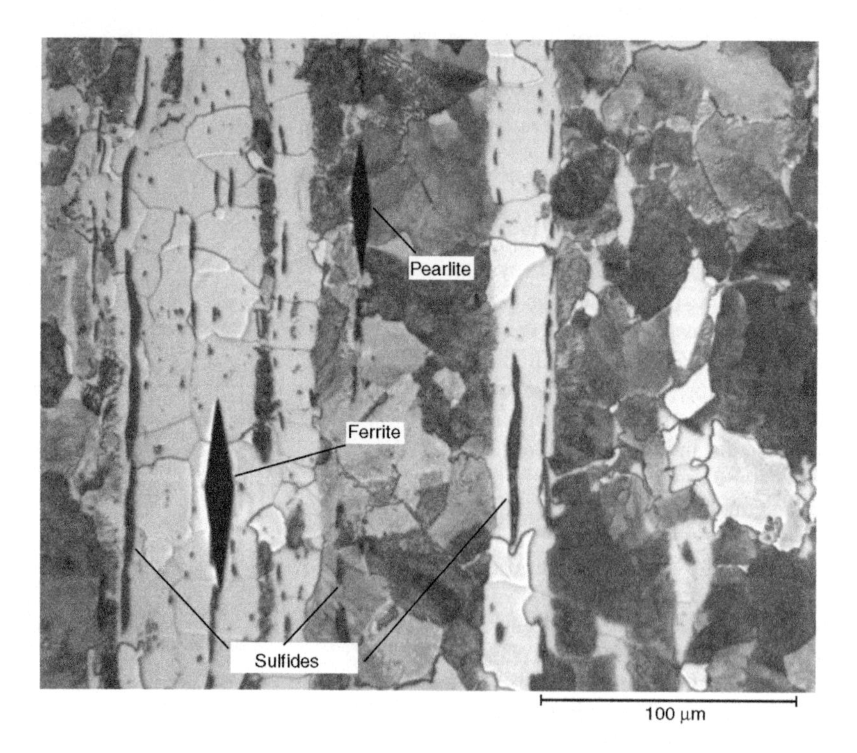

Fig. 5.6 Knoop indentations in ferrite and pearlite of banded 1144 steel. Original magnification 230×

amounts of sulfides, as shown in the figure. The sulfides are ductile at the hot working temperature and become elongated during hot working. The sulfides make the steel machine more easily.) Notice that the Knoop indenter is quite oblong in shape. This allows it to fit into thin layers more easily than a Vickers indent. Therefore, it is particularly useful for microstructures with thin layer morphologies. Hardness is characterized with its own set of Knoop hardness numbers, which can be related to HRC values in tables similar to Table 5.3 (Ref 5.1).

The indentation hardness numbers correlate well with the UTS of steels that have been quenched and tempered. These data are given by the UTS column in Table 5.3, and they are plotted in Fig. 5.7. The correlation becomes less reliable at HRC values above 55, but the figure shows the UTS values that are obtained by extrapolating the curve to HRC values of 60 and 65. The results show that when a steel is hardened into the range of HRC = 60 to 65, the corresponding tensile strengths are extremely high, 2450 and 2850 MPa (351 and 408 ksi) at HRC=60 and 65, respectively. For comparison, the highest tensile strengths of commercial aluminum, copper, and titanium alloys are approximately 580, 1240, and 1330 MPa (83, 177, and 190 ksi), respectively. The highest of these strengths is exceeded by steels hardened to an HRC value of only approximately 43.

Chapter 4 discussed the various microstructures that occur in steel. Table 5.4 presents estimates of the maximum hardness for the various microstructures. Notice that both ferrite and austenite are fairly soft. The hardness of pearlite is controlled by its spacing, and the values of 40 to 43 given in Table 5.4 are for pearlite with the finest spacing that can be obtained on cooling, which was illustrated in Fig. 4.21. Finer spacing can be obtained in pearlitic steels by mechanically deforming this pearlite to thin wire form and winding into cable. Such pearlitic steel cable is used in wire rope, which has tensile strengths to 2200 MPa (320 ksi). Quenched and tempered steel wire can be made to this strength level, but experience has found the pearlitic wire to be tougher, and for that reason, it is the wire of choice for such industrial applications as bridges and crane cables. Lower bainite has hardnesses approaching that of martensite and also finds industrial uses because of its slightly superior toughness to that of quenched and tempered steels. Martensite has the highest strength and hardness of all, and, as shown in Fig. 4.13, the hardness of fresh martensite depends on the %C in the steel. Fresh martensite is rarely used industrially because of its lack of toughness. It is tempered (heated to modest temperatures), which lowers its strength but increases its toughness. Measurement of hardness is a major method of control of the tempering process.

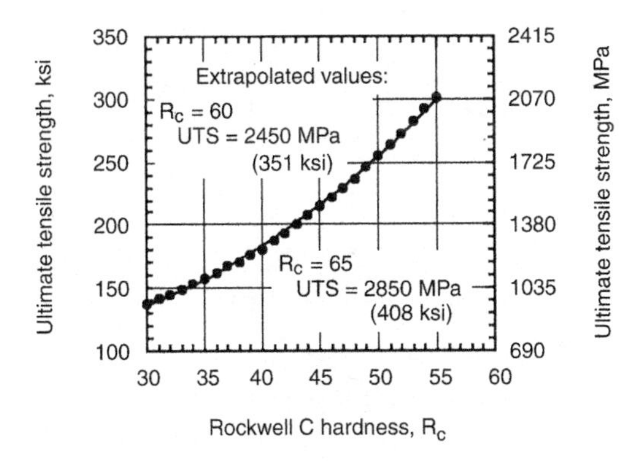

Fig. 5.7 Plot of data from Table 5.3 showing the ultimate tensile stress (UTS) of quenched and tempered steels as a function of Rockwell C hardness (HRC)

The Notched Impact Test

During WWII, the need for higher production rates of warships led to construction using welded steel plate rather than the standard riveted construction. The ships were called liberty ships, and brittle failure in the welded plates of these ships produced spectacular failures, where the entire ship broke in half, with catastrophic results. Cracks began at local points in the welded joints and propagated around the ship, passing from plate to plate and causing failure by a brittle mode, that is, little to no plastic flow. This brittle behavior was not detected by a sudden loss in ductility in the simple tensile test. These disasters led to a wide appreciation of the fact that ductility as measured by the tensile test was not a good

Table 5.4 Hardness of various steel microstructures

Steel microstructure	Maximum HRC values
Ferrite	10–15
Austenite	10–15
Pearlite	40–43
Upper bainite	40–45
Lower bainite	58–60
Martensite	65–66

measure of susceptibility to brittle behavior in complex steel parts.

Ferritic steels, as well as other body-centered cubic (bcc) metals, have the disadvantage of breaking in a brittle fashion at low temperatures. This means, in terms of the ideas of the tensile test presented previously, that the percent elongation at failure is close to zero. As the temperature is lowered, there is a small temperature range over which the bcc metals suddenly begin to fail in this brittle mode. An average temperature of the small range, called the ductile-brittle transition temperature (DBTT), is often chosen to characterize the temperature where the transition occurs. The simple tensile test will detect this transition, but unfortunately, it detects DBTT values well below those that occur in complex steel parts.

The tensile test applies stress in only one direction. In complex steel parts, the applied stress will act in all three possible directions, a situation called a triaxial stress state. The DBTT is raised by a triaxial stress state. A triaxial stress state will develop at the base of a notch when a notched sample is broken in a tensile machine, and such tests are called notched tensile tests. However, it is more useful to break the sample with an impact test, where the load is applied much more rapidly than in a tensile machine, because the combination of the notch geometry and the high load rate produces values of DBTT close to the temperature where brittle failure begins to occur in complex steel parts.

In the United States, the Charpy impact test, shown schematically in Fig. 5.8, is most often used for the impact test. A V-notch is machined into a square bar and placed in the holder. The specimen is broken cleanly in two with a weighted hammer attached to the arm of a pendulum. The pendulum arm is raised to a specified height and then released. It swings down through the sample and swings up on the opposite side. By comparing the rise after breaking the sample to the rise with no sample present, the amount of energy absorbed by the specimen on breakage can be calculated. This energy, usually given in units of joules, J, or foot-pounds force, ft-lbf, is called the Charpy impact energy, or CVN energy. (Note: The Izod impact test is similar to the Charpy. It is used mainly in Europe. The main difference in the Izod test is that the specimen is clamped at one end, and the pendulum strikes the opposite end.)

Figure 5.9 presents Charpy impact data on plain carbon steels that were slow cooled from the austenite region, so that the microstructures of these steels are mixtures of ferrite and pearlite. The Charpy impact energy is very sensitive to %C in the steels. At all compositions, there is a transition from ductile to brittle failure as the temperature drops, but at %C levels of 0.11 and below, the transition is much sharper and the CVN energy is considerably higher for the high-temperature ductile mode of fracture. For steels, a CVN energy value of 20 J (15 ft · lbf) is often taken as the onset value for brittle failure. The dashed line on Fig. 5.9 at 20 J (15 ft · lbf) shows

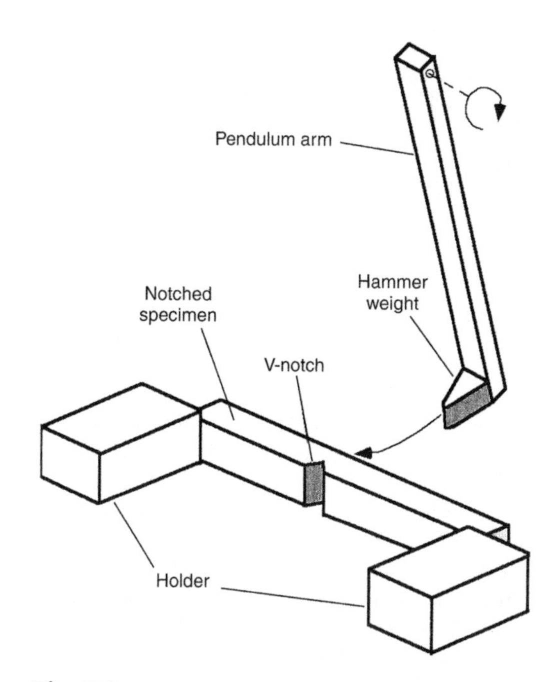

Fig. 5.8 Schematic diagram of the Charpy impact test

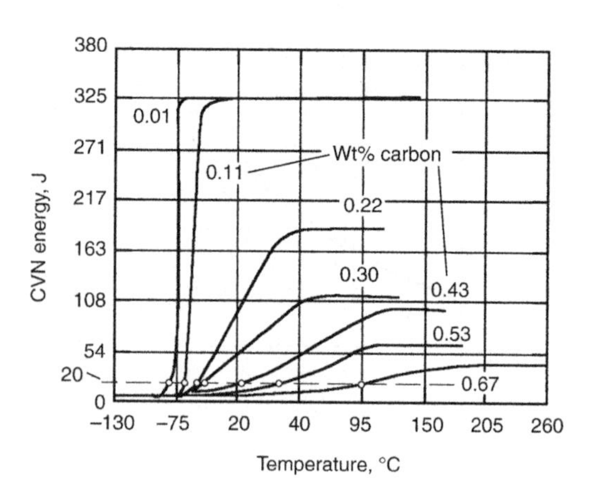

Fig. 5.9 Charpy V-notch (CVN) data on plain carbon steels austenitized at 870°C (1600°F) for 4 h and slow cooled. Source: Ref 5.3

that for this criterion, steels with %C above approximately 0.5 would be considered brittle at room temperature. However, the DBTT predicted by the Charpy test is really only a guide to actual transition temperatures that will occur for a complex piece of steel used in the field. So, the data in Fig. 5.9 do not mean that a 1060 steel will always fail by brittle fracture at room temperature.

It is common to refer to the CVN energy as a measure of the notch toughness, or simply the toughness of the steel. This measure of toughness is much more useful than ductility measurements, such as percent elongation, for evaluating the potential of a steel for brittle failure in service. Toughness and ductility have related but slightly different meanings, and it is high toughness that is desirable in steel parts. The toughness of steels can be improved by the following factors:

1. Minimizing the carbon content
2. Minimizing the grain size
3. Eliminating inclusions, such as the sulfide stringers shown in Fig. 5.6. This is done by using steels with low impurity levels of sulfur, phosphorus, and other elements.
4. Using a microstructure of either quenched and tempered martensite or lower bainite rather than upper bainite or ferrite/pearlite

There is a large difference in the nature of the fracture mechanism when a steel fails in a brittle mode versus a ductile mode. In brittle failure, there is little to no plastic flow of the metal prior to a crack forming and running through the metal, causing it to break into pieces. The metal seems to pull apart along the fracture surface. This sudden pulling apart occurs in one of two ways: (1) the metal separates along grain boundaries, or (2) the metal grains are cleaved in half along certain planes of their crystal structure. These brittle fracture modes are called grain-boundary fracture and cleavage fracture and can be distinguished by looking at the fracture surface in either a low-power microscope or a scanning electron microscope (SEM). Figure 5.10 illustrates a cleavage fracture surface with an SEM micrograph. The characteristic parallel lines are often called a river pattern, and they outline the paths of cleavage. Figure 5.11 presents an SEM image of a fracture surface that is roughly 60% grain-boundary fracture. On over 60% of the surface,

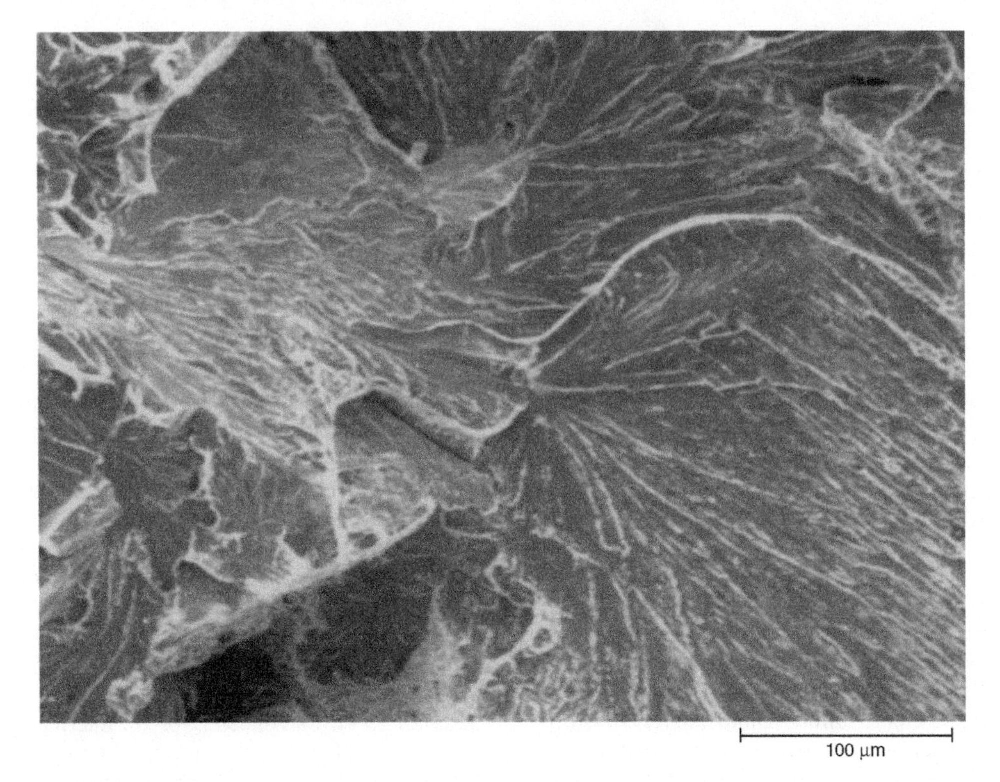

Fig. 5.10 SEM micrograph of a cleavage fracture surface on a 1018 steel. Original magnification 160×

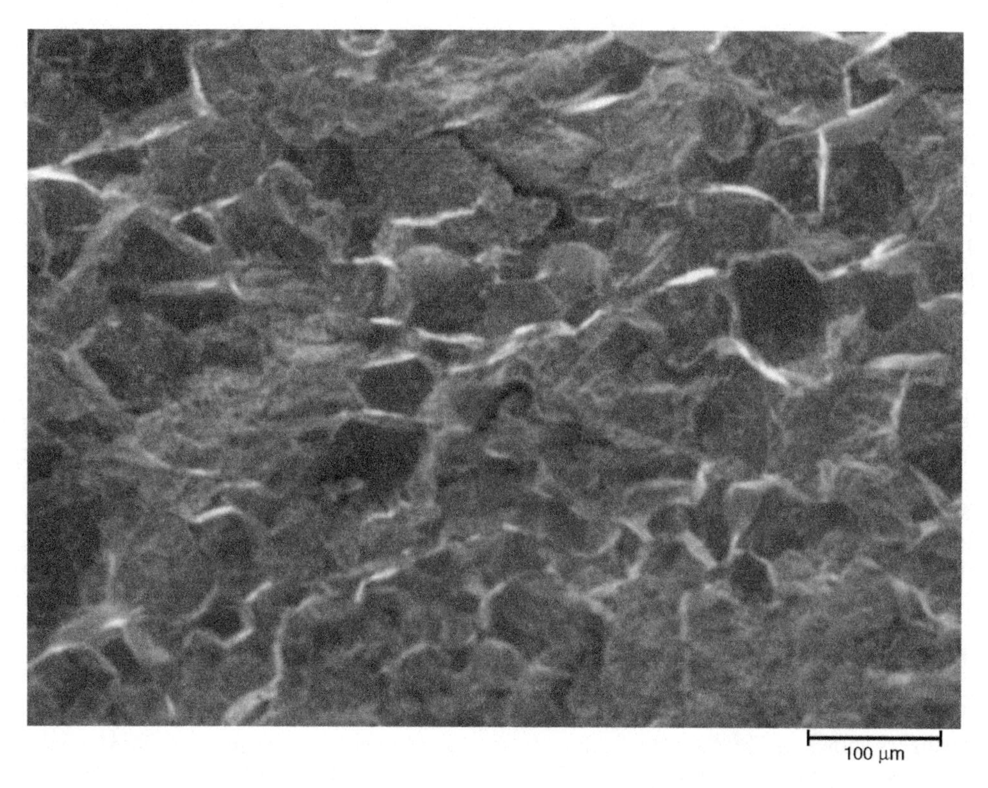

Fig. 5.11 SEM micrograph of a mostly grain-boundary fracture surface on a 1086 steel. Original magnification 95×

one sees the three-dimensional shape of individual grains revealed by failure along the old grain boundaries.

When a steel fails in a ductile mode, there will be some plastic flow of the metal prior to breakage occurring along the fracture surface. This flow causes very tiny voids to form within the metal grains. As the flow continues, these voids grow and coalesce until breakage occurs. The SEM micrographs at fairly high magnifications, such as that shown in Fig. 5.12, show remnants of the individual voids exposed on the fracture surface. The fracture is often called *microvoid coalescence*, for obvious reasons. If the steel contains small foreign particles, they will often be the site where the voids form first, and these particles are found at the bottom of the voids on one or the other of the mating fracture surfaces. The 1018 steel in Fig. 5.12 contained enough small sulfide particles distributed throughout its volume that many of the voids on the fracture surface reveal the sulfide particles that caused them to nucleate during the plastic flow, leading to failure. Unfortunately, it is difficult to see the microvoids on fracture surfaces with optical microscopes, because the depth of field is too shallow to image them at the high magnifications needed.

Grain-boundary fracture surfaces virtually always indicate that brittle failure has occurred, with little to no plastic flow before fracture. Cleavage fracture surfaces sometimes occur with significant plastic flow and, by themselves, do not indicate brittle failure. Microvoid coalescence fracture surfaces indicate that a ductile fracture occurred.

Fatigue Failure and Residual Stresses

A crude but quick way to break a steel rod in two is to make a small notch across the surface with a file or a cold chisel and then bend the rod at the notch. The rod will generally break with a single bend or maybe two. When the rod is bent away from the notch, a tensile stress develops at the base of the notch that is much higher than would be present if the notch were not there, and this leads to fracture starting at the notch. The small radius of curvature at the root of the notch causes a local rise in the stress there. The sharp radius is a stress concentrator or, alternately, a stress intensifier, and the smaller the radius, the higher the stress concentration. Small surface scratches, even those too small to see by eye, can produce stress concentrations during bending,

> **NOTE**
>
> The %C in structural steels that are subject to cold weather conditions is kept as low as possible to avoid embrittlement problems. A fascinating question is whether or not the steel plate used in the *Titanic* may have become embrittled at the −2 °C (29 °F) temperature of the saltwater when it struck the iceberg and sank. Studies of recently recovered pieces of *Titanic* steel plate have not completely settled this question. Charpy tests of the *Titanic* plate reveal CVN energies of only 4 J (3 ft · lbf) at −2 °C (29 °F), compared to values of 325 J (240 ft · lbf) for modern ship plate (Ref 5.4). Modern ship plate is superior due to lower levels of sulfur, phosphorus, and silicate particles, along with smaller grain sizes. However, examination of *Titanic* rivets (Ref 5.5) reveals a high level of slag inclusions, and it is not possible to rule out failure due to popped rivets.

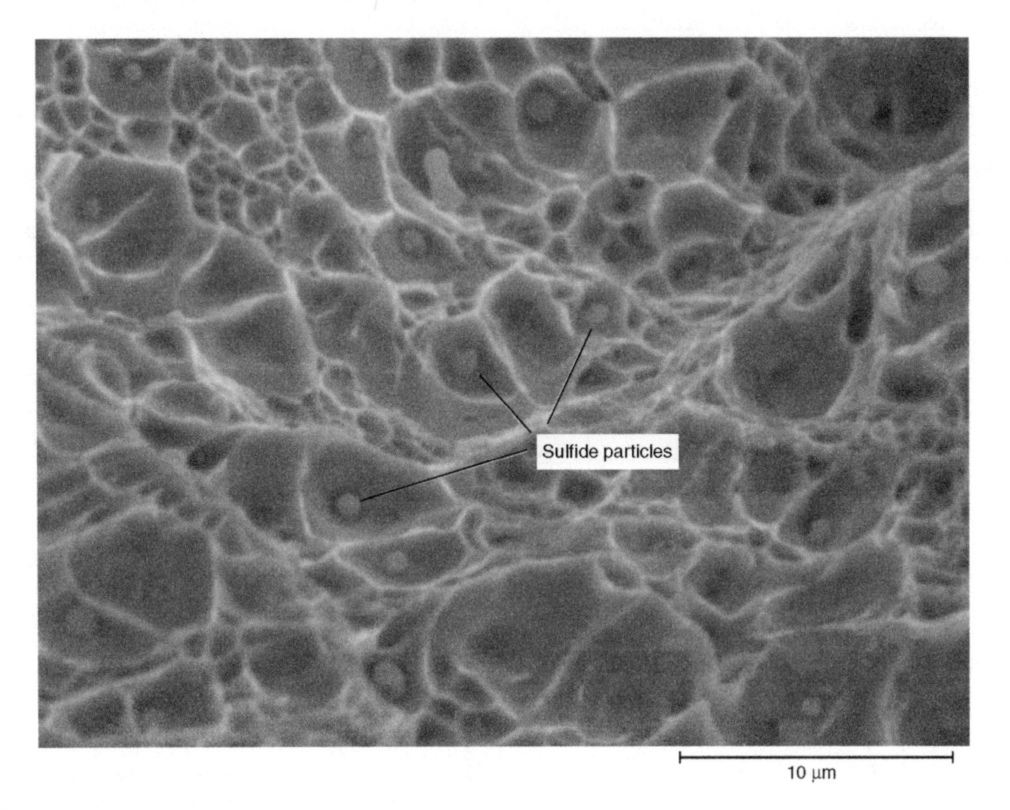

Fig. 5.12 SEM micrograph of a ductile fracture surface on a 1018 steel. Original magnification 2300×

which are localized at the root of the scratch. If sufficiently high, the concentrated stress can exceed the yield stress locally and produce tiny cracks that have the potential to lead to failure.

This idea can help in understanding a type of failure that occurs in metals and is called fatigue failure. Consider an axle supporting a load heavy enough to cause it to bend down slightly at its center point between the wheels, as shown schematically in Fig. 5.13. This bending will cause the metal to be pulled apart at the location *T* and also to be pushed together at the location *C*. This means that the metal will undergo a tensile stress at *T* and a compressive stress at *C*. Therefore, as a point on the axle surface rotates, it will undergo a cyclic stress: tension when it is down and compression when it is up. In a well-designed axle, the maximum tensile stress will be well below the yield stress, and all the strain experienced in the metal surface during rotation will be in the elastic region of the stress-strain diagram, as shown at the bottom of Fig. 5.13. However, suppose there is a very small scratch on the metal surface. If the stress is raised locally at the scratch root to a point above the yield stress, it may generate a crack there. Each time the axle rotates, the crack grows, and eventually, the crack becomes large enough to break the axle, a process known as fatigue failure. Fatigue failures occur in metal parts subjected to cyclic stresses, as often occurs in rotating machinery, valve and coil springs, or vibrating parts, such as airplane wings.

When a steel bar is pulled on with a tensile stress, the atoms of the iron grains are pulled away from each other, so that, on average, the distance between atoms increases. Similarly, if a compressive stress is applied, the average distance between atoms decreases a small amount. Hence, if the metal is stress free, there is some average distance between atoms. A metal rod lying on the bench generally is assumed to be stress free; that is, the atoms are spaced at their stress-free distance. Because the rod is just sitting there, this must be true, but only for the average distance between atoms over the whole rod. It may be that in parts of the rod, the atoms are sitting there with average distances between them greater than their stress-free distance, and in other parts, less than this value. If this is the case, the rod is said to contain residual stresses, some of them tensile and some of them compressive. Suppose, for example, that the outer millimeter of the rod is under compression. Then, below this outer 1 mm (0.04 in.) thick cylindrical region there must be some region where the atoms are under tension. In this case, there is residual stress on the surface that is compressive, with residual tension below. It is the surface where the scratches are that leads to fatigue failure. So, by producing residual compressive stresses on the surface, it is possible to reduce the crack growth during cyclic loading. As shown in Fig. 5.14, the local residual compressive stress at the surface will cause the tensile stress on the rotating axle to decrease, because the cyclic stress produced by rotation is added to the axle, starting at the value of the residual compressive surface stress. Thus, it is very desirable to have residual compressive stresses on metal parts.

There are several ways to produce residual compressive stresses in steels. These involve heat treating and/or mechanical deformation. Consider first the heat treating technique called flame hardening. The surface of a piece of steel heats up much faster than its interior when an intense flame is directed at the surface. The flame heating causes a layer of the steel at the surface to become austenite, while the interior remains ferrite + pearlite. On rapid cooling with

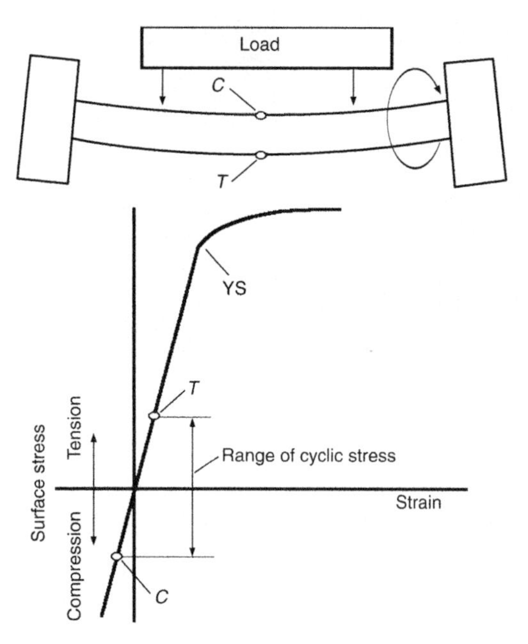

Fig. 5.13 Development of compressive and tensile stress states on the surface of a rotating axle

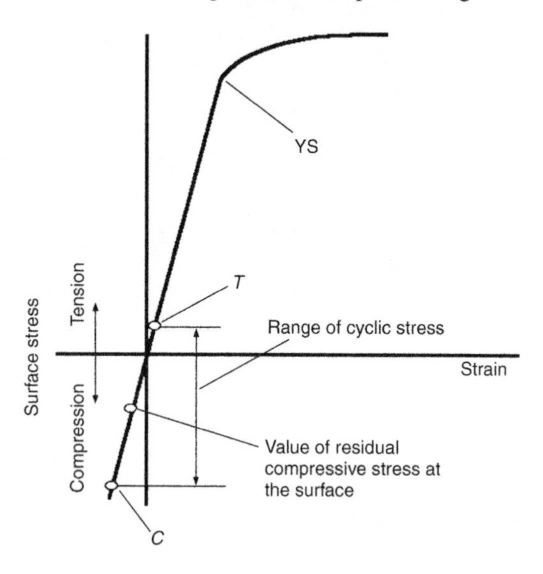

Fig. 5.14 Residual compressive stress on the surface will subtract from any applied tensile stress and reduce the net tensile stress experienced in the piece

a quench of some sort, the surface layer will transform to martensite, while the inner region remains ferrite + pearlite. Martensite has a lower density than the austenite from which it forms. Therefore, the outer layer expands when martensite forms. The inner region resists this expansion, causing the atoms in the outer layer to be forced closer together than they want to be, and a residual compressive stress is formed in the outer martensitic layer. An alternate way of heating the surface layer much faster than the interior is by induction heating. A copper coil is placed at the surface of the steel, and a high- frequency current is passed through the coil. The magnetic field generated by this current induces currents in the steel surface, causing localized surface heating. This technique is widely used in industry. It is so effective that axles used in motor vehicles are routinely induction hardened, not for increased surface wear resistance but simply for improved fatigue life. Figure 5.15 is presented to illustrate the magnitude of the surface stresses that are realized with this technique. It also shows the distribution of the residual stress below the surface.

The famous Japanese samurai swords are a fascinating example of localized martensite formation on the surface of steel playing an important role. When these swords are quenched, they are made to cool faster on the thin cutting edge and form martensite locally along the cutting edge, not back from it. Not only does this produce a hard cutting edge, but the expansion on martensite formation all along the cutting edge forces the blade to curve; the longer the blade, the more the curve, as Fig. 5.16 demonstrates. This effect has been discussed in a recent paper (Ref 5.7).

Residual compressive stresses can also be produced on heat treating, due simply to thermal expansion/contraction effects that occur on heating and cooling. If a round bar is quenched, the outer surface will thermally contract faster than the interior. The interior resists the contraction of the outer surface and pushes it into tension. If the tension is large enough to cause plastic flow in the outer surface, a stress reversal will occur on further cooling, and the result is a residual compressive stress on the outer surface. As with flame hardening and induction hardening, the end result is a desirable residual compressive surface stress. However, if a phase transformation occurs on cooling, things become more complicated. So, for steels, the compressive residual surface stress is assured if the steel is not heated above the A_1 temperature before quenching. Heating above A_1 generates austenite, which complicates matters due to the formation of martensite on quenching. If martensite forms only in an outer surface layer on quenching, then, as explained previously, a surface residual compressive stress is obtained. However, if

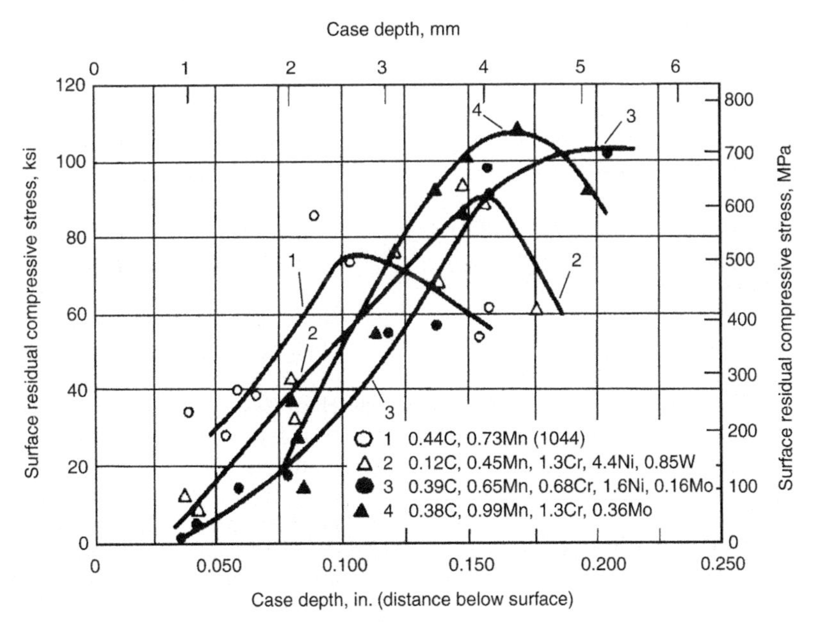

Fig. 5.15 Longitudinal residual surface compressive stresses versus case depth in four induction-hardened steels. Source: Ref 5.6 (Copyright 1963, Society for Experimental Mechanics)

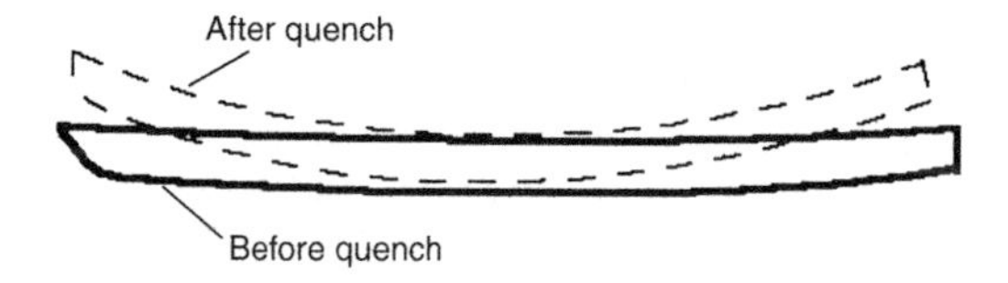

Fig. 5.16 Development of curvature on the cutting edge of a samurai sword caused by martensite formation during quenching

martensite develops all the way to the center of the bar (a condition called through hardening), the result is a residual surface tensile stress, an undesirable situation.

As discussed in Chapter 17, "Surface Hardening Treatment of Steels," there are many surface treatments that are given to steels. The two most common such treatments involve carburizing and nitriding. These treatments are usually done to produce hardened layers on the surface that reduce surface wear rates. These treatments also enhance the formation of residual compressive stresses, which is a fortuitous benefit accompanying the increased surface hardness.

Residual surface compressive stresses can also be generated by mechanical means. Perhaps the most common technique employed in industry is shot peening. The technique involves bombarding the steel surface with small steel balls (shot). This action causes localized plastic flow at the surface, resulting in the desired residual surface compressive stress. Steel toughness is enhanced by surface residual compressive stresses. An example involves the high-strength pearlitic wire used in cables, which was discussed earlier. It was pointed out in the section on impact testing that, in general, pearlitic steel is not as tough as quenched and tempered martensitic steels. Yet, the high-strength pearlitic wire used in bridge and crane cables displays superior toughness compared to quenched and tempered martensitic wires of the same hardness. The improved toughness of the pearlitic wire is thought to arise from the drawing operation used to refine the pearlite spacing, which is necessary to achieve the high strength. Apparently, the wire-drawing operation produces residual surface compressive stresses that result in the improved toughness.

Summary of the Major Ideas in Chapter 5

1. The mechanical properties of metals are often characterized with the tensile test. The test applies a stretching (tensile) force to a rod and measures the fractional length increase (Fig. 5.1). Results are shown as plots of *stress* (force per area of original rod) on the vertical axis against the *strain* (fractional length change) on the horizontal axis. At small strains, the rod behaves elastically; it springs back to its original length when the applied stress is released. Beyond a stress called the *yield stress*, the rod begins to deform plastically, meaning that on release of the applied stress, it has become permanently elongated. By stretching the rod to fracture, one finds a maximum stress on the diagram, and this is called either the *tensile stress* or the *ultimate tensile stress*.
2. In a simple bending test, metals that bend significantly before fracturing are said to be ductile, as opposed to brittle metals that will fracture before significant bending occurs. Good ductility shows up in a tensile test by high values of percent elongation and/or percent reduction in area. Percent elongation is simply 100 times the total fractional length change after fracture in the tensile test. Percent reduction in area is 100 times the fraction that the final area of fracture surface has dropped below the original cross-sectional bar area.
3. Mechanical properties are often summarized by listing yield strength, tensile strength, and elongation, as in Table 5.1. The strength values in the United States are usually expressed as MPa (megapascals), psi (pound per square inch) or ksi (thousands of pounds per square inch). The data in Table 5.1 show that as the carbon content of plain carbon steel, either air cooled (normalized) or as rolled, goes up from 0.2 to 0.95%, the tensile strength increases from approximately 450 to 1000 MPa (65 to 145 ksi), and the percent elongation decreases from approximately 35 to 9%. Virtually always, as the strength in a metal or alloy is increased, a loss of ductility occurs.
4. Hardness tests evaluate mechanical properties by forcing an indenter material, often diamond, into the surface of a metal under a given load and measuring either the depth of penetration (the Rockwell test) or the size of the indent impression (Brinell and Vickers tests). Table 5.2 lists the various tests and the conditions for choosing between them.
5. In quenched and tempered steels, there is a good correlation between the ultimate

tensile strength and hardness. Figure 5.7 presents a graph of this correlation for Rockwell C hardness, called HRC. It is common to heat treat steels to HRC values of 45 and above, which corresponds to tensile strengths of 1480 MPa (215 ksi) and above. Compare this to the strongest commercial alloys of aluminum, copper, and titanium, which are roughly 570, 1220, and 1310 MPa (83, 177, and 190 ksi), respectively. Clearly, steels are much stronger than all of these alloys.

6. All body-centered cubic metals, such as ferritic steel, share the problem that at low temperatures they fracture in a brittle manner, as opposed to their normal ductile fracture at higher temperatures. The ductile-brittle transition temperature (*DBTT*) is defined as the temperature below which the brittle failure occurs. Values of the DBTT can be measured with the tensile test, but they are much lower than where the transition is observed in complex steel parts. A V-notch impact test reveals a DBTT that does correspond fairly well with values experienced in the field. The Charpy V-notch test measures the energy needed to break the notched bar, and data such as those of Fig. 5.9 show that the DBTT in carbon steels is increased as %C increases.
7. It is common to use the term *toughness* as a measure of the ability of a metal to fracture in a nonbrittle manner. There are four important ways that the toughness of a steel may be enhanced: minimize carbon content, minimize grain size, eliminate inclusions, and develop a microstructure of quenched and tempered martensite or lower bainite. The fracture surface can distinguish brittle failure from ductile failure.
8. The iron atoms in a ferritic steel are positioned on the corners and at the center of the body-centered cubes that make up its grains. There is some average distance of separation of atoms in a steel before it is stressed. A *tensile stress* increases this distance, and a *compressive stress* decreases it. Residual stresses occur in steels when there are regions in the steel part where the atoms lie, on average, closer together (compressive residual stress) or further apart (tensile residual stress) than in the unstressed state.
9. Residual stresses can be produced in steels during heat treatments that involve quenching, particularly when martensite forms only at the surface of a quenched steel. The density of martensite is less than that of the austenite from which it forms. This means that when a given volume of austenite transforms to martensite, the volume of the martensite needs to increase. When the martensite forms as a surface layer on a bar, the interior of the bar will restrain its expansion and pull its atoms into compression, thereby creating a residual compressive stress on the surface. It is common to generate surface layers of martensite by flame hardening or induction hardening, which rapidly heats only the surface layers into the austenite temperature region prior to quenching. Figure 5.15 shows that very large surface compressive stresses (on the order of 480 to 690 Mpa, or 70 to 100 ksi) can be produced in steels with induction hardening.
10. Surface compressive stresses are very beneficial to toughness, because they inhibit crack formation and growth at the surface. The compressive stress will subtract from any applied tensile stress and reduce the net tensile stress at crack tips. (The tempered glass in automobile windows is another example of the benefits of producing materials with surface compressive stresses.)
11. Fatigue failure is a type of failure that occurs in metal parts subject to cyclic loading. The applied stress at the surface of a loaded axle will undergo alternating tension and compression on each rotation of the axle (Fig. 5.13). The geometry of a crack tip leads to a concentration of applied stress right at the tip. Thus, the applied surface stress increases locally at the tip of any crack lying on the surface, and this can cause small enlargements of the crack size on each rotation until failure eventually occurs. Such fatigue failure can be reduced significantly by the presence of surface residual stresses, which subtract from applied tensile stresses at the surface of the part.

REFERENCES

5.1. H.E. Boyer, *Hardness Testing,* American Society for Metals, 1985
5.2. G.F. Vander Voort, *Metallography,* McGraw-Hill Book Co., 1984

5.3. J.A. Rinebolt and W.J. Harris, Jr., Effect of Alloying Elements on Notch Toughness of Pearlitic Steels, *Trans. Am. Soc. Met.,* Vol 43, 1951, p 1197
5.4. B.L. Bramfitt, S.J. Lawrence, and H.P. Leighly, Jr., A Perspective on the Quality of Steel Plate from the RMS *Titanic, Iron Steelmaker,* Vol 26, Sept 1999, p 29–40
5.5. T. Foecke, "Metallurgy of the RMS *Titanic,*" NIST-IR 6118, National Institute of Standards and Technology, 1998
5.6. M. Hetenyi, Ed., *Handbook of Experimental Stress Analysis,* John Wiley, 1963, p 459
5.7. W.N. Weins and P. Bleed, Why Is the Japanese Sword Curved?, *Mater. Res. Soc.,* 1991, p 691–701

CHAPTER 6

The Low-Alloy AISI Steels

THERE ARE MANY different chemical elements that occur in the various types of steel, either as direct additives or as impurity elements. In order to understand the chemical symbols used for the various elements, it is helpful to review the periodic table of elements from chemistry or science classes in school. Inside the back cover of this book is a periodic table of elements that has been prepared to emphasize the crystal structures of the many different metal elements that occur in nature. The box in the lower left corner identifies the information for each element. For example, the element iron has chemical symbol Fe, atomic number 26, atomic weight 55.85, density of 7.8 g/cm^3, melting temperature of 1538 °C, and a mixed crystal structure of body-centered cubic/face-centered cubic/body-centered cubic (bcc-fcc-bcc) between room temperature and its melting temperature. Notice that the elements in the central columns of the table are grouped into three groups of columns, with the elements having crystal structures of either bcc; some mixtures of structure, such as bcc and fcc for iron; and fcc. These groups of columns will be of primary interest because they contain most of the alloying elements found in steels.

The discussion to this point has been restricted to plain carbon steels. The code for classifying these steels in the United States was developed by two professional organizations, the American Iron and Steel Institute (AISI) and the Society of Automotive Engineers (SAE). The code number of a particular steel is generally referred to as the AISI-SAE code number, but for the sake of simplicity, it is referred to here simply as the AISI code number. Table 6.1 presents a tabulation of the chemical compositions of a limited number of the plain carbon steels. Notice that the first two digits of the code change from 10 to 11 for the free machining steels and from 10 to either 13 or 15 for high-manganese plain carbon steels. A complete list of the chemical compositions of all the plain carbon steels may be found in Ref 6.1 and 6.2. In addition to the AISI specifications for steels in the United States, ASTM International and the military also specify types of steels. The ASTM International specifications are generally preferred for applications of steels for uses where avoidance of failures is imperative, such as bridges and boilers, and the military specifications are used for military applications. Cross references between AISI and ASTM numbers, as well as with the numbering system used in foreign countries, can be found in Ref 6.2 and 6.3. Steel castings, such as those used in large valves, gears, machinery, and so on, have their own codes, most of which are set by ASTM International. The AISI steel code is used for steel produced by steel mills and then sized by rolling or forging techniques; these are called wrought steels.

Table 6.1 Selection of some plain carbon steels

AISI No.	%C	%Mn	%other
1018	0.18	0.75	...
1020	0.20	0.45	...
1044	0.44	0.45	...
1045	0.45	0.75	...
1060	0.60	0.75	...
1078	0.78	0.45	...
1080	0.80	0.75	...
1095	0.95	0.75	...
Free machining steels			
1141	0.41	1.45	0.11 S
1144	0.44	1.45	0.26 S
Plain carbon manganese steels			
1340	0.40	1.75	...
1518	0.18	1.25	...
1541	0.41	1.45	...

All steels contain a low level of impurity elements that result from the steelmaking process.

These impurity elements can be eliminated from laboratory-prepared steels, where cost is not a problem. However, steel is an industrial metal produced in megatonnage quantities, and economic production processes result in a low level of certain impurity elements present in steel. These elements and their maximum levels traditionally found in steels are shown in Table 6.2. These four elements (aluminum, silicon, phosphorus, and sulfur) appear next to each other in the third row from the top of the periodic table, where they have been shaded to highlight their importance. Improvements in steel-processing techniques in the last two decades of the 1900s have allowed the industry to lower the residual level of sulfur and phosphorus in steels, and present-day steels often run levels of these elements at half or less than those given in Table 6.2.

For reasons to be explained subsequently, it is often advantageous to add additional elements to steel; the added elements are called alloying elements. If the percent of these added elements is fairly low, such as total additions of under 4 wt%, then the steels are called low-alloy steels. Fortunately, the large bulk of alloy steels involve the addition of only three chemical elements over and above the elements in plain carbon steels: chromium, molybdenum, and nickel, which are also shaded in the periodic table. There has been an evolution in the types of alloy steels used in industry, particularly the automotive industry, during the first part of the 20th century, so that at the present time, these alloy steels can be classified into three main categories: molybdenum steels, chromium steels, and triple-alloyed (Mo+Cr+Ni) steels. Table 6.3 lists a select number of the low-alloy steels grouped into these three categories. A complete listing of the low-alloy AISI steel compositions can be found in Ref 6.1 to 6.3.

The following are some interesting points. Note that the chromium steels, 5*xxx*, are the simplest of the steels, in that chromium is the only added alloying element in all of them. The molybdenum steels, 4*xxx*, are more complex, in that they can contain various combinations of all three of the major alloying elements, depending on the second digit of the code. For example, the straight molybdenum steels are coded 40*xx*, and the molybdenum-chromium steels are coded 41*xx*. Notice that there are no nickel steels. The nickel steels lost their popularity after the first half of the 20th century, and large-scale production of the 2*xxx* nickel steels and the 3*xxx* nickel-chromium steels was discontinued in approximately 1964.

There are two main reasons for the widespread use of low-alloy steels, which is discussed later. These are to improve the hardenability of the steel and to improve the toughness of tempered steels at a given level of strength.

Manganese in Steel

The element sulfur is present in all steels as an impurity element, as illustrated in Table 6.2. There are two unique facts about sulfur in iron that lead to a potential for sulfur impurities causing embrittlement of steel when present at even extremely low concentrations: (1) The solubility of sulfur in both austenitic iron and ferritic iron is exceedingly low, being essentially zero, and (2) sulfur forms a chemical compound with iron, iron sulfide (FeS), that melts at 1190 °C (2175 °F). To understand how these two facts combine to produce the embrittlement problem, consider the following analysis. Suppose a steel contains only 0.04 wt% S impurity. Because the solubility of sulfur in the iron grains of either austenite or ferrite is essentially zero, virtually all of the sulfur will be present as the compound FeS at all temperatures. Because hot rolling and forging temperatures are often above 1190 °C (2175 °F), the FeS will be present as a liquid at

Table 6.2 Impurity elements present in all steels

Element	Maximum level, %
S	0.04
P	0.05
Si	0.2
Al	0.04

Table 6.3 Selection of some low-alloy steels

AISI No.	%C	%Mn	%Cr	%Mo	%Ni
Molybdenum steels (4*xxx*)					
4023 (Mo)	0.23	0.80	...	0.25	...
4042	0.42	0.80	...	0.25	...
4130 (Mo-Cr)	0.30	0.50	0.95	0.20	...
4140	0.40	0.87	0.95	0.20	...
4320 (Mo-Cr-Ni)	0.20	0.55	0.50	0.25	1.82
4340	0.40	0.70	0.80	0.25	1.82
4620 (Mo-Ni)	0.20	0.55	...	0.25	1.82
Chromium steels (5*xxx*)					
5046	0.46	0.87	0.27	...	...
5120	0.20	0.80	0.80	...	...
5160	0.60	0.87	0.80	...	...
52100	1.00	0.30	1.45	...	...
Triple-alloyed (8*xxx*)					
8620	0.20	0.80	0.50	0.20	0.55
8640	0.40	0.87	0.50	0.20	0.55

these temperatures. Because of the very low sulfur level, only 0.04 wt%, the corresponding very small amount of liquid would not be expected to have any noticeable effect on the hot working of the steel. However, there is another factor that changes the picture. To understand this factor, the concept of wetting, which is well illustrated by the soldering process, must be understood. A good solder joint at a fitting occurs when the molten solder metal runs into the thin joint between a copper pipe and the fitting being attached to it. If the solder at the joint region is simply melted without applying a flux, the molten solder metal will form a ball shape and not flow into the joint. Applying a flux causes the molten solder to flow into the joint. The liquid metal appears to be sucked into the joint, and, in fact, it does become sucked in by a force called surface tension. Surface tension is a force acting along the plane of a surface. The flux removes an oxide from the copper surface and generates the desired surface tension force at the point where the clean surface contacts the molten solder, causing the solder to be pulled along the cleaned surface.

To obtain a feel for the nature of the surface tension force, consider the inflation of a rubber beach ball. The ball is inflated by increasing the pressure of air inside the ball above room pressure. This causes the rubber to stretch into tension. This stretching tension means that there is a pulling (tensile) force acting in the surface plane of the rubber. This surface tensile force is similar to the surface tension force that pulls the solder along the copper surfaces into the joint between pipe and fitting. It is an actual force that lies in the surface plane where the liquid solder contacts the copper.

Surface tension force in action can also be observed by gently floating a paper clip on the surface of the water in a glass. Note that the water surface is bent down a small amount at the contact point. It is the surface tension force in the water–air surface that holds up the clip. The force acts at right angles to the axis of the clip wire along the bent-down water surface and is directed slightly upward, as shown in Fig. 6.1. It is the upward component of this surface tension

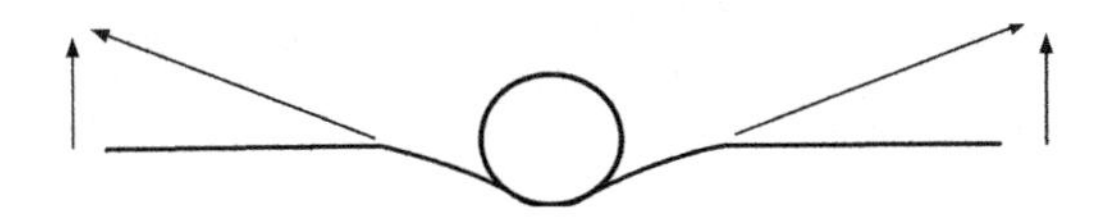

Fig. 6.1 Surface tension forces that allow a wire to float on water

force that is holding up the clip. The addition of soap to water reduces this surface tension force. To illustrate dramatically, dip the tip of a toothpick into a liquid soap typically used in the kitchen. Now, gently touch the soap-coated tip to the water surface well away from the clip. Observe what happens to the floating clip.

This sucking-in phenomenon is often called wetting, for obvious reasons. Now, consider a small spherical ball of liquid FeS lying at a grain boundary in steel at the hot rolling temperature. The grain-boundary surface is similar to the surface on a copper pipe. The grain boundary is a solid-solid surface compared to the solid-vapor surface on the copper pipe. Just as there is a surface tension force where the liquid solder contacts the copper-vapor surface, there is a surface tension force where the molten FeS contacts the grain-boundary surfaces between austenite grains. This force causes the molten FeS to wet the austenite grain boundaries. The FeS is pulled into the grain boundaries as a very thin film of liquid on the grain boundaries, which destroys their strength. Because the thickness of the liquid film is extremely thin, the very small amount of FeS present is adequate to cover quite large fractions of the grain boundaries, causing the steel to break apart by brittle failure (grain-boundary fracture) in the rolling or forging operations. Such an embrittlement occurring during mechanical deformation at high temperatures is called hot shortness. It almost always occurs when a liquid phase forms in metals, but only if the liquid wets the grain boundaries. For example, the lead in leaded brasses becomes molten in the solid brass matrix on heating. However, the lead does not wet the grain boundaries; it forms small spheres in brass and does not embrittle the alloy. The impurity element phosphorus will also cause hot shortness in steels, but only at higher impurity levels than are found for sulfur. The phosphorus forms a low-melting liquid called steadite and can produce hot shortness at phosphorus levels of approximately 0.1 wt% and above.

A primary reason manganese is present in all steels is to control problems with sulfur. Like iron, manganese forms a chemical compound with iron called manganese sulfide. It has the chemical formula MnS and melts much higher than FeS, at 1655 °C (3010 °F). In steels, where both iron and manganese are necessarily present, this compound will have chemical formula (MnFe)S, and the melting point will be reduced somewhat. However, it is still higher than the hot rolling and forging temperature, and it removes the hot shortness problem. It is customary to call the (MnFe)S particles present in steels manganese sulfide, even though they often contain appreciable amounts of iron. The sulfide particles highlighted in the 1144 steel of Fig. 5.6 are examples of such manganese sulfide.

Even though the manganese addition removes the problem of hot shortness, the presence of these particles (usually called inclusions) in steel often can lead to embrittlement problems in room-temperature operations. Because the sulfide inclusions are often ductile at the hot working temperature, they elongate into stringers, as demonstrated in Fig. 5.6. This produces a type of embrittlement in sheet and rod, depending on the direction in which the load is applied. The embrittlement often leads to failure problems in products made from steel plate. Both the longitudinal and transverse directions of the rolling operation are shown on a steel plate in Fig. 6.2, with the inclusions being elongated in the longitudinal (rolling) direction. The orientations of two Charpy impact bars are also shown. Notice that in the transverse bar, the elongated inclusions will run parallel to the base of the V-notch, while in the longitudinal bar, they will run at right angles to the base of the V-notch. Brittle failure occurs by cracks being opened up by the triaxial stresses generated at the base of the V-notch. Now, consider the effects of the elongated inclusions. When the inclusions lie parallel to the V-notch base, it is possible to have an inclusion lying along the entire base of the V-notch. However, when the inclusions lie at right angles to the base of the V-notch, an inclusion will pass the base of the notch at only one point. Hence, the inclusions will enhance crack formation much more effectively for the transverse bar orientations, where they lie parallel to the base of the V-notch. Charpy data on

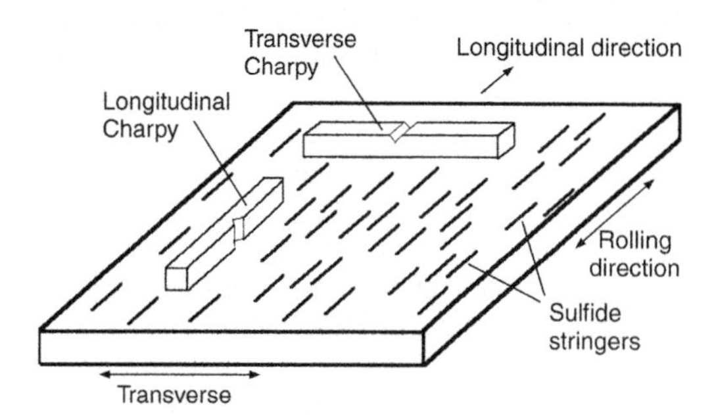

Fig. 6.2 Orientation of the longitudinal and transverse Charpy specimens cut from steel sheet with sulfide stringer inclusions

rolled sheet containing elongated sulfide inclusions give impact energies of approximately 60 J (44 ft·lbf) for longitudinal bars and only 20 J (15 ft·lbf) for transverse bars. The data provide dramatic evidence illustrating how elongated inclusions reduce the transverse toughness of wrought steels.

The manganese sulfides are called inclusions. Inclusions are generally produced by impurities in the steel. Residual oxygen and nitrogen impurities generate inclusions of oxide and nitride particles. In general, these inclusions are brittle and do not elongate at the hot working temperature. Unless their volume fraction becomes excessive, they do not lead to dramatic reductions of toughness.

Effect of Alloying Elements on the Iron-Carbon Phase Diagram

The phase diagrams for steel presented in Chapter 3, such as Fig. 3.5, are all pure iron-carbon phase diagrams. When manganese and the various alloying elements are added to iron+carbon alloys, the three important lines on the phase diagram, A_3, A_1, and A_{cm}, will shift a little. If only one addition is considered, such as the Fe-C-Mn compositions of plain carbon steels, the phase diagram is now called a ternary phase diagram. The study of ternary phase diagrams is considerably more complex than binary phase diagrams. The iron-carbon diagram of Fig. 3.5 involves only two elements and is called a binary phase diagram. Details are not presented here on ternary phase diagrams, but Fig. 6.3 is presented to give a flavor of the complexity. This diagram is for a hypothetical ternary alloy of Fe + C + *X*, where *X* is the third element addition. The diagram refers to alloys restricted to 2 wt% of the *X* element, and it shows how the three important lines of the iron-carbon diagram are shifted by the addition of element *X*. The A_3 and A_{cm} lines are only shifted up or down relative to their position on the binary iron-carbon diagram, but the A_1 line is split into two lines, labeled A_1(L) and A_1(U) in Fig. 6.3. For low-alloy steels, the temperature difference between these two lines is usually small and can be neglected. In plain carbon steels, the shift of the three important lines that occurs because of the manganese present is small enough that the pure iron-carbon diagram presents a good approximation for these steels. However, for the low-alloy steels, there are other elements present in addition to the manganese, and the effects can become noticeable, although still fairly small. To simplify things, the split of the A_1 is ignored, and the shift of the lines is characterized by determining how the added elements shift the temperature of the A_1 and the composition of the pearlite point.

The results are fairly simple: the elements manganese and nickel shift A_1 down, and chromium and molybdenum shift it up. Reference 6.4 presents an equation for the A_1 temperature that illustrates these effects:

$$A_1(\text{in °C}) = 727 - 10.7(\%\text{Mn}) - 16.9(\%\text{Ni}) + 16.9(\%\text{Cr}) + 29.1(\%\text{Si}) \qquad \text{(Eq 6.1)}$$

The equation is only valid for low levels of the elements shown but applies to plain carbon and low-alloy steels. The element molybdenum does not appear because it has no effect at levels below 0.5 wt% (Table 6.3). To illustrate the use of the equation, consider a 1018 steel, which Tables 6.1 and 6.2 show contains 0.75 wt% Mn and 0.2 wt% Si. Plugging these values into the equation gives an A_1 temperature of 725 °C (1340 °F), which was the value used in Fig. 3.9 and 3.10. All four elements shift the pearlite composition to values lower than 0.77 wt% C. All of the various other elements that are considered later with tool steels raise A_1 and also drop the pearlite composition.

As a first example, consider the steel 5160, which is the alloy of choice for most leaf springs used widely in motor vehicles. Comparing this steel to the plain carbon steel 1060 (Tables 6.1 and 6.3), it is evident that the main difference is the addition of 0.80 wt% Cr. The added chromium shifts the pearlite composition down to approximately 0.66% C for this alloy. Therefore, 5160 is closer to the pearlite composition than 1060, and the increase in the volume fraction pearlite formation as cooling rates are increased will be greater for 5160 than in 1060.

As a second example, consider the widely used bearing steel 52100, which has a carbon content of 1.0% (larger than 0.77% C and called hypereutectoid) and a chromium content of 1.45 wt%. The common heat treatment for this steel differs from that of most low-alloy steels in that the austenitization temperature lies *below* the A_{cm} line. The solid lines in Fig. 6.4 show the positions of the A_1, A_3, and A_{cm} lines for the pure iron-carbon diagram, and the dashed lines are an estimate of their positions in 52100 steel. The recommended austenitization temperature for 52100 (Ref 6.2, p 429) is 845 °C (1560 °F), which is shown on the diagram. Notice that at this temperature, a pure Fe+1% C steel would be fully austenite, because it would lie above the A_{cm} for pure iron-carbon alloys. However, in 52100 steel, the composition point lies well below its A_{cm} line, and, as such, the steel will contain particles of cementite at the austenitization temperature. On quenching, these particles simply remain in the steel, and when the austenite transforms to martensite, the particles become surrounded by martensite rather than austenite. The cementite particles that are present in the 52100 bearing steel after quenching and tempering are very small in size. Figure 6.5 presents a high-magnification scanning electron

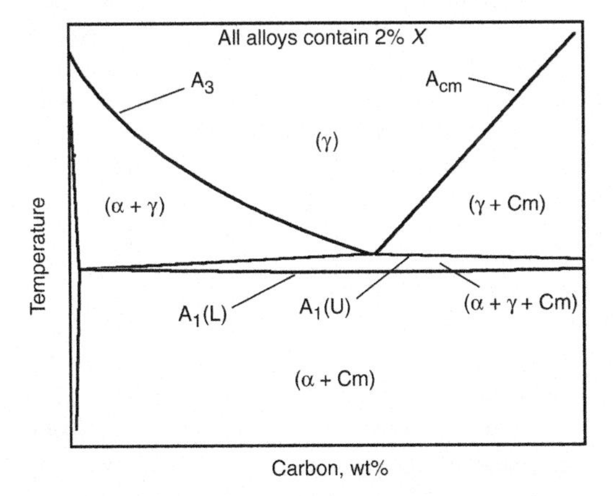

Fig. 6.3 Ternary Fe-C-X phase diagram at 2 wt% X, where X represents a steel alloying element such as molybdenum, chromium, or nickel. Alloying causes the A_3 and A_{cm} lines to shift and the A_1 line to split into A_1(L) and A_1(U)

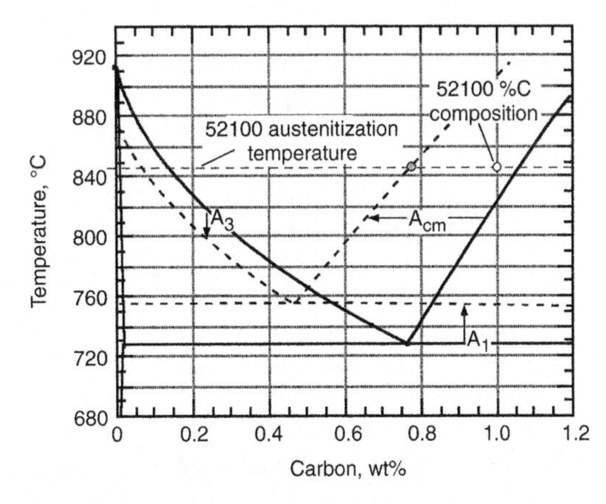

Fig. 6.4 Dashed lines are estimates of new positions of the A_1, A_3, and A_{cm} lines in 52100 steel, a hypereutectic chromium steel. Solid lines represent A_1, A_3, and A_{cm} in the binary iron-carbon phase diagram.

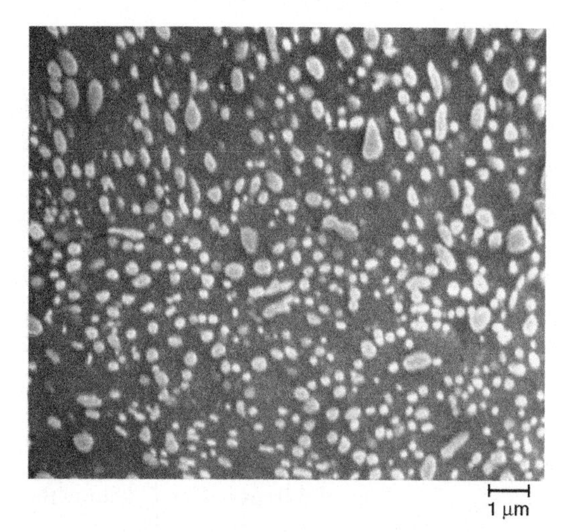

Fig. 6.5 Cementite particles present in 52100 steel following standard quenching and tempering. Original magnification: 5000×

microscope micrograph showing a typical distribution of the particles. The small particles in this micrograph average a diameter of just under 0.2 μm (8 × 10^{-6} in.) and are too small to be resolved in an optical microscope. Smaller-sized particles are preferred, because some mechanical properties of 52100 are improved as these particles are made smaller. This steel is discussed further in Chapter 11, "Austenitization."

Summary of the Major Ideas in Chapter 6

1. The major code used to classify the various steels in the United States was developed by the American Iron and Steel Institute (AISI), and all the common wrought steels have compositions specified by their AISI number. As illustrated in Table 6.1, the AISI numbers of the plain carbon steels are in the form 1*XXX*, where the last two digits tell the carbon composition, and the second digit tells if excess sulfur or manganese has been added. Manganese is present in all steels. The amount of manganese present in a steel can only be found in a table of compositions, because the AISI code does not contain this information.
2. Understanding the composition of steels requires a minimal familiarity with the chemical symbols used to represent the various alloying and impurity elements present in steels. The symbols of all the elements of nature are given in the periodic table of elements (see inside back cover), which is taught in high school chemistry classes. The important alloying elements and impurity elements in steel are shaded on the periodic table.
3. The common alloy steels have one or more of the three elements chromium, molybdenum, or nickel added, and these steels are referred to as the low-alloy AISI steels. Table 6.3 lists the composition of the most important of these steels. The steels may be grouped into three types: molybdenum steels (4*xxx*), chromium steels (5*xxx*), and triple-alloyed steels (8*xxx*). Again, the last two digits tell the %C in the steel. As seen in Table 6.3, the meaning of the second digit varies in an irregular fashion.
4. Manganese is present in all steels to overcome problems with sulfur embrittlement. Sulfur is an impurity atom in steels that cannot be economically removed. It forms a compound with iron, iron sulfide (FeS), that is molten at the hot rolling temperatures of steel. The molten FeS wets the austenite grain boundaries and leads to brittle grain-boundary fracture during hot deformation, a phenomenon known as *hot shortness*. The addition of manganese replaces the FeS compound with a manganese sulfide compound (MnS) that is not molten at the hot rolling temperature and cheaply overcomes the hot shortness problem.
5. The small particles of manganese sulfide become elongated in wrought steels and form stringers lying in the direction of deformation, as illustrated in the resulfurized 1144 steel of Fig. 5.6. Manganese sulfide stringers are a type of *inclusion* in the steel. They have little effect on the toughness of the steel for deformation in its longitudinal direction (the direction of the stringers) but can dramatically reduce toughness for deformation in the transverse direction, that is, deformation at right angles to the stringers.
6. The addition of manganese and the chromium, molybdenum, and nickel alloying elements change the position of the A_3, A_1, and A_{cm} transformation lines from their locations on the binary iron-carbon phase diagram. Equation 6.1 shows how the A_1 line is lowered by manganese, raised by chromium and nickel, and not affected by molybdenum at its levels in the 4*xxx* steels. All of these additions reduce the

composition of the pearlite point below 0.77%. In general, these shifts are small and often can be ignored in designing heat treatments. However, as alloying content increases, they can be significant, as is illustrated for the heat treatment of 52100 steel.

REFERENCES

6.1. Classification and Designation of Carbon and Low-Alloy Steels, *Properties and Selection: Irons, Steels, and High-Performance Alloys*, Vol 1, *Metals Handbook*, 10th ed., ASM International, 1990, p 141

6.2. Practices and Procedures for Irons and Steels, *Heat Treater's Guide*, 2nd ed., ASM International, 1995

6.3. *Worldwide Guide to Equivalent Irons and Steels*, 2nd ed., ASM International, Nov 1987

6.4. K.W. Andrews, *J. Iron Steel Inst.* (London), Vol 203, 1965, p 721

CHAPTER 7

Diffusion—A Mechanism for Atom Migration within a Metal

MODERN ARCHEOLOGY teaches that our earliest ancestors evolved in East Africa and then spread very slowly to the rest of the world by a process called migration or diffusion. Figure 7.1 indicates the direction of population flow out of Africa into Europe, Asia, and India. The lower right of the figure illustrates the two main requirements for diffusion. First, the direction of flow is from high density to low density, from populated regions to unpopulated regions. This results in what is known as a gradient in the density of people, with the flow going from the dense area of the gradient to the less dense area. Second, the moving species, in this case man, must have mobility that produces a back-and-forth movement. Because there are more people in the high-density regions, on average, more people will jump from high-density regions to low-density regions, which causes the net flow down the gradient.

Now, consider the experiment shown in Fig. 7.2. The end of a bar of pure iron is packed in charcoal and heated in a furnace to 925 °C (1700 °F). At this temperature, the iron is austenite, and the phase diagram shows that it is capable of dissolving approximately 1.3% C. However, because the bar is pure iron, it contains little to no carbon, while the charcoal at its surface contains close to 100% C. Hence, the carbon atoms are expected to move into the pure iron bar by the mechanism of diffusion, if the atoms are able to jump back and forth fast

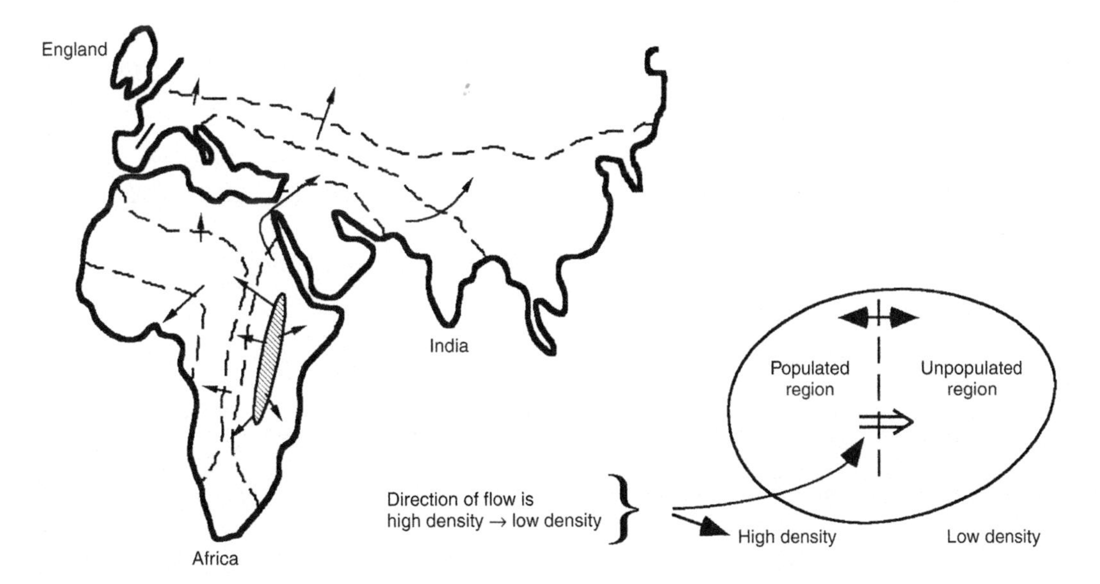

Fig. 7.1 Migration (diffusion) of early man out of East Africa

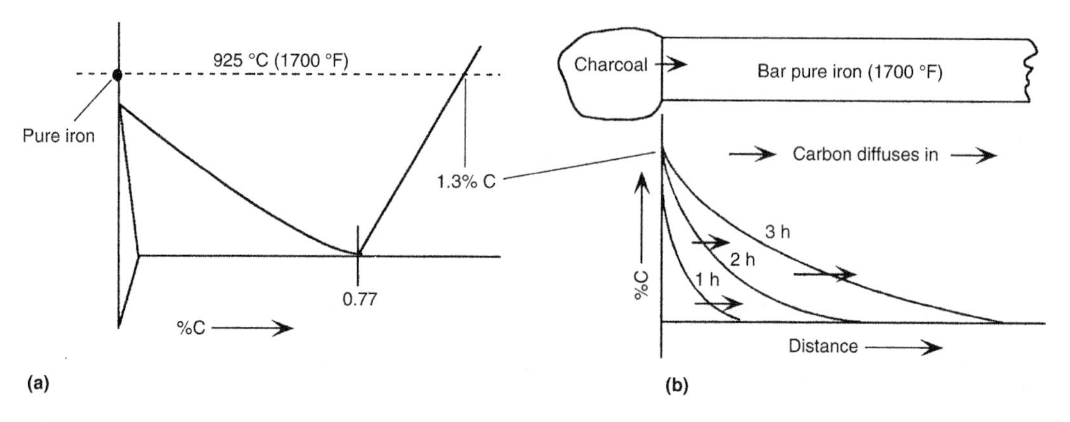

Fig. 7.2 (a) The iron-carbon phase diagram, indicating that iron can dissolve up to 1.3% C at 925 °C (1700 °F). (b) The diffusion of carbon into pure iron. As the carbon migrates into no-carbon regions of the bar, it continues to be absorbed from the charcoal at the surface

enough. Notice in the periodic table in the inside back cover that the atomic number of a carbon atom is much less than that of an iron atom, only 12 versus 26. This means that the carbon atom is much smaller than an iron atom. Mainly for this reason, a carbon atom dissolved in austenite will be in the holes in between the iron atoms. Figure 3.2 shows the iron atoms as spheres on the corners and face centers of the face-centered cubic (fcc) cubes that make up the crystals of the austenite grains. The small, open circles locate the centers of the holes between the iron atoms. If a carbon atom wants to move, it must jump from one hole (usually called a site) to another between the bigger iron atoms. Thermal energy associated with increased temperature causes atoms to vibrate back and forth. As temperature increases, the magnitude of the thermal vibrations increases, and it becomes increasingly probable that a given carbon atom will jump to a nearby empty site.

Experiments show the following results. In austenite at room temperature, it takes years for a carbon atom to jump to a nearest site, but this changes tremendously on heating. At 925 °C (1700 °F), a carbon atom in an austenite iron grain will jump back and forth between neighboring holes between the iron atoms of the fcc crystal at nearly unimaginably high rates, approximately 1.8 billion times per second. For this reason, diffusion of the carbon atoms will occur out of the graphite into the pure iron at rates that are extremely sensitive to the temperature. At 925 °C (1700 °F), it is possible to diffuse the carbon atoms well into the pure iron bar. The carbon concentration increases to the maximum possible value in austenite, 1.3%, at the charcoal interface, and, as shown in Fig. 7.2, with increasing time, more carbon diffuses into the interior. In the carbon-iron case, the two basic requirements for diffusion are also met: (1) The carbon atoms have a high mobility (they rapidly jump back and forth), and (2) the carbon atoms move from high-carbon compositions on the left to low-carbon compositions on the right. Here, the flow is down a concentration gradient, which is similar to the population density gradient.

Figure 7.3 illustrates that it is also possible to remove carbon atoms from steel by the process of diffusion. The figure shows a bar of 1095 steel heated to 925 °C (1700 °F), with only its left end exposed to air. The oxygen in the air will react with the carbon atoms at the surface and form gaseous molecules of carbon dioxide (CO_2), which will then drift away from the surface, removing carbon from the left end of the

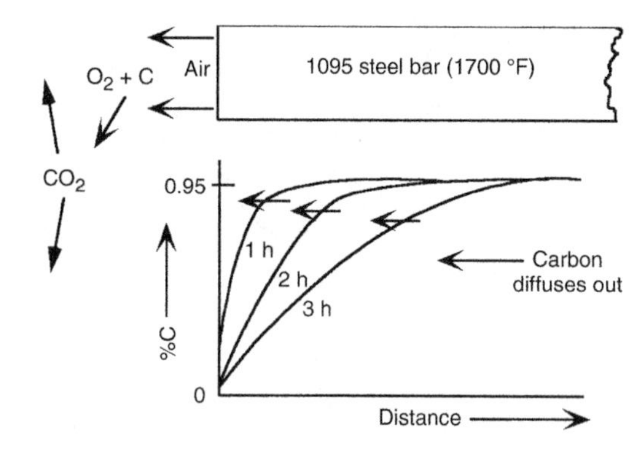

Fig. 7.3 Decarburization removes carbon from a 1095 steel bar by diffusion. A surface reaction between the steel and the oxygen to form CO_2 depletes the surface of carbon atoms. Carbon atoms migrate from the interior of the bar to the carbon-depleted surface

bar. This will cause the %C at the left end of the bar to decrease, and, as the carbon diffuses out of the bar, the %C is reduced in the bar, as shown in the figure. This process is called decarburization, and it can be a big problem for steels heated in air. To avoid the problem, it is necessary to control the composition of the gas surrounding the steel during heating. Methods of doing this are discussed later.

The rate at which diffusion occurs is characterized by a parameter, D, called the diffusion coefficient. To a good approximation, one may calculate how long it will take a carbon atom to diffuse a distance, d, by dividing the distance squared by six times D:

$$\text{Time (s)} = \frac{d \times d}{6 \times D} \qquad \text{(Eq 7.1)}$$

The units of D used here are microns squared per second (µm²/s). To calculate the time in seconds needed to convert the distance of interest, d, to microns, it is necessary to square it and divide by six times D. To illustrate the very strong effect of temperature on diffusion, the two plots in Fig. 7.4 and 7.5 are presented. Figure 7.4 is a plot showing how much the diffusion coefficient of carbon in austenite, D_c, changes as the temperature rises from 820 to 1150 °C (1500 to 2100 °F). This 330 °C (600 °F) increase in temperature raises D_c from approximately 5 to 150 µm²/s, a factor of 30 (or 2900%—a very large change!). Figure 7.5 uses Eq 7.1 to calculate how long it will take a carbon atom to diffuse a distance of 1 mm in austenite (1 mm = 1000 µm = 0.04 in.). Again, the strong temperature dependence is obvious, as the time drops from 9.3 h at 816 °C (1500 °F) to only 18 min at 1150 °C (2100 °F). Most heat treating is done in the 815 to 870 °C (1500 to 1600 °F) temperature range, while most forging is done in the 1090 to 1200 °C (2000 to 2200 °F) range.

In addition to being very sensitive to temperature, the time to diffuse a distance, d, is surprisingly sensitive to the length of the diffusion distance. Equation 7.1 shows that if the distance, d, is reduced to ½ of the original value, the time decreases by a factor of 4, and for a reduction to ¼ of the distance, it decreases by a factor of 16. Hence, when forging steel at 1150 °C (2100 °F), the time to diffuse a carbon atom 0.25 mm (250 µm, or 0.01 in.) is 18/16 min, or roughly only 1 min, compared to the 18 min needed to diffuse 1 mm (0.04 in.).

A proper understanding of diffusion is important when, for example, fabricating steel blades. The alternating layers in a pattern-welded Damascus blade are spaced at distances on the order of 50 to 100 µm (2 to 4 mils, or 0.002 to 0.004 in.). Blacksmiths often assume that if one of the original layers is a high-carbon steel and the other a low-carbon steel, the final blade will consist of high- and low-carbon layers. Suppose the forging is being done at a typical forging temperature of 1150 °C (2100 °F). Figure 7.4 shows that the value of D is 150 µm²/s (240×10^{-9} in.²/s), and Eq 7.1 gives the time to diffuse the carbon

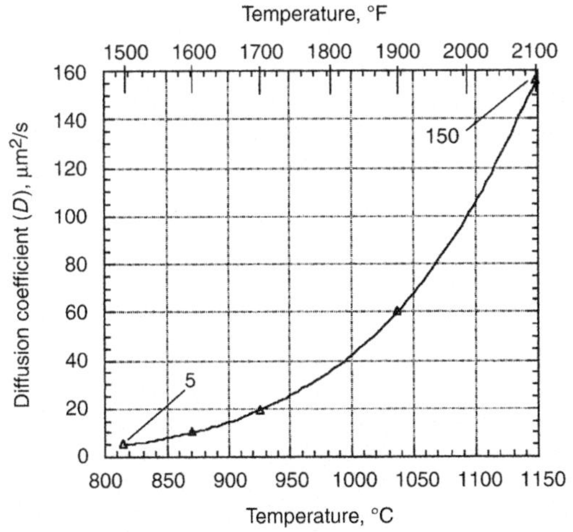

Fig. 7.4 Temperature dependence of the diffusion coefficient of carbon in austenite

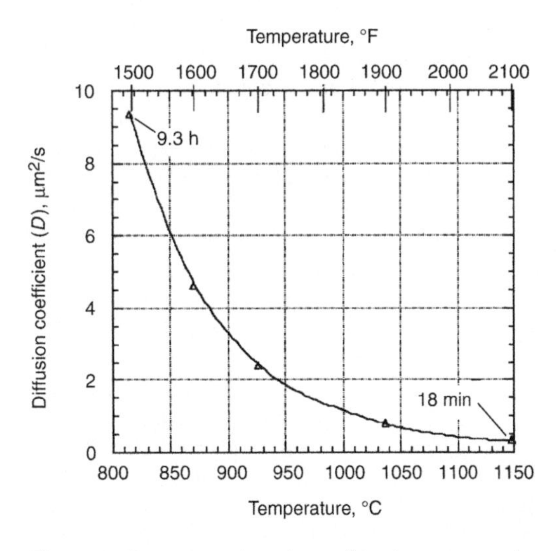

Fig. 7.5 Temperature dependence of the time necessary for a carbon atom to diffuse a distance of 1 mm (0.04 in.) in austenite

atoms over a spacing of 100 μm (0.004 in.) as $t = (100 \times 100)/(6 \times 150) = 11$ s. This result shows that diffusion will homogenize the steel during the forging operation, so that both layers have the same carbon composition. For more on this topic, see Ref 7.1.

Notice in the periodic table inside the back cover that the size of the three major alloying elements (molybdenum, chromium, and nickel) as well as the element manganese is approximately the same size as an iron atom or larger (compare their atomic numbers). This means that they will have a difficult time dissolving in austenite, because they are too big to fit in between the iron atoms. When these atoms dissolve in austenite, they do so by replacing the iron atoms at the corners or faces of the fcc crystals. How can one of these alloying atoms diffuse in the fcc austenite grains? A carbon atom has an easy time diffusing, because all it has to do is jump into one of the neighboring holes between the iron atoms that is not occupied by another carbon atom, and there are many such empty holes available. However, an alloying element lying in a face center of a given cube cannot jump into one of the holes between the iron atoms, because the holes are too small. Also, they generally will not be able to jump onto a nearby corner site of the cube, because those sites usually will have iron atoms on them. However, not all corner sites are occupied by iron atoms, and when there is not an atom on a corner site, it is called a vacancy site.

A very small fraction of the corner sites are vacant, and this fraction increases very rapidly as temperature increases. Consequently, the alloying elements are able to diffuse in austenite by jumping between vacant sites, and the diffusion rates increase rapidly with temperature. The rates of diffusion, however, are very much lower than the rates for carbon diffusion. The diffusion coefficients of manganese and the three major alloying elements (molybdenum, chromium, and nickel) have been measured in austenite. They are roughly equivalent to one another and thousands to millions of times smaller (depending on temperature) than the carbon diffusion coefficient (Ref 7.2). To illustrate how very small their diffusion rates are compared to carbon, Fig. 7.6 presents diffusion times for molybdenum atoms in austenite. The times needed for molybdenum to diffuse 1 mm (0.04 in.) at 820 and 1150 °C (1500 and 2100 °F) are 2600 and 1.6 years, respectively. As seen in Fig. 7.5, carbon diffuses the same distance (1 mm, or 0.04 in.) in only 9.3 h and 18 min at the same respective temperatures. These results show that when steel is heat treated, the manganese and the three major alloying elements will not move significantly, whereas the carbon atoms will move significantly. It is possible to make the larger atoms move smaller distances in reasonably short times, but only by heating them at very high temperatures for long times. For example, a molybdenum atom will diffuse 50 μm (0.002 in.) when austenite is held at 1200 °C (2200 °F) for 8 h. However, the corresponding diffusion time for a carbon atom at the same temperature is only 1.7 s.

Reference 7.2 summarizes diffusion coefficient measurements of the important alloying elements in steel in graphical form. For each alloying element, the equations for the temperature dependence can be determined. For chromium, $D = 0.17 \exp(-61300/RT)$; molybdenum, $D = 1.8 \exp(-68200/RT)$; manganese, $D = 1.1 \exp(-68200/RT)$; and for nickel, $D = 0.51 \exp(-68200/RT)$, where R = 1.987, T is temperature in degrees Kelvin, and the units of D are centimeters squared per second (cm²/s). A good approximation for the diffusion coefficient of carbon in austenite is $D = 0.12 \exp(-32000/RT)$ cm²/s. From these equations, one can calculate how many times bigger the diffusion coefficient of carbon is than the alloying elements at any temperature. Some results are given in Table 7.1. The numbers illustrate why carbon diffuses very much faster than alloying elements in austenite.

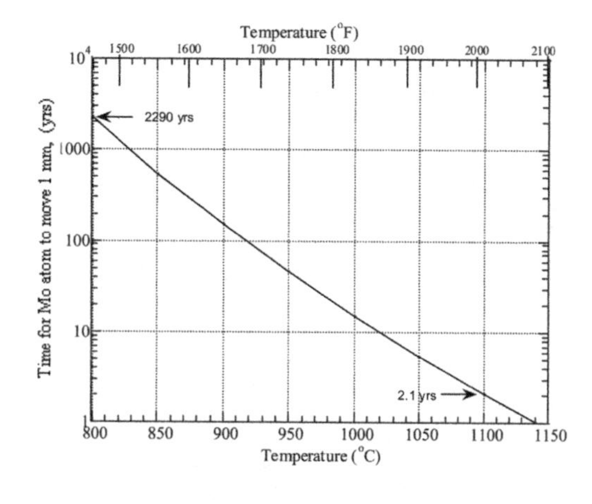

Fig. 7.6 Temperature dependence of the time necessary for a molybdenum atom to diffuse 1 mm (0.04 in.) in austenite

Table 7.1 Ratio of diffusion coefficient values of carbon in austenite to that of four major alloying elements in austenite at three temperatures

Element	800 °C (1470 °F)	1000 °C (1830 °F)	1200 °C (2190 °F)
Cr	724,000	82,000	16,900
Mo	1,740,000	119,000	16,800
Mn	2,840,000	194,000	27,500
Ni	6,140,000	419,000	59,400

Source: Ref 7.2

Carburizing and Decarburizing

The carburizing process is shown in Fig. 7.2. Because charcoal is a solid form of carbon, it may be expected that the carbon atoms are transported from the charcoal to the iron surface at the interface where the charcoal makes contact with the iron. However, such is not the case. If pure carbon powder instead of charcoal is used, no carburization occurs. Charcoal, being the remains of burned wood, contains hydrocarbons that vaporize and produce a gas containing carbon monoxide (CO). The carbon atoms are carried by gas molecules of carbon monoxide.

Suppose a piece of steel is heated in a charcoal fire (forge) the way blacksmiths have done for centuries. Air is a mixture of approximately 20 vol% oxygen (molecules of O_2) and 80 vol% nitrogen (molecules of N_2). When air is pumped into the charcoal, the oxygen burns and makes the fire hot. The reaction of the O_2 in the air with the carbon in the charcoal can be written as:

$$2O_2 + 3C \Rightarrow CO_2 \text{ (carbon dioxide)} + 2CO \text{ (carbon monoxide)} \qquad \text{(Eq 7.2)}$$

The chemical reaction expressed in Eq 7.2 says that two molecules of oxygen react with three atoms of carbon to produce two different gases, that is, one molecule of carbon dioxide and two molecules of carbon monoxide. The nitrogen of the air does not react during the burning; after the reaction, the gases in the charcoal bed still contain mostly N_2, approximately 72%. The remaining 28% is a mixture of CO_2+CO. In general, if the mixture contains an abundance of CO_2, the surface of the steel bar will be decarburized and maybe even oxidized. Being oxidized means that a layer of iron oxide (FeO) forms on the surface; that is, an oxide scale forms on the surface. However, if the mixture contains an abundance of CO, the steel may be carburized, and it is even possible that cementite or graphite will form on the surface. These ideas may be understood by consideration of the following chemical reaction equation:

$$CO_2 + [C]_S \underset{\text{Decarburize}}{\overset{\text{Carburize}}{\rightleftarrows}} 2CO \qquad \text{(Eq 7.3)}$$

The symbol $[C]_S$ refers to a carbon atom dissolved in the steel. If the reaction proceeds to the right (decarburizing), Eq 7.3 indicates that one molecule of carbon dioxide will react with an atom of dissolved carbon lying at the surface, and two molecules of carbon monoxide will be generated. These carbon monoxide molecules then drift off, and the net result is removal of carbon from the steel surface, that is, decarburization. However, if the reaction proceeds to the left (carburizing), Eq 7.3 indicates that two molecules of carbon monoxide decompose at the steel surface, dissolving an atom of carbon into the steel and generating a CO_2 molecule that will float off. Therefore, the relative amount of CO versus CO_2 in the gas determines whether the steel will be carburized, decarburized, or oxidized. As the CO_2 content increases, the reaction is biased to the right, and the likelihood of decarburization and oxide scale formation is increased.

The branch of science called thermodynamics that evolved near the end of the 19th century provides the tools to calculate the actual percentages of CO_2 needed to promote decarburization. The calculations require measurements of what are known as equilibrium constants of reactions and activity coefficients of the alloying elements in steel. Using these measurements, Fig. 7.7 was determined for the three steels 1020, 1060, and 1095, assuming they contained 0.75% Mn. The figure makes it possible to predict the vol%CO_2 on these steels that produces either carburization or decarburization. For example, suppose a 1095 steel is heated to 1000 °C (1830 °F). The 1095 curve of Fig. 7.7 intersects the 1000 °C (1830 °F) temperature line at approximately 0.068% CO_2. Therefore, if the %CO_2 in the gas surrounding the 1095 steel is less than 0.068%, the steel will carburize, but if it is above this value, it will decarburize. If the gas contains exactly 0.068% CO_2, there will be no addition or depletion of carbon from its surface. A gas composition that, for a given temperature, neither carburizes nor decarburizes is called a neutral gas, and, in this example, the 0.068% CO_2 value is the neutral CO_2 composition for a 1095 steel at 1000 °C (1830 °F). Figure 7.7 shows that the neutral CO_2

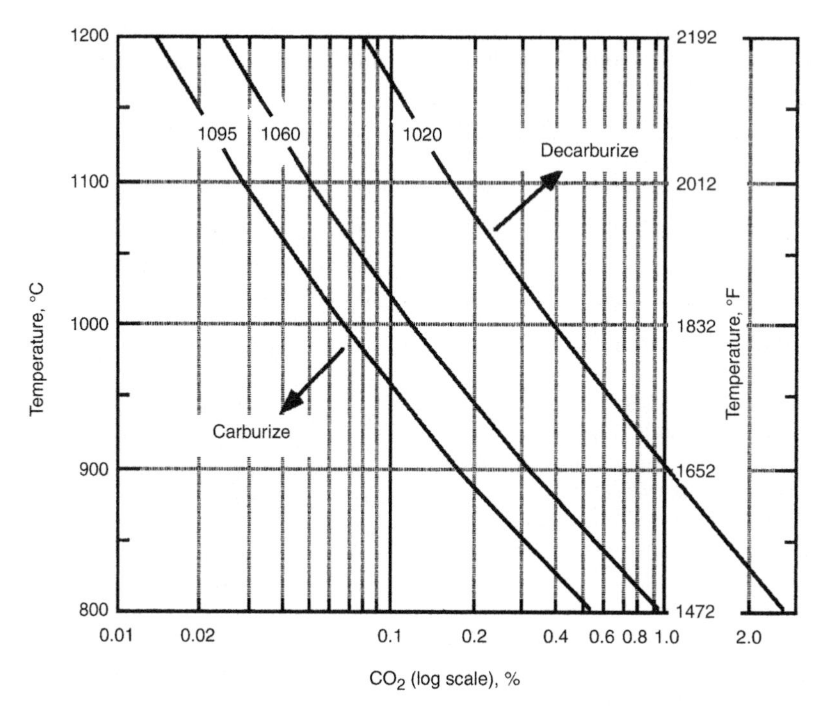

Fig. 7.7 The %CO_2 above which decarburizing occurs and below which carburizing occurs for the three steels 1095, 1060, and 1020

composition increases as the carbon content of the steel decreases, going from 0.068% C for a 0.95% C steel (1095) to 0.4% CO_2 for a 0.2% C steel (1020) at 1000 °C (1830 °F). For any of the steels, as the %CO_2 exceeds the neutral gas composition by large amounts, an oxide scale (FeO) will eventually form on the steel surface. The curve for the formation of FeO scale lies well to the right boundary of the diagram shown in Fig. 7.7 and is on the order of 5% CO_2 at 1000 °C (1830 °F).

The heat treating industry in the United States uses a gas called endothermic gas (often referred to as just endo gas) to control the CO_2 levels in furnaces to produce either neutral or carburizing gases for steel heat treating and carburizing. The gas is generated by burning natural gas (methane, or CH_4) in a very hot catalytic converter. The endo gas contains roughly 40 vol% hydrogen (H_2) + 40 vol% nitrogen (N_2), and the remaining is a mixture of CO and CO_2. The curves in Fig. 7.7 apply to its use. The %CO_2 in the gas is controlled by changing the ratio of air to natural gas fed to the gas generator, which allows the gas to easily switch from a neutral gas for heat treatments to a carburizing gas for carburizing treatments. The presence of the H_2 does two notable things. First, it makes the gas explosive at lower temperatures, so great care is needed with its use. Second, it speeds up the rate of carbon transport from the gas-to-steel surface so that carburizing rates are increased, particularly at short times.

Many bladesmiths use furnaces that are heated by combustion of air-natural gas or air-propane. Both natural gas (CH_4) and propane (C_3H_8) are hydrocarbon gases containing hydrogen and carbon. Combustion with air will produce gas mixtures of CO, CO_2, and H_2, and the %CO_2 in the combustion gas will control whether or not decarburization and/or oxide scale will form on the steel being heated. Proper design of the combustion chamber, combined with control of the air-gas ratio, can minimize these problems.

Although present-day bladesmiths generally use gas-fired furnaces (forges), some still use old-style, traditional forges, which produce heat by air flowing through a bed of hot charcoal. In a charcoal forge, the CO_2/CO ratio is controlled by the air flow rate through the hot bed. High flow rates produce a higher temperature but increase the CO_2/CO ratio, which can decarburize or even oxidize the steel. Lower air flow rates drop the temperature but can carburize iron and steel. These facts explain why only the more clever blacksmiths of old were able to consistently carburize bloomery iron to levels that produced strong steels (see historical note in Chapter 3, "Steel and the Iron-Carbon Phase Diagram").

Summary of the Major Ideas in Chapter 7

1. Diffusion is a process that allows carbon atoms to migrate into solid iron. A carbon atom dissolves in iron by fitting into the holes (interstices) lying between the much larger iron atoms. As the temperature is raised, the carbon atoms begin to vibrate so much that they will often jump to the nearest empty hole. There will be net migration of carbon atoms in the direction with more empty holes. Thus, diffusion causes atom migration from the regions where it is plentiful into regions where it is scant.
2. Any foreign atom dissolved in iron can move by diffusion. Migration is always directed from high concentrations of the foreign atoms to low concentrations. If charcoal is placed on the surface of an iron bar and heated, the carbon atoms in the charcoal will diffuse into the iron bar, diffusing from high carbon concentrations to low carbon concentrations. Similarly, if a steel bar is heated in air, carbon atoms are removed from its surface by oxidation to form carbon dioxide molecules. The concentration of carbon near the surface drops, and more carbon atoms will diffuse toward the surface and be removed, thus resulting in a decarburized layer near the surface.
3. The rate of diffusion is characterized by a number called the diffusion coefficient, D. Using Eq 7.1, it is possible to estimate how long it will take an atom to diffuse an arbitrary distance, d.
4. Rates of diffusion are strongly dependent on temperature. Experiments have measured how much the diffusion coefficient, D, increases as the temperature is raised. Figure 7.4 gives the diffusion coefficient of carbon in austenite versus temperature. Figure 7.5 shows that it takes a carbon atom 9.3 h to diffuse a distance of 1 mm (0.04 in.) at 815 °C (1500 °F) but only 18 min at 1150 °C (2100 °F). To move a carbon atom the distance between the bands of a forged Damascus blade (a distance of approximately 100 μm, or 0.004 in., or 4 mils) at a forging temperature of 1150 °C (2100 °F) takes only 11 s.
5. The important alloying elements in steel, chromium, molybdenum, and nickel, as well as the manganese present in all steels have sizes very similar to iron atoms. Their large size prevents them from fitting into the small interstices (holes) between the atoms of solid iron. Thus, it is much more difficult for these atoms to migrate by diffusion in steels, and this is reflected in their much lower diffusion coefficients. For example, Fig. 7.6 shows that it takes a molybdenum atom approximately 2600 years to migrate 1 mm in austenite at 815 °C (1500 °F), compared to only 9.3 h for a carbon atom.
6. The large difference in diffusion rates for carbon versus the common alloying elements is important in the heat treatment of steel. It allows the carbon atoms to be moved around without significantly changing the distribution of the alloying elements.
7. Two important processes in the heat treating of steel that are dependent on diffusion of carbon atoms in iron are *carburizing* and *decarburizing*. Carburizing requires adding carbon atoms to the surface of an iron or steel bar, and decarburizing requires removing carbon atoms from the surface of a steel bar. The addition and removal of carbon at the bar surface is accomplished by a reaction involving gaseous carbon monoxide (CO) and carbon dioxide (CO_2). Carbon atoms diffuse into the bar as more carbon atoms are added to the surface by carburizing. Similarly, carbon atoms diffuse out from the interior of the bar as they are removed from the surface by decarburizing.
8. When charcoal burns, a mixture of CO and CO_2 gases is produced, as shown by the chemical reaction in Eq 7.2. These gases are also produced as reaction products in methane- (natural gas) or propane-fired furnaces. If the fraction of the CO_2 is larger than approximately 1%, the hot gas will decarburize steel. However, if the $\%CO_2$ is reduced adequately, the hot gases will carburize the steel. It is possible to calculate the $\%CO_2$ that produces a neutral gas, that is, a gas composition where there is no carburizing or decarburizing. As shown in Fig. 7.7, the $\%CO_2$ in a neutral gas depends on both the temperature and the amount of carbon in the steel.

9. Many industrial steel parts (such as gear teeth and axle journals) are given a carburization heat treatment, which locally increases the hardness near the surface over a small distance of approximately 1 mm (0.04 in.). Low-carbon steels, such as 1020 or 8620, are often used. After quenching and tempering, there is a high-carbon martensite on the surface and either a low-carbon martensite or a bainitic/pearlitic interior. As shown in Fig. 4.13, the high-carbon surface will be much harder than the low-carbon interior. Thus, a part with a very high hardness having excellent wear resistance combined with an interior of high toughness can be produced. Carburizing is discussed further in Chapter 17, "Surface Hardening Treatment of Steel."

REFERENCES

7.1. J.D. Verhoeven and H.F. Clark, Carbon Diffusion Between the Layers in Modern Pattern Welded Damascus Blades, *Mater. Charact.*, Vol 41, 1998, p 183–191

7.2. J. Fridberg, L.E. Torndahl, and M. Hillert, Diffusion in Iron, *Jernkontorets Ann.*, Vol 153, 1969, p 263–276

CHAPTER 8

Control of Grain Size by Heat Treatment and Forging

AS MENTIONED in Chapter 5, "Mechanical Properties," it is important to keep the grain size of steels as small as possible in order to improve the toughness of the steel. Wrought or cast steels will arrive with a certain grain size. This grain size can be changed significantly by further heat treatment or forging operations, and it is important to understand what controls these changes.

The shape and some of the properties of grains in steel and other metals can be modeled surprisingly well with soap bubbles. A very simple kitchen experiment can give a feel for the geometry of grains in metals. In the bottom of a small-diameter clear glass, place several drops of a liquid dishwashing soap, and then add enough water to make a layer approximately 3 mm (⅛ in.) deep. After gently mixing with the end of a straw, blow vigorously through the straw until an array of soap bubbles fills the entire glass. The surface tension forces in the liquid soap film are similar to the surface tension forces in the solid-solid surfaces of metal grain boundaries, so the shape and growth of the soap grains simulate metal grains very well. The soap grains provide an excellent model of the three-dimensional shape of metal grains. For example, there is a seven-sided soap grain contacting the glass surface at the lower right of Fig. 8.1(a) (the sides are labeled 1 through 7), and the third dimension of this grain can be seen below the glass surface. Metallographic pictures of grains, such as Fig. 1.1, show only a two-dimensional section of the grain, and the soap grains help to visualize how these grains generally extend below the surface. Grain growth can be modeled by placing the tip of the straw at the center of the glass and gently blowing inward to expand the bubble positioned at the end of the straw. Perhaps a more pleasant way to model metal grains is to rapidly empty a bottle of beer, leaving behind an array of bubbles, as illustrated in Fig. 8.1(b).

Grain Size

As seen in the Fig. 8.1 bubble models, grains are irregularly shaped structures, which makes it difficult to easily characterize their size. ASTM International adapted a method for measuring grain size in 1947 that is widely used for characterizing grain size in steels; it is discussed in Ref 8.1. A grain size number, G, is defined by the following equations:

$$n = 2^{G-1} \quad \text{(Eq 8.1)}$$

Solving for G yields:

$$G = 1 + \frac{\log n}{\log 2} \quad \text{(Eq 8.2)}$$

where n is defined as the number of grains per square inch (645 mm^2) on a micrograph having a magnification of 100×. Figure 1.1 is at a magnification of 100×, so by counting the number of grains shown and measuring the area of the circular micrograph, the grain size number, G, from Eq 8.2 can be calculated. Because a large fraction of the grains in this micrograph contact the edge and are therefore not fully represented, some error will be introduced. Corrections for such problems are discussed in Ref 8.1.

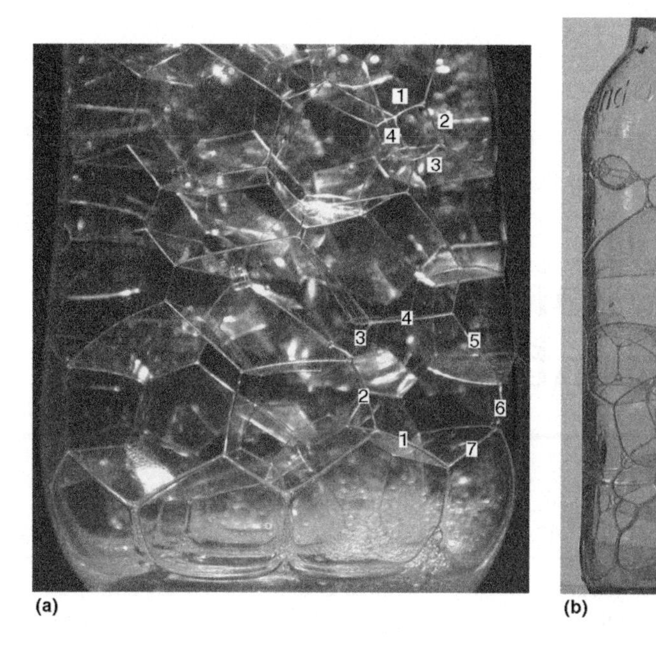

Fig. 8.1 (a) Soap grains and (b) beer bubbles simulating metal grains

Table 8.1 ASTM grain size numbers versus average grain diameter

ASTM No.	Average diameter, µm	Relative size
−1	510	Very coarse
0	360	
1	250	Coarse
2	180	
3	125	
4	90	Medium
5	65	
6	45	
7	32	Fine
8	22	
9	16	
10	11	Very fine
11	8.0	
12	5.6	
13	4.0	Ultrafine
14	2.8	
15	2.0	

Table 8.1 presents a list of the ASTM grain size number, *G*, along with a corresponding average grain diameter given in microns (1 µm = 0.001 mm = 4×10^{-5} in.). The column on the right side of the table indicates general guidelines for characterizing sizes. The very coarse sizes are only encountered in large ingots or in steels held near their melting points for long times. The finest sizes are only encountered in steels given special treatments that inhibit grain growth, some of which are discussed later.

Grain Growth

As austenite is heated to higher temperatures or held for longer times at temperature, the average grain size increases. This process of grain growth occurs by smaller grains shrinking in size until they disappear and larger grains growing in size, the net effect being an increase in the size of the average grain. It is the surface tension force in the austenite-austenite grain boundaries that causes grain growth.

The process may be understood by considering the soap bubble grains of Fig. 8.1(a), which also grow by the same mechanism. Figure 8.2 presents an analysis of a force balance on a short segment of grain boundaries snipped out of a small and a large grain. The magnitude and direction of the surface tension force is represented by the arrows labeled "T," and this force is pulling out from both ends of the segments, as shown. (The grain boundary is like a piece of stretched rubber, so a segment must have a force pulling out from each end.) This surface tension force always acts in the plane of the boundary. Therefore, the lower curvature of the big grain results in the T-force being directed more vertically in Fig. 8.2. The arrow labeled "T_h" is the horizontal component of the T-force. This is the fraction of the surface tension force that is directed in the horizontal direction. Note that there are two such forces acting to the right on each segment. However, movements of the soap grain boundaries caused by the forces are too slow to be visible. This indicates that the force of $2T_h$ is not able to pull the grain-boundary segment to the right. The reason is that a pressure difference develops across the boundary that balances this force. The air pressure on the concave (in) side of the boundary is called P_{cc}, and on the convex (out) side, P_{cv}. The air pressure on the concave side increases over that on the convex side by the amount needed to balance the surface tension force to the right, $2T_h$. (This pressure increase is similar to the air pressure causing a balloon to blow up; it is higher on the convex side, or inside, of the balloon.) Remember, pressure is force per unit area, so the pressure difference, $P_{cc}-P_{cv}$, is a force per area to the left, and it adjusts to balance the surface tension force going to the right. Therefore, the pressure is always higher on the concave side of curved boundaries. In soap bubble grains, this higher pressure on the concave side causes the air molecules to diffuse through the soap to the low-pressure convex side. So, the soap bubble grain boundaries always move slowly toward the concave side as the air diffuses to the convex side.

Note that the force $2T_h$ is larger for the smaller grains. This means that the small grains will shrink faster due to a larger pressure difference across them. In fact, very large grains will be convex and will always tend to absorb their neighbors by grain growth. Note that side 2 of the labeled seven-sided grain of Fig. 8.1(a) is concave toward the neighboring five-sided grain to its left. Hence, boundary 2 will tend to move into the five-sided grain, causing it to shrink and the larger seven-sided grain to grow. The four-sided grain labeled at the top right of Fig. 8.1(a) illustrates the concave geometry that causes the smaller grains to shrink.

This same type of action occurs in the grain growth of austenite. It is a little unusual to talk about a pressure in a solid, but the analogy holds. The higher pressures on the concave side of the austenite grain boundaries cause the iron atoms to jump across the boundaries from concave to convex sides, thereby causing large grains to grow at the expense of small grains. This motion of the iron atoms occurs by a jumping process similar to the impurity diffusion discussed in Chapter 7, "Diffusion—A Mechanism for Atom Migration within a Metal." Like diffusion, grain growth is very sensitive to temperature, going faster at higher temperatures, where thermal energy is causing larger vibrations of the atoms. The grain growth of austenite is very sensitive to the austenitizing temperature, and that is why it is important not to austenitize at temperatures any higher than needed to homogenize the austenite. Figure 8.3 presents data on the austenite grain size of 1060 steels held for 6 min and 2 h at increasing austenitization temperatures. Note that raising the temperature of the 6 min hold from 1400 to 1700 °F (760 to 930 °C) nearly triples the grain size, from 33 to 94 μm (0.001 to 0.004 in.). Like most metallurgy rate processes, grain growth is more sensitive to temperature than time. For example, increasing the austenitization hold time at 930 °C (1700 °F) from 6 to 120 min, a factor of 20, increases the

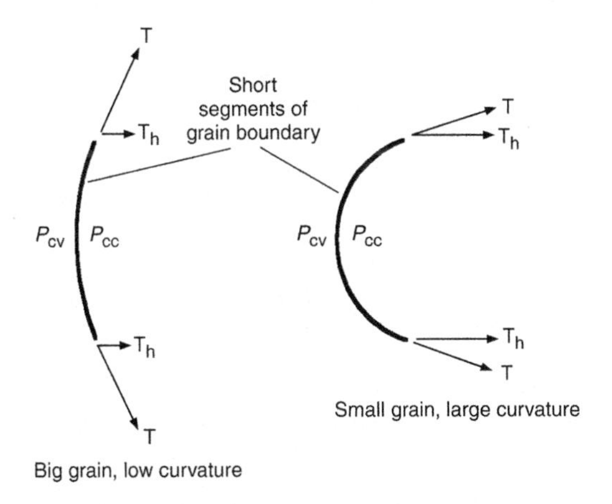

Fig. 8.2 Surface tension forces acting on grain-boundary segments produce pressure differences across the boundaries

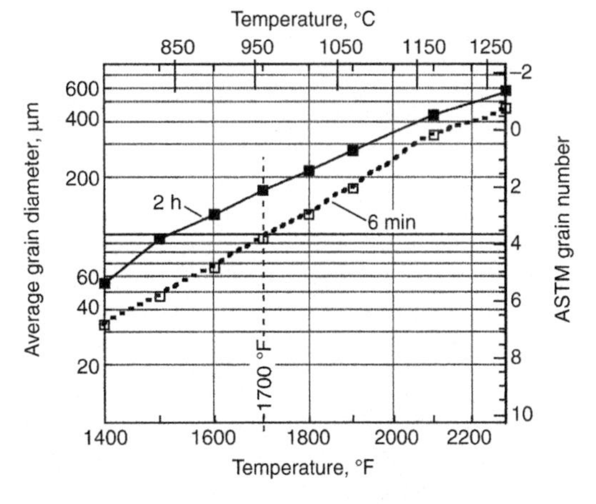

Fig. 8.3 Average grain diameter (alternatively, ASTM grain number) as a function of austenitization temperature for 1060 steel austenitized for 6 min and 2 h. Source: Ref 8.2

grain size from 94 to 174 μm (0.004 to 0.007 in.), a factor of only 1.85.

New Grains Formed by Phase Transformation

For simplicity, consider first a pearlitic steel. At room temperature, pearlite will have an average grain size. When it is heated above its A_{c_1} temperature, austenite grains begin to form. The new austenite grains start to form on the old pearlite grain boundaries, as shown schematically in Fig. 8.4. After a short time, all the old pearlite grains are replaced with a completely new set of austenite grains. The new austenite grains are at their smallest size immediately after the pearlite is consumed and before significant grain growth occurs as the temperature rises and time proceeds. There are two factors that contribute to the formation of the smallest possible initial austenite grain size:

- Faster heating rates increase the pearlite-to-austenite transformation rate and cause the austenite grains to nucleate closer together, leading to small grain size.
- Smaller original pearlite grains produce smaller austenite grains.

When the austenite is cooled to below the A_{r_1} temperature, a whole new set of pearlite grains is formed, and the same two factors—rate of transformation and size of the prior grains—control the size of the new grains. By simply heating and cooling through the transformation temperature, three different sets of grains are involved (original pearlite, austenite, and new pearlite). When dealing with hypoeutectoid steels, the same ideas apply, only now it is necessary to heat above the A_{c_3} temperature before 100% austenite is formed. A simple technique to reduce the grain size of a ferrite-pearlite steel is to heat it into the austenite region, holding the maximum temperature as low as possible and the time as short as possible, and then rapidly cool to room temperature as fast as possible without forming bainite (assuming it is not desired). Because smaller initial grain size promotes smaller final grain size, repeating this cycle several times will develop fine grain size.

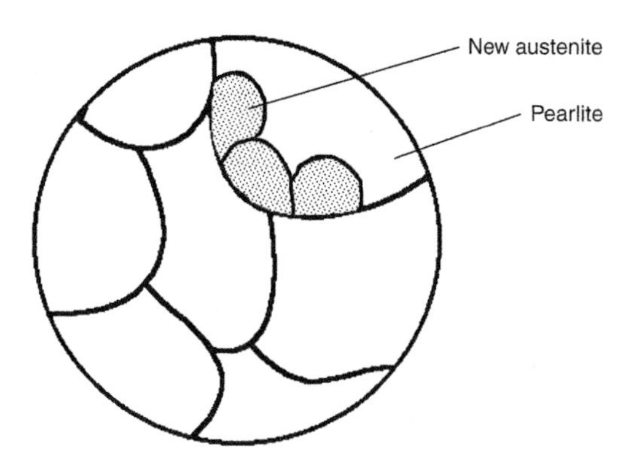

Fig. 8.4 Three new austenite grains forming along an old pearlite grain boundary

When heat treating to form martensite, toughness is also enhanced by fine-grained austenite, because it results in a finer lath or plate size in the martensite. Again, the same ideas apply. Rapid heating and repeated heating and cooling cycles produce smaller (finer) martensite microstructures. Grange (Ref 8.3) demonstrated the beneficial effect of small austenite grain size on the mechanical properties of 8640 steel. Grain sizes in the ultrafine range of ASTM No. 13 to 15 were achieved by a four-cycle process, where the steel was austenitized in a molten lead bath for approximately 10 s, cooled to room temperature, cold worked, and then cycled again. A series of similar experiments were performed on three steels to examine the effectiveness of thermal cycling alone; no cold working was employed. The steels were heated by immersion in a salt pot. Initially, the steels were austenitized for 15 min at 900 °C (1650 °F) and oil quenched in rapidly stirred oil. Then, the steels were given three thermal cycles consisting of a 4 min austenitization in 790 °C (1450 °F) salt and a quench in rapidly stirred oil. The grain sizes were measured with the same technique described by Grange (Ref 8.3), and the ASTM numbers before and after the three-cycle treatment are given in Table 8.2, showing that ultrafine grain sizes were obtained.

Figure 8.5 presents micrographs of the martensite structures found in 1086 steel before and after thermal cycling. The composition of this steel is in the range where the martensite is expected to be a mixture of lath and plate morphologies. In the uncycled, coarser-grained sample (Fig. 8.5a), dark plates are visible in a matrix

Table 8.2 ASTM grain size number of austenite before and after the three-cycle heat treatment

	ASTM No.	
Steel	Initial	After cycling
1045	9	14
1086	11	15
5150	8.5	14

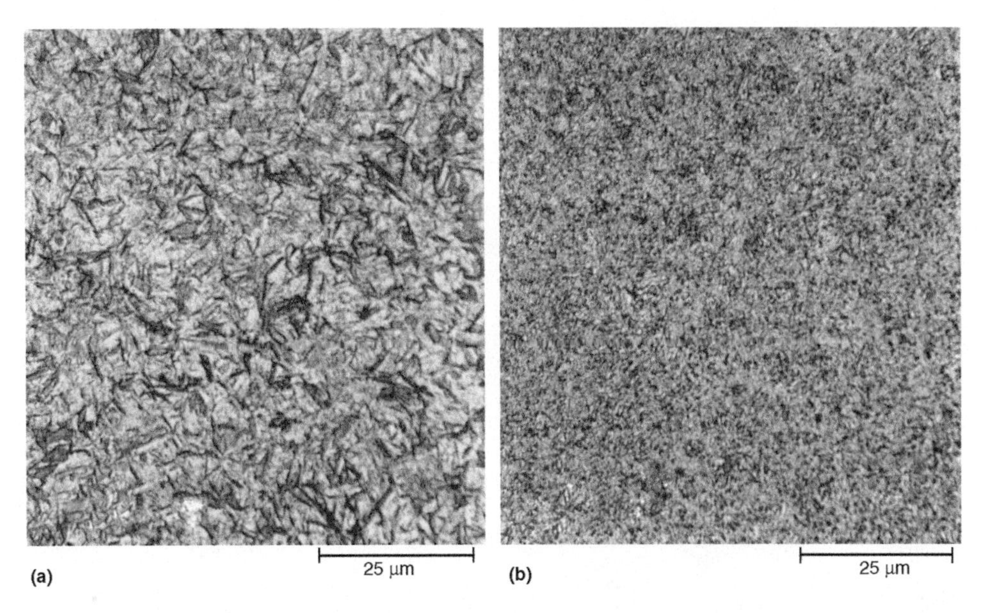

Fig. 8.5 Thermal cycling produces a finer martensite structure in 1086 steel. (a) Uncycled sample (grain size: ASTM No. 11). (b) Cycled sample (grain size: ASTM No. 15). Original magnification: 800×

of the lath structure. However, when a finer-grained austenite is produced by thermal cycling (Fig. 8.5b), the final martensite structure is also finer, and the plates are not easily identified.

New Grains Formed by Recrystallization

When a metal is plastically deformed, the flow of the metal matrix generates defects in the crystal structure. These defects (called dislocations) are planes of atoms that are shifted off the sites they occupied in the unit cell prior to the deformation, and the energy of the crystal is increased in the defect locations. This means that as the metal is made to plastically flow, energy is being stored in the crystals in the form of defects. More flow leads to a higher defect density and more stored energy. If a metal is plastically deformed by forging or rolling, the grains in the metal will align into elongated, pancaked shapes in the direction of the metal flow. The thinner grains of Fig. 8.6(a) show elongated grains in a 1010 steel that was cold rolled 90%. This terminology of 90% cold work means that the cross-sectional area of the metal was reduced 90% by mechanical deformation at room temperature. In rolling deformation, the slab width changes little during reduction, so a 90% reduction in area is the same as a 90% reduction in thickness. These elongated grains will contain a high defect density. On heating, a new set of grains will form that contain a low defect density, because the heat treatment allows the atoms of the grains to begin to move into a more desirable low-energy configuration. Figure 8.6(b) shows the new grain set after approximately 80% of the original structure has been replaced, and Fig. 8.6(a) shows the new grain set after approximately 10% replacement. The formation of this new set of low-defect-density grains is called *recrystallization*. The new grains are all individual crystals, and the name conveys the idea of an old set of crystals (the deformed grains) transforming into a new set.

In metals such as steel, it is possible to obtain new sets of grains by the two different techniques described here: heating to produce a phase transformation, or deforming and heating to produce recrystallization. The first technique works for steel because of the body-centered cubic (bcc) to face-centered cubic (fcc) crystal structure change that occurs in iron at the A temperature. In metals and alloys of copper and aluminum, however, there is usually no A-type transformation possible, because these metals retain the fcc crystal structure from room temperature to the melting point. Therefore, in these metals, recrystallization is the only way to produce a new set of grains.

For recrystallization to occur, the cold-worked metal must be heated to a temperature

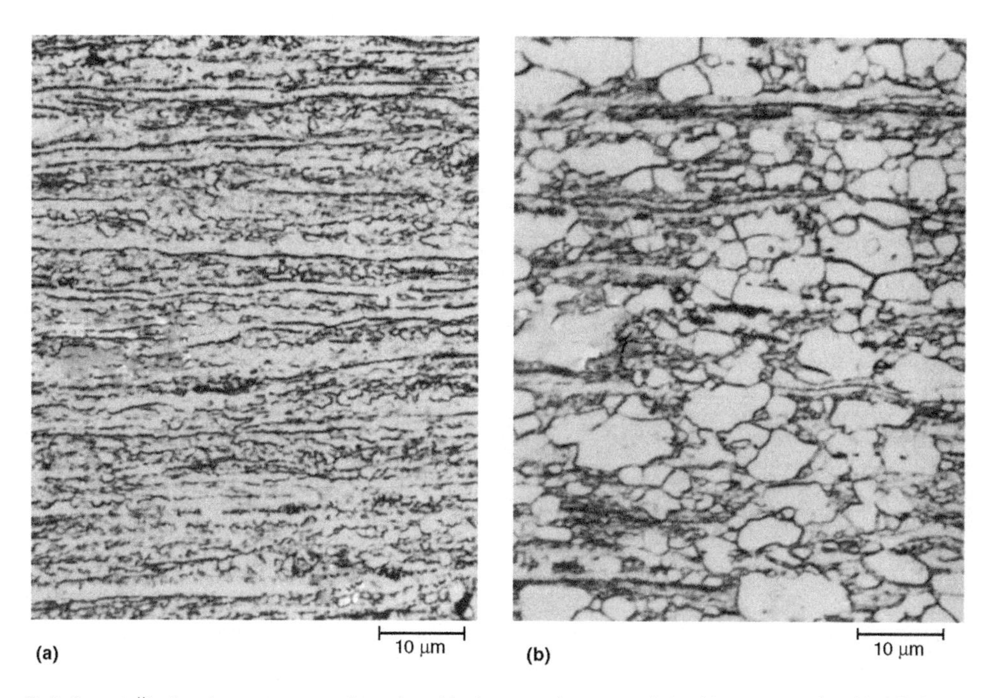

Fig. 8.6 Recrystallization increases proportionately with time, as demonstrated in this 1010 steel cold rolled 90% and annealed at 550 °C (1025 °F). (a) 2 min, 10% recrystallized. (b) 15 min, 80% recrystallized. Source: Ref 8.4

called the *recrystallization temperature*. The recrystallization temperature depends on the amount of prior deformation that was applied. More deformation increases the defect density and lowers the recrystallization temperature. However, as shown in Fig. 8.7, the drop in recrystallization temperature stops after approximately a 50% reduction. This is because the increase in defect density reaches a maximum value for deformations larger than approximately 50%. Therefore, when referring to the recrystallization temperature of a given alloy, it is assumed that the deformation is approximately 50% or more. Another consideration is the time for recrystallization. Recrystallization occurs faster at higher temperatures. Data such as that plotted in Fig. 8.7 generally are presented for the 1 h recrystallization temperature. That temperature is defined as the temperature where recrystallization is complete in approximately 1 h. An important practical question to ask when trying to produce recrystallization is how much cold work is needed to cause recrystallization to occur. According to Fig. 8.7, low-carbon steels will recrystallize with only approximately 5% cold work, but the temperature needed is significantly higher than is required with larger amounts of cold work. In general, the recrystallization temperature of a steel refers

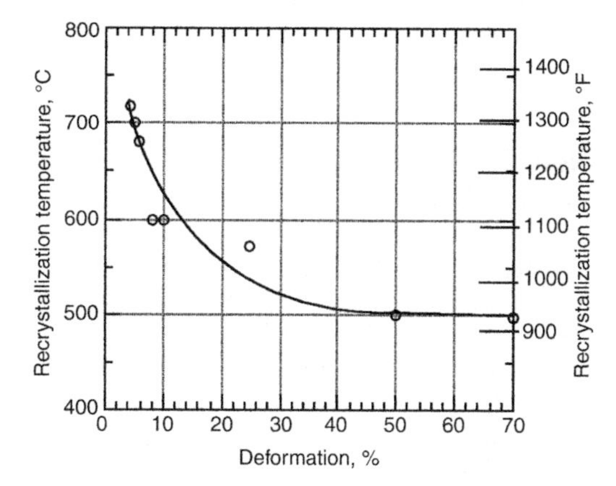

Fig. 8.7 Recrystallization temperature of low-carbon steels as a function of the amount of prior cold work. Source: Ref 8.5

to the temperature at which recrystallization occurs in one hour after a deformation of 50% or more.

When forging austenite, the metal is necessarily at a high temperature. Because the deformation is being applied rapidly in the forging operation, defects can accumulate in the deformed austenite grains and cause recrystallization to occur during the reduction that occurs on forging. The same is true of hot rolling. This type of recrystallization is often called dynamic recrystallization, in contrast to static recrystallization, which occurs after mechanical flow has stopped.

Dynamic recrystallization of austenite does several things. The reason metals work harden at low temperatures is because the increased defect density makes the metal stronger. During dynamic recrystallization, work hardening does not occur, because the new recrystallized grains have a low defect density. Therefore, the worked metal remains soft and pliable, due in large part to the dynamic recrystallization. Also, dynamic recrystallization during hot forging allows the continuous generation of new sets of austenite grains with each forging cycle, which prevents excessive grain growth at the high temperatures (approximately 1200 °C, or 2200 °F) where forging and rolling of austenite is usually done.

The temperature at which a metal recrystallizes depends on its melting point. For example, tin melts at a low temperature, 232 °C (450 °F). Alloys such as soft solder (usually a tin-lead or tin-silver alloy) can recrystallize at room temperature. Recrystallization effects can be demonstrated by bending a thick copper wire back and forth. It will harden at the bend and break there. However, a wire from a roll of soft solder can be bent almost indefinitely, with no hardening or breaking. The soft solder is recrystallizing at room temperature, producing new low-defect grains that are soft, while the copper retains its original grains, and they progressively become harder and more brittle as the defect density increases with bending.

Effect of Alloying Elements

Figure 8.8 presents a portion of the periodic table that surrounds iron and includes several elements that are sometimes added to steels in addition to the four major alloying elements of manganese, chromium, molybdenum, and nickel. Some elements tend to be much more reactive than other elements. Such reactive elements will combine with the elements carbon, nitrogen, oxygen, and sulfur to form carbides, nitrides, oxides, and sulfides. In addition to carbon, steels always contain a low level of nitrogen, oxygen, and sulfur. Consequently, it is common to have carbides, nitrides, oxides, and sulfides present in steel. These are chemical compounds that melt at very high temperatures and tend to be very hard and brittle. As discussed previously, cementite is a hard, brittle iron carbide (chemical formula Fe_3C). However, the cementite in steel is in a form that is generally not brittle, because of its very small size and because it is surrounded by ductile ferrite. This is true for the cementite in pearlite (remember, pearlite contains only approximately 11% cementite and is present as

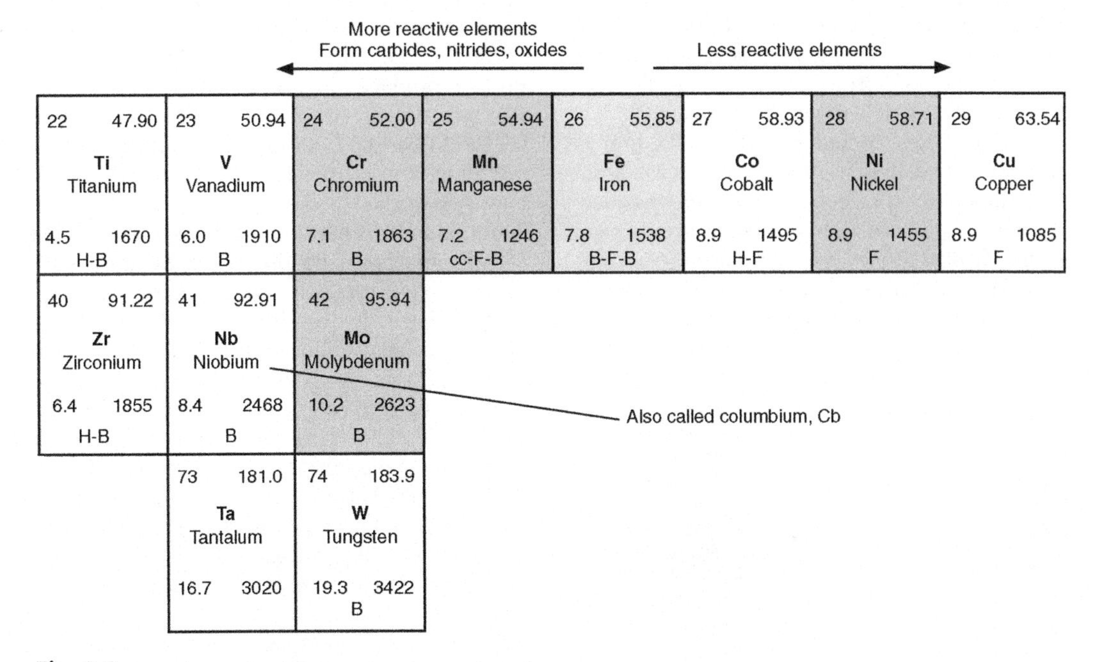

Fig. 8.8 General reactivity of the important elements in steels

very thin plates), and it is true for the very small cementite particles in hypereutectoid steels, such as those shown in Fig. 6.4 in 52100 steel.*

The arrows at the top of Fig. 8.8 show that the elements to the left of iron tend to be chemically reactive elements, while those to the right are less reactive. Of the four common alloying elements, chromium and molybdenum are carbide-forming elements and, in general, will prefer to segregate into the cementite constituents of steel. Nickel is not a carbide-forming element and will prefer to segregate into the ferrite. Manganese tends to be in between these groups. The elements on the left—titanium, zirconium, vanadium, niobium, tantalum, and tungsten—form very stable carbides (except for carbon, tantalum carbide has the highest known melting point, 4000 °C, or 7200 °F). There are three other important reactive elements in steel: silicon, aluminum, and boron. The steelmaking process produces low levels of silicon and aluminum in all steels, and boron is sometimes an additive. The oxides, nitrides, and sulfides formed by all of the reactive elements in steel are generally present as small particles called *inclusions*. These inclusions are usually present at very small volume percentages and, in some cases, are detrimental to toughness, as discussed in Chapter 6 for sulfide inclusions. At other times, however, the inclusions can offer beneficial effects, especially regarding their effect on grain growth of austenite.

Adding alloying elements to steel can produce a significant reduction in the rate of grain growth of austenite. The alloying elements produce this reduction in grain growth rates by two different mechanisms: particle drag and solute drag.

Particle Drag. To understand how the importance of particle drag was discovered, it is helpful to be familiar with steel ingot-making practice. Until only approximately two decades ago, steel in the United States was produced predominately by solidification of molten steel in large individual ingots. (The major tonnage of steel is now produced by continuous casting. It is often referred to as strand cast steel, because the production line facility is called a strand.) The oxygen content in the molten steel was generally high enough that, on cooling in the ingot mold, the oxygen combined with the carbon in the steel to form bubbles of carbon monoxide (CO) gas. This hot gas boiled up out of the surface of the ingot, producing a fireworks-type display. These ingots were called *rimming* ingots, because they had a soft ferrite rim at the surface. Rimming ingots were only useful for low-carbon nonalloy steels, which, incidentally, represented a large fraction of steel production, because rimming ingots were used for the sheet material in such items as car bodies and appliance bodies. To avoid oxidizing the alloying elements in alloy steel, the oxygen level in the ingot needs to be reduced to the point where the CO boil does not occur. These ingots were called *killed ingots*. To reduce the oxygen content of killed ingots, small amounts of silicon and/or aluminum were added to the ingot. These reactive elements tied up the oxygen as silicon oxides and aluminum oxides, thereby lowering the amount of oxygen dissolved in the molten steel and stopping the CO boil. In the first half of the 20th century, it was realized that steels that were aluminum killed had a much superior resistance to grain growth than steels that were silicon killed. The grain growth data of Fig. 8.3 were taken on a silicon-killed 1060 steel. These data are presented again in Fig. 8.9 and compared to grain growth data for an Al+Si-killed 1060 steel. Note the dramatic reduction in grain growth produced by the increase of the aluminum content from only 0.006 (steel 6) to 0.02% Al (steel 7). All modern strand cast steels are aluminum-killed steels, sometimes referred to as AK steels.

For many decades, it was not known what caused this dramatic difference in grain size. Finally, in the 1960s, it was shown that the effect was produced by the formation of very small particles of aluminum nitride (AlN) that formed in the steel as it cooled down. The small particles exert a drag force on the growing grain boundaries that is called particle drag. The small AlN particles have sizes on the order of 0.01 μm (4×10^{-7} in., requiring electron microscopes to see). On heating, they coarsen in size, and when they become larger than some critical value, they are no longer able to pin the austenite grain boundaries. In steel 7 in Fig. 8.9, the critical size was achieved at a temperature of approximately 930 °C (1700 °F). Above this temperature, grain

*In the world of science, the term *nanomaterial* has become a current buzzword for possible exciting new materials. The pearlite spacing in an air-cooled 52100 steel is approximately 0.075 μm, which is 75 nanometers (nm). The cementite plate thickness is just 11% of this value, which shows that these plates are only approximately 8 nm thick. The pearlite spacing for drawn pearlitic wire used for cables is reduced by the drawing operation, and cementite plate thicknesses of 1 nm are obtained for maximum drawing. Pearlite is an example of a nanomaterial that has been widely used, long before this buzzword came along.

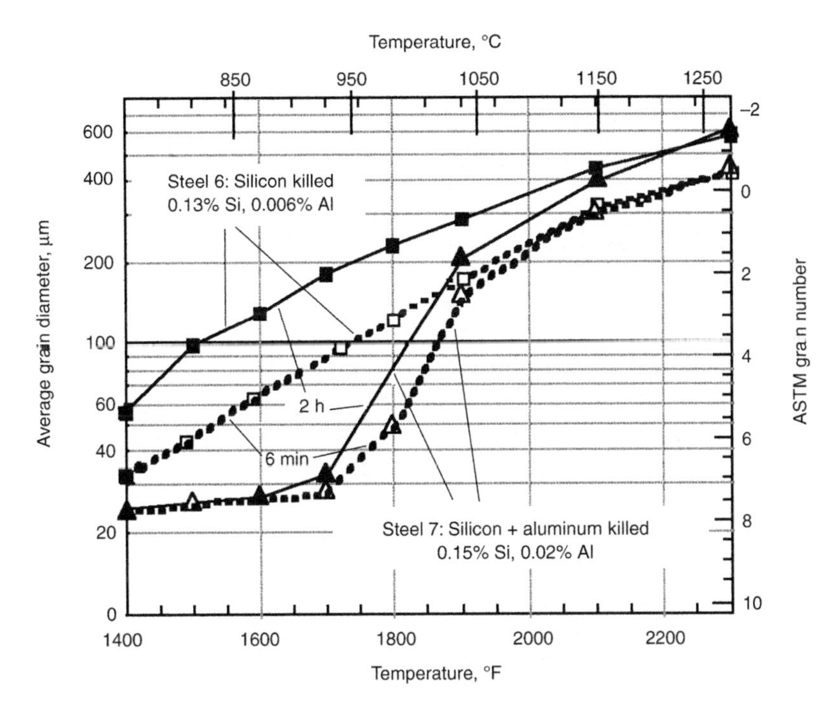

Fig. 8.9 Average grain diameter of austenite versus temperature in two 1060 steels austenitized for 6 min and 2 h. Grain growth is suppressed in the aluminum killed steel. Source: Ref 8.2

growth accelerated until grain size was the same as for steels free of particles.

Experiments have shown that particle drag is optimized by:

1. Minimizing the size of the particles
2. Choosing particles that resist coarsening or dissolving to high temperatures
3. Maximizing the volume fraction of particles

Generally, the higher the melting point of a particle, the less tendency it will have to coarsen. The reactive elements in Fig. 8.8 are ideal for forming particles, and because the highest-melting-point metals, such as niobium (also called columbium in the United States), tantalum, and tungsten, produce the highest-melting carbides, they resist coarsening up to high temperatures. Figure 8.10 presents data showing the effectiveness of very small additions of niobium on grain growth for a 1040 steel. Note that the onset of accelerated grain growth is increased from 930 °C (1700 °F) for the aluminum nitride particles shown in Fig. 8.9 to just over 1100 °C (2000 °F). The niobium reacts with both the carbon and nitrogen in the steel and forms particles that are called niobium carbonitrides. As Fig. 8.10 shows, these particles are able to effectively pin the austenite grain boundaries at temperatures up to 1100 °C (2000 °F).

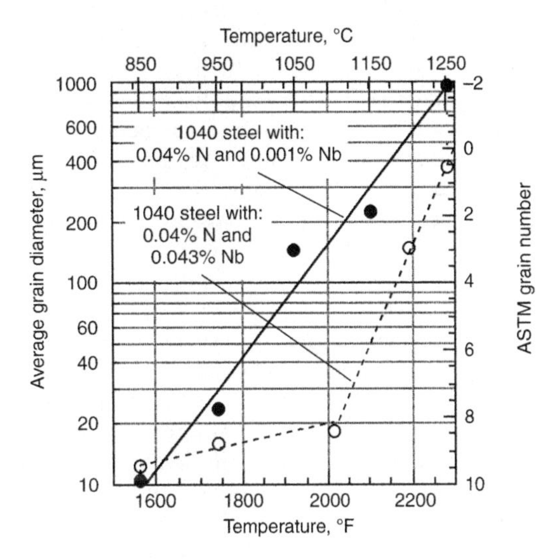

Fig. 8.10 One-hour grain growth is suppressed by adding very small amounts of niobium to a 1040 steel. Source: Ref 8.6

The discovery that particle drag inhibits grain growth played an important role in the development of a new class of steels. These are structural steels called high-strength low-alloy (HSLA) steels, and they are non-heat-treated low-carbon steels. The carbon content is approximately 0.15 to 0.25% C, and, for this low level, they produce relatively high strengths in the hot-rolled condition, with tensile strengths of 415 to 700 MPa (60 to 100 ksi). Grain refinement is aided by the addition of one or more of the reactive metals, niobium, vanadium, titanium, or aluminum. Because these elements are added at very low percentages, on the order of 0.04 to 0.1%, the steels are sometimes called microalloyed steels.

The small particles reduce austenite grain growth and also increase the recrystallization temperature of austenite. This combination allows the steel mill to produce a fine-grained ferrite by a process called controlled rolling. The process first generates a fine-grained austenite and then rolls it at a relatively low temperature and at the highest reduction that will not recrystallize the austenite. The high-density-defect structure remaining in the deformed austenite grains causes the ferrite grains that form in them on cooling below A_{r_1} to be very fine. The resulting ferrite-pearlite or ferrite-bainite steels have very fine grain sizes and higher strength and toughness than plain carbon structural steels.

Solute Drag. Any element that is dissolved in a phase, such as carbon dissolved in fcc iron or nitrogen dissolved in bcc iron, is called a *solute*. When alloying elements such as chromium, nickel, and molybdenum are dissolved in iron, they are too big to fit into the holes between the iron atoms, and they must replace the iron atoms at their preferred sites on the fcc or bcc lattice sites. These solute atoms are not the correct size to easily fit into the vacant iron lattice sites, and therefore, they will introduce a local energy increase by pushing or pulling on the neighboring iron atoms. However, if a solute atom is near a grain boundary, it can reduce this energy increase significantly by hopping onto the grain-boundary surface. The reason for this is that the grain-boundary surface is a region where the iron atoms do not match up very well from one side to the other. Thus, there is more open space along this boundary than elsewhere, and the solute atoms will fit in there with less pushing or pulling of neighboring iron atoms. Consequently, the localized energy increase around the solute atoms will decrease if the atom sits on the grain boundary, as shown schematically at the top of Fig. 8.11. Because of this energy decrease, the solute atoms will prefer to lie on the grain boundary. If the planar concentration of solute atoms is plotted (solute atoms per area parallel to the grain boundary), the solute concentration will be highest on the grain

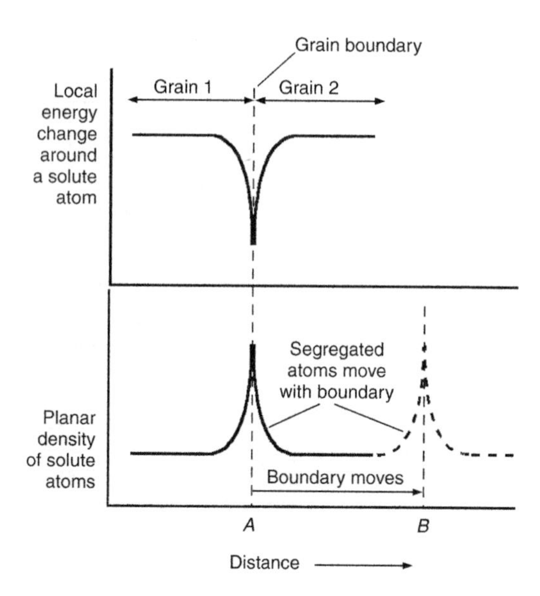

Fig. 8.11 Segregation of solute atoms to grain boundaries and resultant solute drag

boundary, as shown at the bottom of Fig. 8.11. This phenomenon is called grain-boundary segregation, and it is largest for the atoms that do not fit well into the vacant lattice sites.

Now, consider what must happen when a grain boundary with atoms segregated on it tries to move. The sketch at the bottom of Fig. 8.11 shows that when such a boundary moves from location *A* to location *B*, the atoms that have segregated onto the boundary must move along with the boundary as it migrates from *A* to *B*. However, these atoms can only move from position *A* to *B* by the process of diffusion. Also, because these atoms are too large to fit into the holes between the iron atoms of the bcc or fcc lattice sites, they will have very low diffusion coefficients (compare Fig. 7.6 to Fig. 7.5). Therefore, as the grain boundary moves, it must drag the solute atoms along with it, and the solute atoms will exert a drag force on the boundary and slow down its migration rate. This is the effect known as grain-boundary drag.

The four major alloying elements, manganese, chromium, molybdenum, and nickel, are not particularly effective at forming fine particles that will give as strong a particle drag effect as niobium, titanium, vanadium, and aluminum. However, they will produce a grain-boundary drag effect and can reduce the rates of grain growth. The solute drag effect in alloyed steels is illustrated in Fig. 8.12. The data show that grain growth in an 8620 steel is significantly

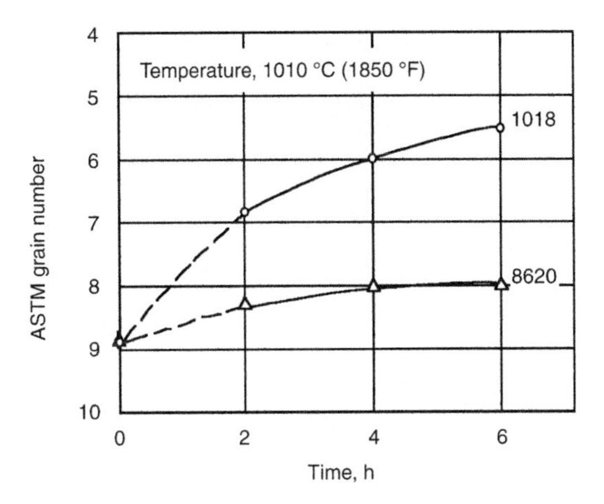

Fig. 8.12 Grain growth in a plain carbon 1018 steel versus a triple-alloyed 8620 steel at 1010 °C (1850 °F). The alloying elements cause a grain-boundary drag effect and inhibit grain growth. Source: Ref 8.7

lower than in a 1018 steel. The 8620 steel is a triple-alloyed steel containing low levels of chromium, molybdenum, and nickel (Table 6.3) and is commonly selected for applications requiring carburization. To reduce carburizing times and save energy costs, it is desirable to carburize at as high a temperature as possible. A problem in doing this with plain carbon steels, such as 1018, is that excessive austenite grain growth may occur. The data of Fig. 8.12 show that the grain growth rate for 8620 steel is less than for 1018 steel, which should eliminate excessive austenite grain growth during carburization.

Summary of the Major Ideas in Chapter 8

1. The grains of steel (first discussed with Fig. 1.2) have three-dimensional shapes that can be modeled by soap bubbles, as shown in Fig. 8.1(a). Similarly, they are modeled by the bubbles that form in a beer bottle after rapidly emptying the bottle (Fig. 8.1b).
2. Grain size is characterized by the *ASTM grain size number*. Large numbers, such as 10, indicate small grains, and small numbers, such as 2, indicate large grains. See Table 8.1 for details.
3. As austenite is heated to higher temperatures, the grain size increases in a process called *grain growth*. Grain growth occurs when surface tension forces act in the grain boundaries and cause small grains to shrink and large ones to grow. It also can be modeled by soap bubbles and can be visualized by studying the growth of bubbles in a beer bottle after rapidly emptying its contents. Grain growth increases very rapidly as temperature is raised and less rapidly with time held at a constant temperature (Fig. 8.3).
4. The toughness of steel is improved as grain size becomes smaller. For this reason, it is important to heat treat at as low a temperature as possible to reduce grain growth of the austenite.
5. A steel grain size can be dramatically reduced by cyclic heat treatments, where the temperature cycles above the A_{c_3} (and/or the A_{c_1}) and below the A_{r_1} values. With each heating ramp, a whole new set of austenite grains forms, and in each cooling ramp, a new set of ferrite/pearlite grains forms. The reduction in grain size is improved by using rapid heating and cooling rates, such as is possible with salt pot quenching. It is also enhanced by keeping the maximum temperature as low as possible. Ultrafine steel grain sizes are easily obtained (compare Tables 8.2 and 8.1).
6. A new set of grains also can be formed by the process of *recrystallization*. The steel must be fairly heavily deformed and then heated. The deformation introduces defects (planes of atoms displaced from their low-energy positions) into the grains, thereby raising their energy level. Subsequent heating produces a new set of defect-free grains, because they have a lower energy content. The new strain-free grains form at a temperature called the recrystallization temperature. The recrystallization temperature depends on the amount of prior deformation for steel, as shown in Fig. 8.7.
7. When forging austenite at high temperatures, recrystallization often occurs during forging, because the metal is deformed in the forging dies in a process called *dynamic recrystallization*. The defects that form in steel grains during deformation cause the grains to become harder, stronger, and brittle. The new strain-free grains formed by recrystallization are softer and ductile. Thus, dynamic recrystallization allows austenite to be deformed by large amounts without becoming brittle and, at the same time, refines (reduces) grain size.
8. Alloying elements in steel can be subdivided into chemically reactive elements to the left of iron in the periodic table (Fig. 8.8)

and nonreactive elements to the right of iron in the table. The reactive elements will form compounds, which appear as discrete particles in the steel. These elements react with carbon, nitrogen, oxygen, and sulfur in the steel to form carbides, nitrides, oxides, and sulfides that are sometimes called inclusions.

9. The addition of alloying elements can produce dramatic reductions in the grain growth of austenite. This reduction is produced by different effects (particle drag, solute drag) in the reactive versus nonreactive alloying elements.
10. If the particles formed by the reactive elements can be made very small, they will reduce austenite grain growth (an effect known as particle drag) until the temperature is high enough to either dissolve them or make them grow to larger sizes (coarsen). The addition of minute quantities (such as 0.04%) of the reactive elements lying in the two left columns in Fig. 8.8 is very effective in reducing grain growth and is the basis of the relatively new high-strength low-alloy (HSLA) steels (Fig. 8.10).
11. By the process called solute drag, the less reactive elements, such as the common alloying elements nickel, manganese, chromium, and molybdenum, produce a reduction in grain growth. These alloying elements segregate to the austenite grain boundaries, where they become trapped and are dragged along with the boundaries as they move, thereby slowing grain growth. This effect is illustrated in Fig. 8.12 by comparing grain growth at 1010 °C (1850 °F) in a plain carbon steel and a Cr-Mo-Ni low-alloy steel, both having approximately 0.2% C.

REFERENCES

8.1. G.F. Vandervoort, *Metallography, Principles and Practice*, McGraw-Hill, 1984
8.2. O.O. Miller, Influence of Austenitizing Time and Temperature on Austenite Grain Size of Steel, *Trans. Am. Soc. Met.*, Vol 43, 1951, p 261–287
8.3. R.A. Grange, Strengthening Steel by Austenite Grain Refinement, *Trans. Am. Soc. Met.*, Vol 59, 1966, p 26–48
8.4. *Atlas of Microstructures of Industrial Alloys*, Vol 7, *Metals Handbook*, 8th ed., American Society for Metals, 1972, p 10
8.5. H.F. Kaiser and H.F. Taylor, *Trans. Am. Soc. Met.*, Vol 27, 1939, p 256
8.6. R. Coladas et al., Austenite Grain Growth in Medium and High Carbon Steels Microalloyed with Niobium, *Met. Sci.*, Nov 1977, p 509–516
8.7. G.O. Ratliff and W.H. Samuelson, High Temperature Carburization Is Routine Process at Shore Metal, *Met. Prog.*, Vol 108, Sep 1975, p 75–78

CHAPTER 9

Hardenability of Steel

THE PROFESSIONAL materials information organization, ASM International (formerly the American Society for Metals) has published an excellent handbook (Ref 9.1) on the heat treating of steels. This handbook is referred to fairly extensively in this chapter and the next. The subject of this chapter, hardenability, is often confused with hardness, and one intent of this chapter is to distinguish the difference between these two concepts.

The general process of hardening a steel (for example, a round bar) involves three steps:

1. *Austenitization:* The steel is heated to form austenite (usually 100% austenite but not always). The austenitization temperatures vary with the carbon composition of the steel and are shown in approximations in Fig. 4.23. Austenitization is discussed in more detail in Chapter 11.
2. *Quenching:* The hot steel is rapidly cooled by immersion in water or oil. Quenching techniques are discussed further in Chapter 12.
3. *Tempering:* The steel is heated to a low temperature to remove brittleness. The tempering process is discussed further in Chapter 10.

The steel will be hardened after the second step, but the tempering step is included in the list because it is virtually always employed in the heat treatment of hardened steels.

In order to fully harden the steel bar, the austenite must transform to martensite at all points on the surface to the center of the bar. Consider again the microstructures of the quenched steels shown in Fig. 4.19 and 4.20. As explained in Chapter 3, "Steel and the Iron-Carbon Phase Diagram," the reason that pearlite and bainite formed on the old austenite grain boundaries is that the temperature of the austenite did not fall below the M_s temperature fast enough to avoid their formation. Also, when pearlite or bainite form in austenite, the volume they occupy cannot transform to martensite. To fully transform austenite to martensite, the cooling rate on quenching must be above some critical value throughout the austenite. Now, think about how the temperature will drop in a round bar that is quenched in water. The rate at which the temperature drops will be fastest at the surface and slowest at the center. Hence, it is possible that the outer regions of the bar will be fully martensitic, but the central regions will have pearlite and bainite mixed in with the martensite. If this happens, the hardness of the bar will be lower in the central regions. Such a situation is shown for a 1060 bar in Fig. 9.1. The figure shows the hardness of an oil-quenched 25 mm

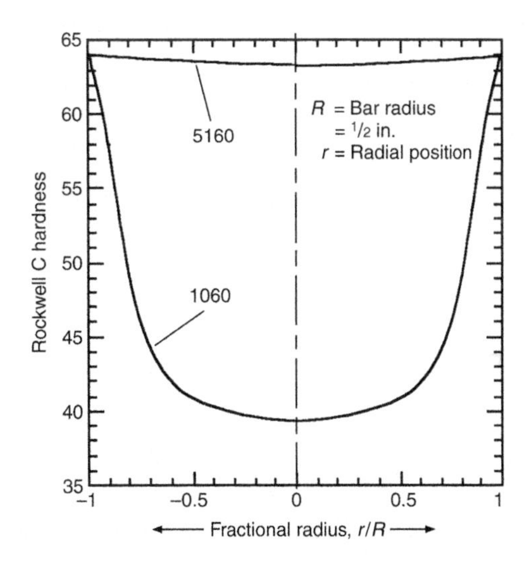

Fig. 9.1 Rockwell hardness versus radius for 25 mm (1 in.) diameter bars of oil-quenched 1060 and 5160 steels

(1 in.) diameter bar at positions from the center ($r = 0$) to the surface ($r = \frac{1}{2}$ in., or $r/R = 1$, where r is any radial position, and R is the outside radius of the bar).

Now, consider a 5160 steel. Tables 6.1 and 6.3 show that the only significant difference in the composition of 1060 and 5160 is the addition of 0.8% Cr in 5160. Figure 9.1 shows that the as-quenched hardness of the 5160 steel does not decrease significantly below the surface of the bar; that is, the bar has been through hardened. The 5160 steel has martensite all the way to its center, but the 1060 has pearlite and bainite in the central regions, causing the drop in hardness. Both bars were cooled at the same rate. This means that the chromium present in 5160 must have made the formation of pearlite and bainite more difficult. The results also illustrate another important point regarding the effect of alloying elements. At the surface, both bars are fully martensitic, and both bars have the same hardness. This means that the chromium addition in 5160 has not changed the hardness of the fresh martensite.

Figure 4.13 presents the hardness of fresh martensite versus the %C in steel. This curve applies to both plain carbon and alloy steels, which means that alloying elements do not affect the hardness of martensite. So, alloy steels are not harder because of the alloying elements making the martensite harder. When as-quenched alloy steels are harder than plain carbon steels with the same carbon composition, it is because the alloying elements are preventing formation of the softer pearlite and bainite. The chromium addition has improved the *hardenability* of the steel, as can be seen by comparing the hardness profiles of 1060 and 5160 in Fig. 9.1. So, hardenability is a measure of how deep below the surface a quenched steel can be made fully martensitic. On the other hand, hardness is a measure of the resistance to penetration that the steel offers to an indenter point forced into its surface under a given load.

From the previous discussion, it should be clear that to improve the hardenability of a steel, it is necessary to slow down the formation of the constituents that can form in austenite prior to martensite and prevent its formation. These product constituents can be any of the three shown in Fig. 4.23 at temperatures above the M_s temperature: ferrite, pearlite, or bainite. Two experimental techniques have evolved for characterizing the rate of formation of these constituents: isothermal transformation (IT) and continuous transformation (CT). Each technique has associated diagrams that are called IT diagrams and CT diagrams. (Note: Some people continue to prefer the use of the older names, time-temperature transformation (TTT) in place of IT and continuous cooling transformation (CCT) in place of CT.)

Isothermal Transformation Diagrams

As previously discussed in Chapter 4, "The Various Microstructures of Room-Temperature Steel," the IT technique involves the use of thin samples that are quenched from the austenitizing temperature to a temperature of interest, and the transformation of the austenite is allowed to occur isothermally at this temperature. Figure 9.2 is used to explain how an IT diagram is constructed. After the quenched sample rapidly cools to the isothermal temperature, the austenite will eventually begin to transform to one of the three product constituents. Consider a pearlite

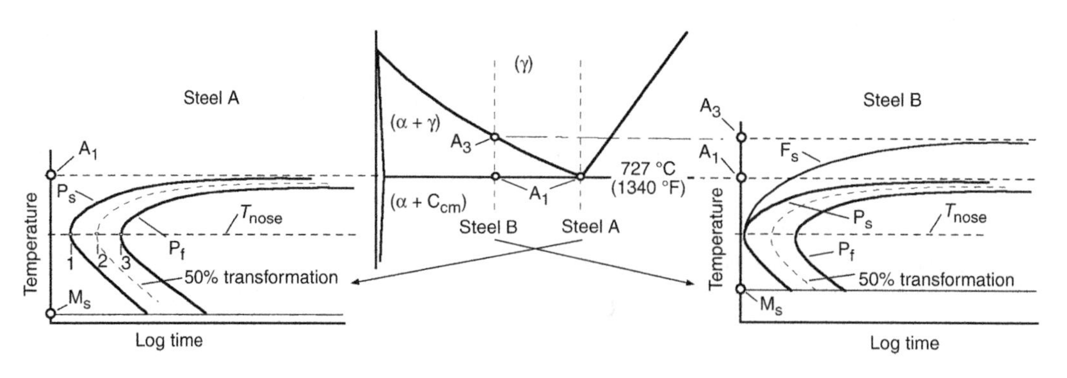

Fig. 9.2 Iron-carbon diagram (center figure) with %C compositions of steel A and steel B. The corresponding isothermal transformation diagrams for the two steels are shown at the sides

steel, such as steel A in Fig. 9.2. After the sample has been quenched to the temperature labeled T_{nose} for the time labeled "1" on the IT diagram at the left, the austenite will start to form pearlite. This point is labeled P_s, for pearlite start. After time 3, the austenite has transformed completely to pearlite, and so this point is labeled P_f, for pearlite finish. Point 2, not always shown on IT diagrams, indicates the time when 50% of the austenite has transformed.

The experimental determinations of these times is often made using a device called a dilatometer, which measures the length change of the sample during the hold time. Austenite has a smaller volume per atom than any of the three product constituents, so the sample will begin to expand at P_s and will stop expanding at P_f. By quenching to different temperatures, the P_s and P_f times for a range of temperatures can be determined. Plotting the P_s and P_f times on a log scale against temperature results in the IT diagram. Note that the P_s curve has a minimum time value at the temperature labeled T_{nose} in Fig. 9.2. This point is called the nose of the P_s curve for obvious reasons. If T_{nose} times are less than a few seconds, the steel will have poor hardenability, because it will be very difficult to cool the interior of a bar fast enough to avoid pearlite formation.

Pearlite cannot form unless austenite is cooled below A_1. So, it would be expected that the more austenite is undercooled below A_1, the faster pearlite will form. The IT diagram for the pearlite steel A shows that, initially, such is the case. However, as the steel is cooled below the T_{nose} temperature, it takes longer for the pearlite to start forming as the undercooling below A_1 is increased. The reason for this result is diffusion. In order for pearlite to form in austenite, the carbon atoms have to rearrange themselves from the random distribution in austenite to the special distribution in pearlite. As discussed in Chapter 3 and shown in Fig. 3.7 and 3.8, pearlite consists of plates of ferrite (%C = 0.02) and plates of cementite (%C = 6.67). The mechanism that redistributes the carbon from the uniform composition of 0.77% in austenite to the 0.02/6.67% mixture of plates in pearlite is diffusion. However, as discussed in Chapter 7, diffusion is very sensitive to temperature. The rapid decrease in the carbon diffusion coefficient as the temperature drops eventually wins out and increases pearlite formation start times as the temperature drops below T_{nose}.

The IT diagrams always show the A_1 and the A_3 or A_{cm} temperatures on the temperature axis (except for pearlitic steels, where only A_1 is needed). Steel B is a hypoeutectoid steel, because its composition is below the 0.77% eutectoid (pearlite) value. Hypoeutectoid steels will have A_3 values lying above the pearlite temperature, A_1, by increasing amounts as %C drops. Therefore, these steels will have a temperature range on the IT diagram that lies between A_3 and A_1, as shown in the IT diagram for steel B at the right of Fig. 9.2. By looking at the phase diagram at the center of Fig. 9.2, it is clear that in this A_1 to A_3 temperature region, only ferrite (α) can form from the austenite (γ). On the IT diagram, there is a line labeled F_s, which locates the times where ferrite begins to form from the austenite, the ferrite start line. This line never extends above the A_3 temperature, because ferrite cannot form above that temperature. However, ferrite can form below the A_1 temperature, and the curve illustrates that, for this example, it can form at temperatures nearly down to the nose temperature. According to Fig. 4.23, the lowest temperature where ferrite can form decreases as the %C decreases from the 0.77% pearlite composition in hypoeutectoid steels.

The actual IT diagrams for three different AISI plain carbon steels that appear in Ref 9.2 and other ASM Handbooks are shown in Fig. 9.3. The diagrams illustrate several points. First, most published diagrams do not label the F_s, P_s, or B_s curves (the start curves for ferrite, pearlite, and bainite), nor do they usually display where pearlite and bainite form. They include the labels, A, F, and C for austenite, ferrite, and cementite (or carbide). The dilatometer that locates the lines on these diagrams cannot distinguish between the three product constituents, and so, it is left to the reader to figure out whether a given line is an F_s, P_s, or B_s line. For plain carbon steels, a reasonable estimate can be made using the information in Fig. 4.23. For example, this figure predicts that in a 1080 steel, pearlite will form from A_1 down to approximately 550 °C (1020 °F), and bainite from there down to the M_s temperature. Therefore, on the IT diagram for 1080 (at the bottom of Fig. 9.3), the start curve would be the P_s for temperatures from A_1 down to approximately 550 °C (1020 °F), and at lower temperatures, the curve would be the B_s down to the M_s temperature.

There are three important effects that occur as the %C is increased in plain carbon steels, as illustrated in Fig. 9.3. First, as the %C increases,

NOTE

(1) The ASM International diagrams label the A_3 temperature as A_f, and the A_1 temperature is labeled A_s. This nomenclature was suggested by R.A. Grange in 1961 for its similarity to the M_s and M_f nomenclature (Ref 9.3). It is used in the various ASM publications, but virtually no one else has adopted this nomenclature, and it is not used here except on copied ASM diagrams. Andrews (Ref 6.4) showed that $A_f = A_3$ and that $A_s = A_1$. (2) It is not uncommon to encounter IT and CT diagrams that use the labels Ac_1 and Ac_3 instead of A_1 and A_3. This is done because the A_1 and A_3 temperatures were measured in heating experiments done at very low heating rates, and resultant values are estimates of A_1 and A_3, thus the use of the "c" subscript (Chapter 3). Many authors use the nomenclature Ae_1 and Ae_3, but as noted in Chapter 3, in this book the "e" is understood, so that $Ae_1 = A_1$ and $Ae_3 = A_3$.

the curves shift to the right, that is, to longer times. Notice that the nose time for the start curves rises from below 0.5 s on the two lower-carbon steels (1021 and 1060) to 0.8 s on the 1080 steel. This shifting of the start curves to longer times with increased %C means that the hardenability of plain carbon steels increases as the %C increases, at least up to the pearlite composition, 0.77% C. The 1060 curve illustrates why the hardness drops so fast below the surface in Fig. 9.1. To avoid pearlite formation and obtain 100% martensite in this steel, Fig. 9.3 shows that the temperature must fall below 540 °C (1000 °F) in roughly 0.3 s. It is only near the quench surface that the temperature can fall fast enough to avoid pearlite formation.

Second, as the %C increases, the M_s temperature decreases, as illustrated in Fig. 4.17 and 4.23. The IT diagrams in Fig. 9.3 show this effect, because the M_s temperatures are shown next to the temperature axis. Note that the M_{50} and M_{90} temperatures are also labeled. These are the temperatures at which 50 and 90% martensite have formed.

Third, as the %C increases, the size of the A + F region decreases and vanishes for the 1080 steel, where A_3 has collapsed to A_1. This effect was explained with the hypothetical IT diagrams of Fig. 9.2. The Fig. 9.2 diagrams also show the first and second effects.

To improve hardenability, it is necessary to force the various start curves that appear on the IT diagrams to longer times. Physically, this means that it must be made more difficult for the three product constituents (ferrite, pearlite, and bainite) to form in the austenite. There are two principal ways in which this can be effected: increasing grain size and adding alloying elements.

Effect of Grain Size on Hardenability. It was pointed out previously that the product constituents virtually always form on the austenite grain boundaries. The grain-boundary area depends on grain size. A larger grain size will reduce the amount of grain-boundary area per unit volume, which will shift the start curves to longer times and improve hardenability. Hence, the position of the curves on IT diagrams depends on the austenite grain size. For this reason, as shown in Fig. 9.3, published IT diagrams always specify the grain size.

Effect of Alloying Elements on Hardenability. As discussed in Chapter 8, "Control of Grain Size by Heat Treatment and Forging," reactive alloying elements such as chromium and molybdenum will form carbides. Consequently, these elements will prefer to go into the cementite (carbide) part of pearlite or bainite when they form. When a given volume of austenite transforms into pearlite or bainite in a plain carbon steel, the carbon atoms must rearrange from their uniform distribution in the original austenite volume. In the transformed volume, there must be virtually no carbon (0.02%) in the ferrite part and 6.7% C in the cementite part. This rearrangement is accomplished by diffusion. Similarly, during transformation in the alloy, the chromium and manganese atoms must rearrange from uniform distributions in the old austenite to higher levels in the carbides that form and lower levels in the ferrite that forms. However, the diffusional rearrangement of the alloying elements is more difficult than it is for carbon because of their tremendously smaller diffusion coefficients, as illustrated in Table 7.1. Therefore, their presence makes it more difficult for pearlite and bainite to form, and the P_s and B_s curves are shifted to the right (longer times) on the IT diagram. All elemental additions to steel, except cobalt, shift the start curves for ferrite, pearlite, and bainite to longer times.

The reason that nonreactive elements such as nickel slow the formation rates of pearlite and bainite is related to their effect on the phase diagram, and is a little too complicated to be explained easily. However, the net result is simple to remember: virtually all alloying additions in steel slow the formation of the constituents that form on cooling austenite. Figure 9.4 presents a comparison of the IT diagrams for 1060 and 5160 steels, which shows the effect of the alloying addition of chromium. Note from Tables 6.1 and 6.3 that 5160 is basically a 1060 steel with 0.8% Cr added. Figure 9.4 shows that the addition of this small amount of chromium has a significant effect on the position of the start curves on the IT diagram. Even though the grain size of the 5160 steel is smaller, the nose of the IT diagram is shifted to approximately 5 s on the 5160 steel versus less than 0.5 s on the 1060 steel.

Hardenability Demonstration. A simple and instructive experiment illustrates the dependence of hardenability on both grain size and composition. Two steels, 1086 and 5150, were forged into the tapered shape of a knife blade, as shown in Fig. 9.5. The 25 mm (1.0 in.) wide blades tapered from a thickness of 6 to 0.03 mm ($\frac{1}{4}$ to $\frac{1}{32}$ in.). Samples of each steel were prepared with a fine and course grain size, using the cyclic heat treatment discussed in Chapter 8 to give the grain sizes shown in Table 8.2. The steels were quenched into agitated room-temperature oil, sectioned in half, polished, etched, and examined in a microscope. Three samples were 100% martensite, both the large- and small-grained 5150 steels and the large-grained 1086 steel. However, the small-grained 1086 steel was a mixture of martensite and fine pearlite with a quite interesting distribution of martensite. Because the cooling rate should be fastest at the surface, the martensite may be expected to form first along the surface, as shown in Fig. 9.5(a). However, it was found to form at the thin end, with a small amount at the back corners, as shown in Fig. 9.5(b).

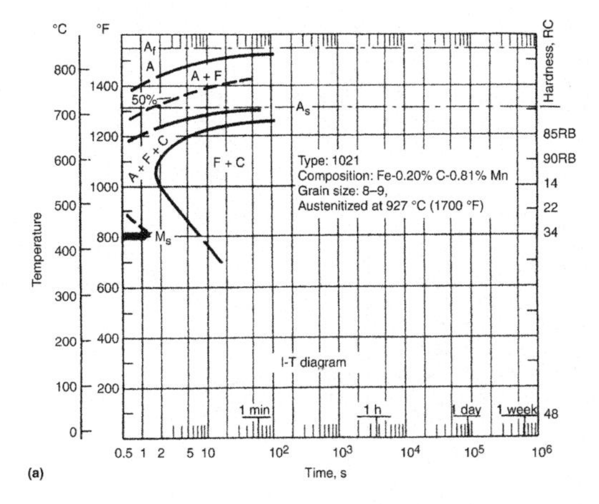

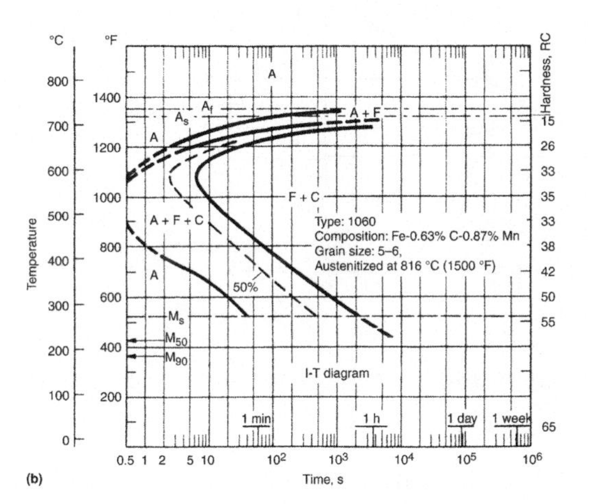

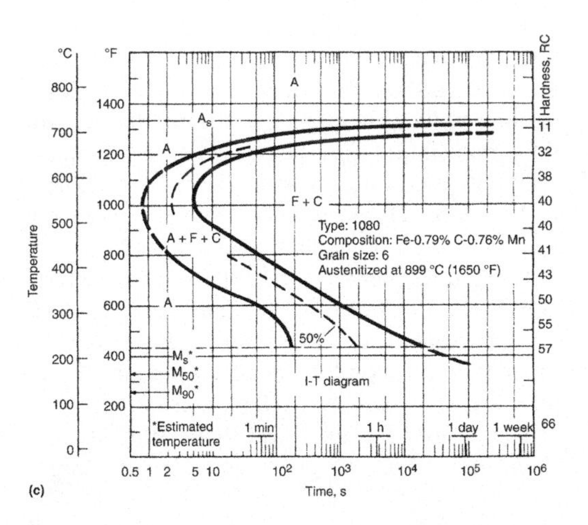

Fig. 9.3 Isothermal transformation diagrams for plain carbon steels 1021, 1060, and 1080 showing the effects of increasing %C. Source: Ref 9.2

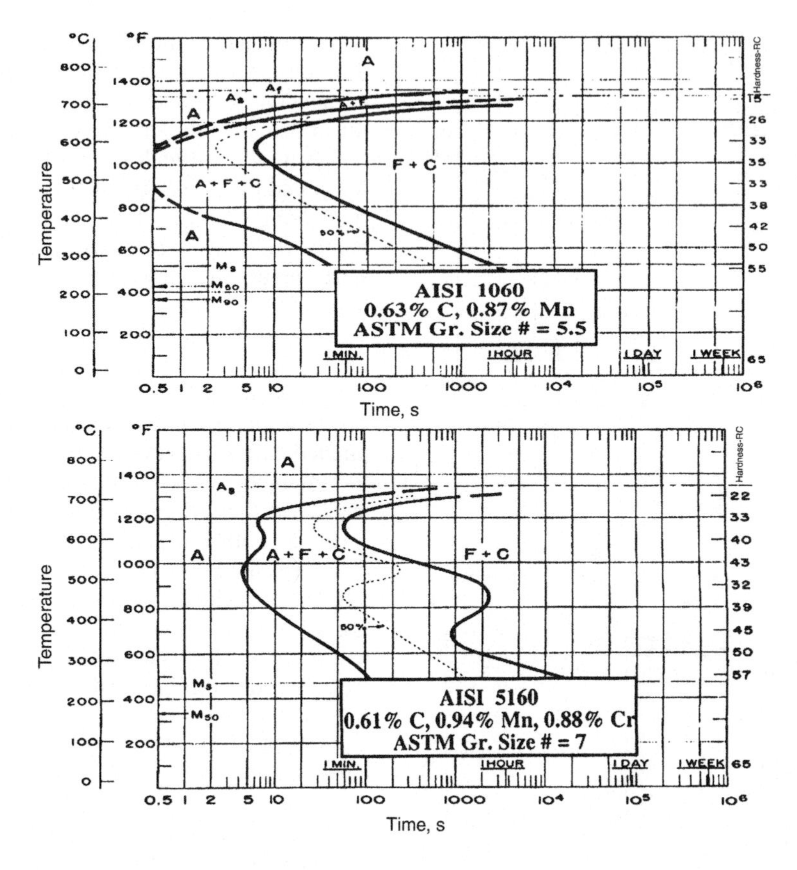

Fig. 9.4 Isothermal transformation diagrams for 1060 and 5160 steels. Alloying with chromium (5160) increases the transformation times. Source: Ref 9.2

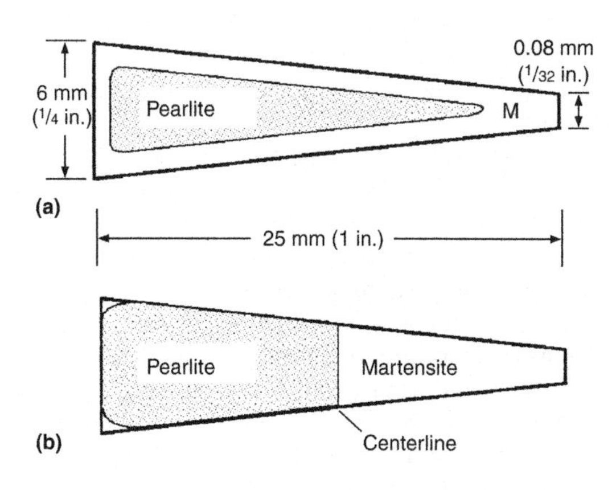

Fig. 9.5 Two possible martensite-pearlite configurations of quenched blades. (a) Expected martensite (M) formation. (b) Actual martensite formation

The fact that it does not form preferentially along the surface means that, for this tapered knife blade geometry quenched in agitated oil, the cooling rate at the centerline of the blade is nearly the same as at the surface for thicknesses up to 6 mm (¼ in.). (Such behavior is predicted for thin samples and is explained by what is called film heat-transfer control.)

The samples were further studied by making HRC hardness measurements at 0.06 mm (1/16 in.) intervals on the transverse blade sections, starting at the thin edge and progressing toward the spine (fat edge). The results of these measurements are plotted versus the blade thickness in Fig. 9.6. (Hardness measurements were not made on blade thicknesses less than 2 mm, or 0.08 in., due to end effects. Microhardness tests confirmed, however, that the hardness remained constant at blade thicknesses from 2 mm, or 0.08 in., to the thin end of the blades, where the thickness was approximately 0.8 mm, or 0.03 in.)

The results for the 1086 blade dramatically show the effect of grain size. In the large-grained sample (1086, grain size = 11), the blade is fully martensitic (HRC = 65) at all thicknesses. However, in the fine-grained, thermally cycled blade (1086, grain size = 15),

100% martensite is only able to a form out to a blade thickness of approximately 2 mm (0.09 in.). Beyond thicknesses of approximately 4 mm (0.15 in.), the blade has no martensite and consists of all bainite or pearlite. On the other hand, the 5160 steel is much less affected by grain size variation. It appears that there is a small loss of martensite due to the fine grains only at thicknesses of greater than approximately 4 mm (0.18 in.). Figure 9.6 illustrates why low-alloy AISI steels are desirable for heat treated parts in industrial sizes. Fine-grained parts are needed for improved toughness, and

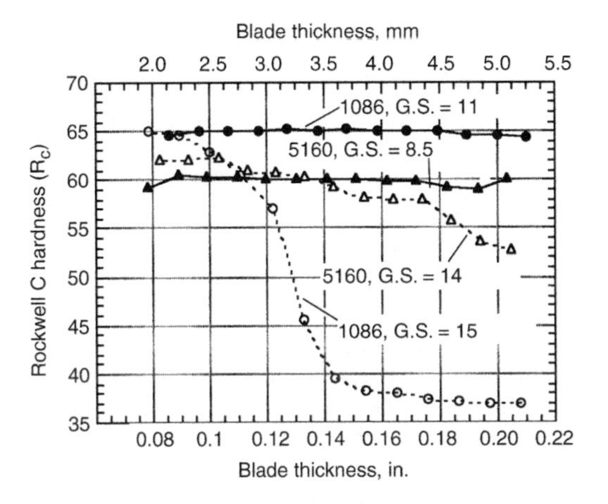

Fig. 9.6 Dependence of Rockwell C hardness on blade thickness. G.S., grain size

the increased hardenability of the alloy steels allows much deeper hardening than is possible with fine-grained plain carbon steels.

The IT diagram for 1080 steel given in Fig. 9.3 and replotted in Fig. 9.7 makes possible a detailed explanation for the martensite formation. Assume the IT curves apply to the small-grained 1086, grain size 15 steel. Hypothetical cooling curves are shown as dashed lines for positions in the blade at thicknesses of 0.09, 0.125, and 0.15 in. (2, 3, and 4 mm). Because 0.09 in. (2 mm) is the largest thickness on the knife blade showing 100% martensite, its cooling curve must be just to the left of the nose of the P_s curve. Because 0.15 in. (4 mm) is the smallest thickness on the knife blade showing no martensite, its cooling curve must be just to the right of the nose of the P_f curve. Also, because the thickness at 0.125 in. (3 mm) corresponds to a hardness 50% of the way between no martensite and all martensite (midway between R_c = 65 and 38 in Fig. 9.6), its cooling curve should pass through the nose of the 50% transformation curve. What happens in the case of course-grained pearlite? The cooling curves will not change their shapes or positions, because cooling rate is not affected by grain size. However, the position of the P_s and P_f curves depends on grain size, and the larger grain size must shift the nose of the pearlite start curve to the right of the cooling curve for the thickest part of the blade, because no pearlite formed in the large-grained 1086 steel.

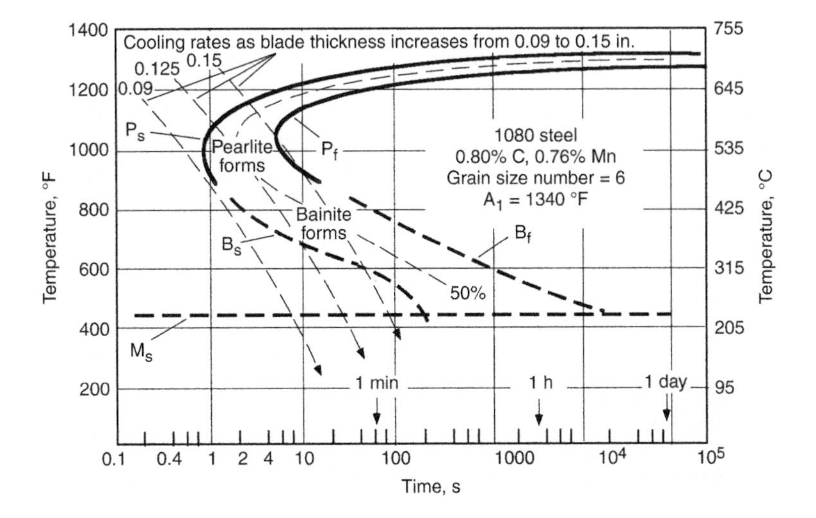

Fig. 9.7 The 1080 isothermal transformation diagram of Fig. 9.3 explains the structure variation found in the 1086 blade. By following the track of the cooling rate lines for a given thickness, it is possible to predict which steel constituents will emerge

Continuous Transformation (CT) Diagrams

The aforementioned analysis using the IT curve in Fig. 9.7 explains the fundamental ideas behind the variation of martensite formation in the 1086 blades, but it can only serve as a qualitative guide. The reason is that IT curves only apply to isothermally cooled steels. In other words, they apply only to steels that are quenched so fast that they remain fully austenite at the temperatures shown on the diagram until the first product constituent (ferrite, pearlite, or bainite) forms in the austenite. Figure 9.8 is a sketch of the P_s curve for 1080 steel and shows the cooling path that must be followed in order to use an IT diagram. It is a very rapid drop followed by an isothermal hold. In the blade experiment, this did not happen, because of a combination of the size of the sample and the slowness of the oil quench. Instead, the cooling curve for this experiment corresponds to the dashed line in Fig. 9.8. Experiments have shown that when a sample is cooled along a continuous cooling curve, such as the dashed line in Fig. 9.8, the first pearlite will not form at point 1 but will be delayed and form later at point 2. Hence, the P_s curve on a CT diagram will not have the same position as on an IT diagram. The CT curves are shifted down and to the right. This means that the previous analysis shown on the IT diagram of Fig. 9.7 can only be an approximation, because the cooling curves superimposed on that IT diagram should have been superimposed on a CT diagram.

Unless working with very thin samples and very fast quench conditions, the CT rather than IT diagrams should be used. Examples of CT diagrams are presented in Fig. 9.9 for two steels having compositions close to 0.4% C. The CT

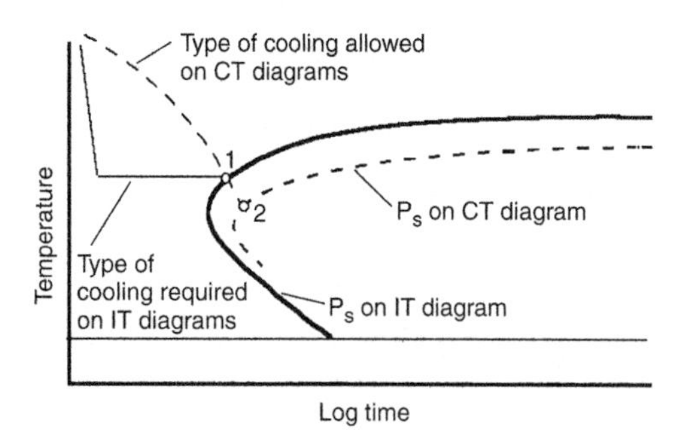

Fig. 9.8 Differences in cooling rate patterns result in differences between isothermal transformation (IT) and continuous transformation (CT) diagrams

diagrams show cooling curves directly on them. The downward-sloped lines in Fig. 9.9 with fractions such as ¼ and ½ are the cooling curves. The shape of the cooling curves depends on how the sample was cooled and will, in general, vary, depending on the source of the CT diagram. The cooling curves in Fig. 9.9 are for samples cooled by the Jominy technique, which is described in the next section, and the fractions in inches on the cooling curves refer to the distance from the quenched end on the Jominy bars.

Notice that in the 5140 diagram, the F_s, P_s, B_s, and M_s curves are labeled. By comparing the two diagrams in Fig. 9.9, these same curves for the 1036 steel can be identified. Even though the 1036 steel has considerably more manganese in it than most plain carbon steels (see composition in upper right corner of figure and compare to Table 6.1), its hardenability is significantly lower than the 5140 steel. The nose of the F_s curve is only approximately 2 s in the 1036 compared to approximately 7 s in the 5140.

The diagrams also show dashed lines with percentage numbers labeling them. These lines are rough estimates of the percentage of the various constituents that will be contained in a steel cooled along one of the cooling curves. For example, consider the cooling curve marked ¾ on the 5140 diagram. This cooling curve crosses the 1% line just below the P_s line at approximately 30 s. Notice that the cooling curve had crossed the F_s curve at approximately 20 s, so at this cooling rate, the steel started to form pearlite after slightly less than 1% ferrite had formed. The cooling curve crosses the B_s curve approximately midway between the 10 and 50% line, so bainite started to form after approximately $(50 - 10)/2 = 20\%$ pearlite had formed. It crosses the M_s line just beyond the 90% line, so approximately 9% martensite will form. In summary, a steel cooled at this rate will result in an approximate mixture of at least 1% ferrite, 20% pearlite, 9% martensite, and the remainder bainite. By similar reasoning, it can be seen that a steel cooled along the curve labeled ⅜ will result in approximately at least 3% ferrite, no pearlite, at least 20% bainite, and the remainder martensite.

The CT diagrams are presented in more formats than IT diagrams. To illustrate this idea, Fig. 9.10 presents an alternate type of CT diagram for a 0.38% C steel, with the composition given at the top of the diagram. The cooling curves for this diagram do not correspond to

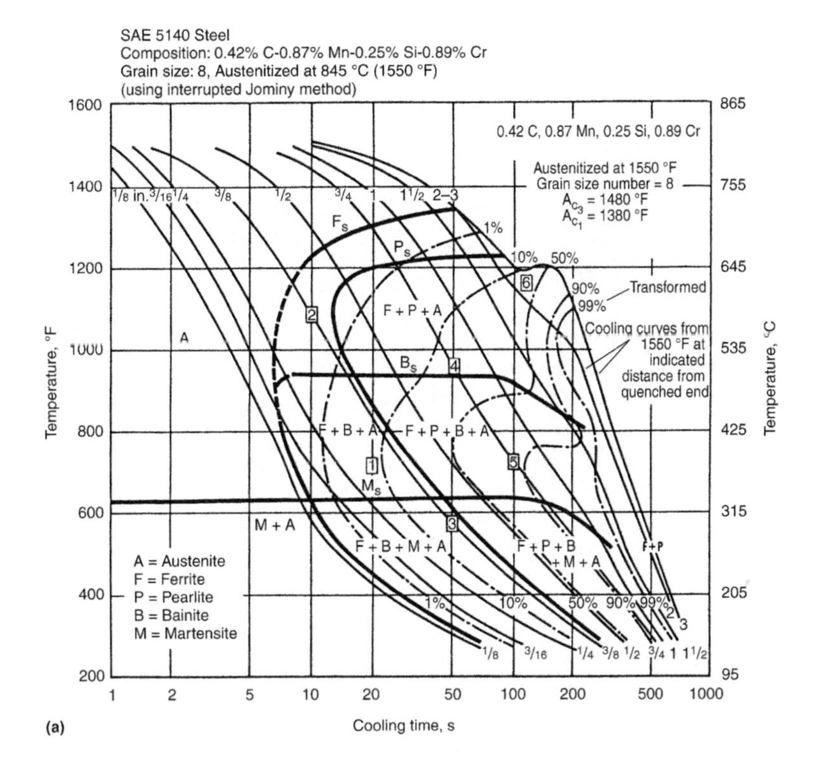

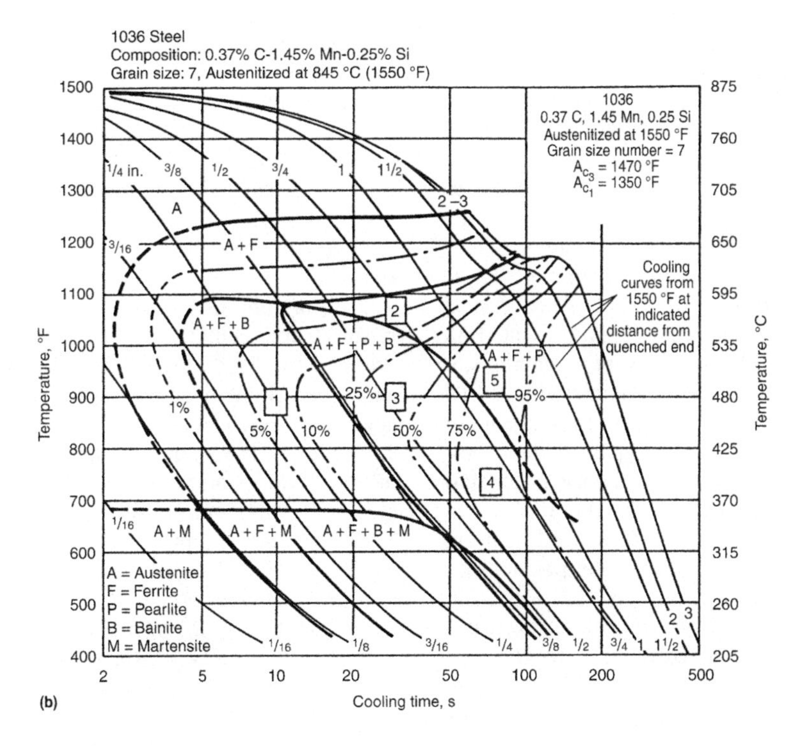

Fig. 9.9 Continuous transformation curves for steels 1036 and 5140 having approximately 0.4% C. Source: Ref 9.2

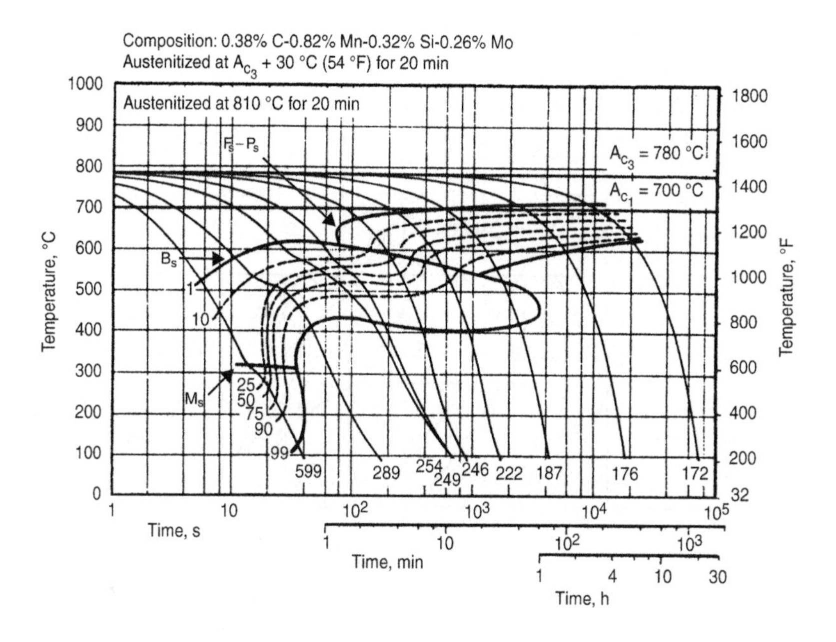

Fig. 9.10 Alternate-style continuous transformation curve that is often presented. Source: Ref 9.2

positions on the Jominy bar but to locations in the specimen from which this particular diagram was made. The number at the lower end of each cooling curve is the room-temperature diamond pyramid hardness found for that cooling curve. The dashed lines give the estimated percent of the austenite that has transformed. The start curves are not labeled on the original diagram, but labels have been included here after consulting the original reference. No distinction was made between the ferrite and pearlite start, so the upper start curve is a combined F_s-P_s curve.

There is a third style of CT diagram that is presented in Ref 9.2 without explanation. These diagrams were produced by the British Steel Corporation, and an example is presented in Fig. 9.11 for a 4130 steel. With these diagrams, it is possible to determine the microstructures that will be present only at the center of round bars of various diameters for three types of quenches: water, oil, and air. To illustrate the use of the diagram, consider three bars, each 50 mm (2 in.) in diameter. There are vertical downward-pointing arrows at the 50 mm (2 in.) diameter positions on the bottom horizontal scales for the water, oil, and air quenches. By following these lines upward, the predicted microstructures for the center of the bars can be determined. For example, consider the water-quenched arrow. It passes through the martensite-bainite transition just below the 90% transformation line, so the center of this steel would be predicted to contain approximately 93% bainite and 7% martensite. Similarly, the structure at the center of a 50 mm (2 in.) air-cooled bar is estimated to be approximately 40% ferrite, 53% pearlite, and 7% bainite.

Figure 9.11 allows for easy determination of the maximum bar diameter that can be through hardened, that is, consisting of 100% martensite at the center (see bold upward-pointing arrow indicating largest diameter with no bainite). For a water quench, the maximum diameter would be just less than 20 mm (0.8 in.), while for an oil quench, it would be just less than 15 mm (0.6 in.).

The diagram also specifies equivalent cooling rates at the center after its temperature has dropped to 750 °C (1380 °F). This information can be applied to any arbitrary quench if the cooling rate at 750 °C (1380 °F) is known. For example, at the moment that the center of a 50 mm (2 in.) water-quenched sample is 750 °C (1380 °F), the cooling rate at the center is just less than 2000 °C/min. (Note: References 9.1 and 9.4 have mislabeled the cooling rate ft/min.)

There are many sources of IT and CT diagrams in the literature. ASM International has published a book titled *Atlas of Time-Temperature Diagrams for Irons and Steels* that brings most of these together (Ref 9.2) and is the most complete source of IT and CT diagrams. Table 9.1 is a summary of its contents.

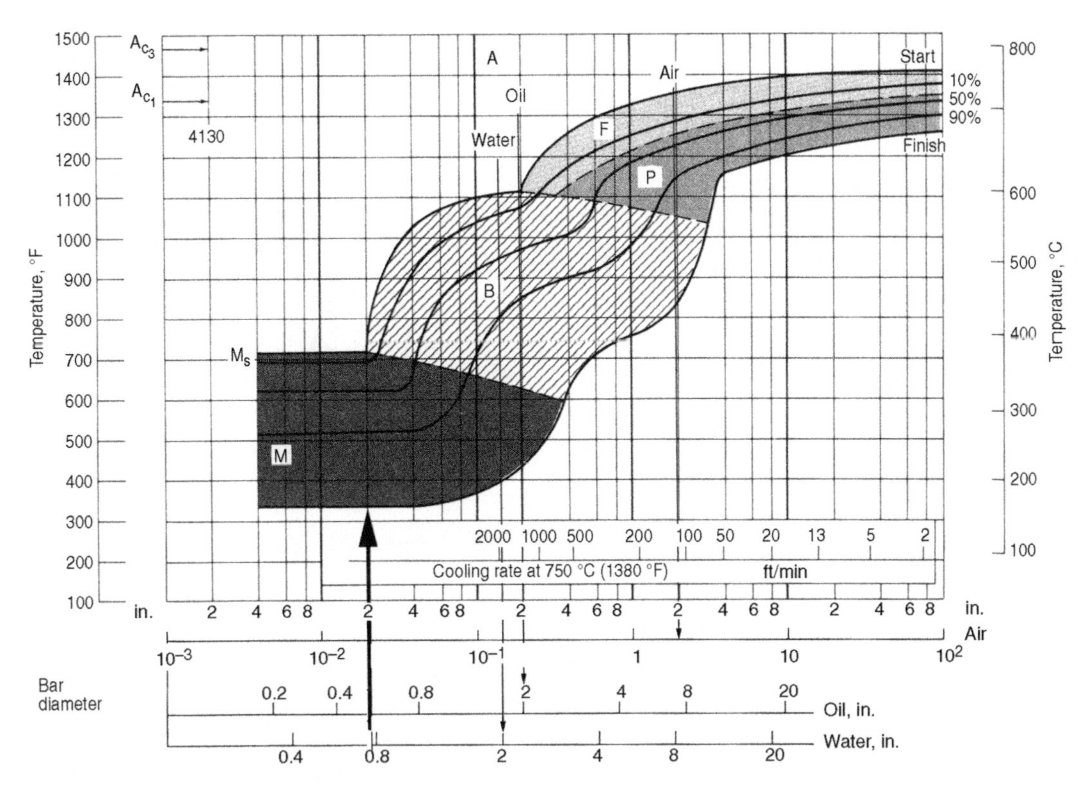

Fig. 9.11 Alternate continuous transformation diagram that gives microstructures at the centers of bars of varying diameter subjected to quenches in air, oil, or water. Source: Ref 9.4

Table 9.1 Summary of transformation diagrams given in Ref 9.2

Steels	Pages	Type of diagram[a]
United States	3–51	IT
British	53–113	IT
German	115–161	IT and CT
French	163–220	IT and CT
Molybdenum steels	243–296	CT (W.W. Cias, Climax Molybdenum Co.)
Vanadium steels	297–369	CT (M. Atkins, *Atlas CT Diagrams,* American Society for Metals,1980)
British engineering steels	373–452	CT (M. Atkins, *Atlas CT Diagrams,* American Society for Metals,1980)
Other steels	453–520	IT
Additional steels	521–607	CT in old 1977 ASM book (Ref 9.5)

(a) IT, isothermal transformation; CT, continuous transformation

The Jominy End Quench

Walter Jominy, a University of Michigan graduate working at Chrysler in the 1930s, developed a test that has become widely adopted for evaluating the hardenability of steels. The test is basically a controlled heat-transfer experiment in which a round bar 100 mm (4 in.) long with a 25 mm (1 in.) diameter is heated to a temperature in the austenite range and placed in a jig that sprays water on one end, thereby cooling the bar directionally from the quenched end (Fig. 9.12a). The test has been standardized by ASTM International as ASTM A 255, which can be found in the ASTM book of standards available in most technical libraries. To ensure consistency, the test is conducted with a specified bar temperature depending on %C level, a specified rate of water flow, and a specified diameter of the water stream. After cooling, small flat surfaces are machined onto opposite sides of the bar with a surface grinder in the geometry shown in Fig. 9.12(b). The bar is then placed in a hardness tester, and HRC measurements are made at $\frac{1}{16}$ in. intervals along the bar, starting at the quenched end and extending to the opposite end. Measurements are made on both sides, and the average at each location is used to generate a hardness profile, such as that shown in Fig. 9.13 for a 1080 steel.

The distance along the bar is often called the J_D position, and in the United States, a value of $J_D = 1$ indicates a position of 1.6 mm ($\frac{1}{16}$ in.), but it is necessary to be alert to the particular

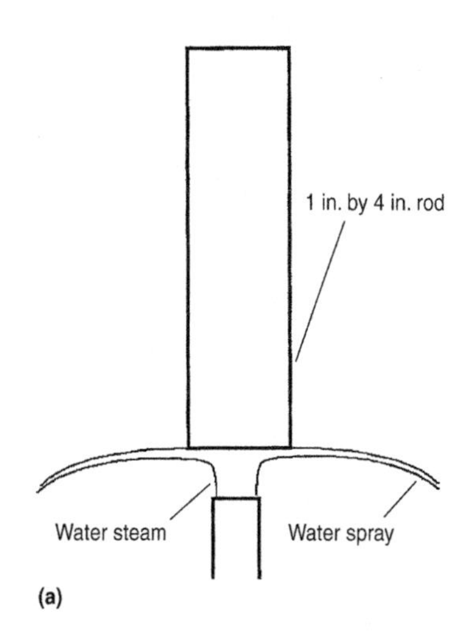

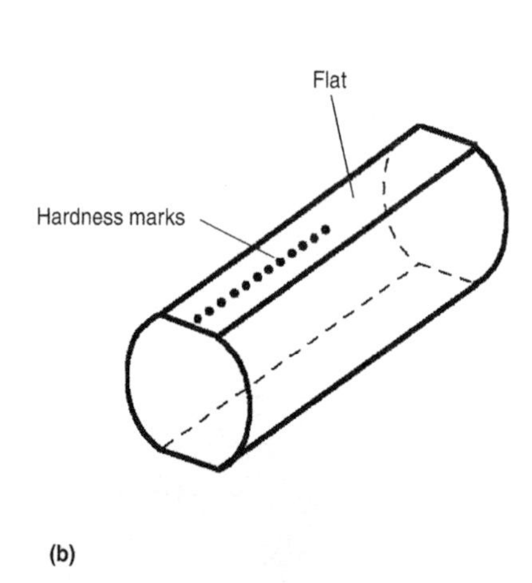

Fig. 9.12 (a) Experimental setup for the Jominy end quench test. (b) Machined flat sides (flats) and hardness measurement locations on quenched bar

units being used. In non-U.S. countries, distance is generally in units of millimeters. The cooling curves that are superimposed on the CT diagram in Fig. 9.9 refer to specific J_D values. For example, the curve labeled $\frac{3}{4}$ gives the cooling curve for a position located 19 mm ($^3/_4$ in.) from the quenched end of the J-bar, that is, $J_D = 12$. One of the things that makes the Jominy test useful is that cooling rates of low-alloy AISI steels with different chemical compositions as well as with different grain sizes are essentially the same. This is because the alloying element additions do not change the thermal conductivity significantly in these steels, nor does grain size variation. Consequently, the various cooling curves shown in Fig. 9.9 will be the same for different steels. The only thing that changes from steel to steel is the position of the individual start curves. As the alloy composition is increased and the grain size is increased, those curves shift to the right.

The Jominy bar provides a map of all the microstructures that appear in a continuously cooled sample. At the quenched end, the structure will virtually always be 100% martensite. Moving in from the quenched end, the cooling rates decrease; at some point, nonmartensitic constituents such as bainite, pearlite, or ferrite will appear, and the hardness will start to drop. This effect is shown in Fig. 9.13 for the case of a 1080 steel. The Jominy data are given by the curve marked "Hardness". The percent martensite curve shows the amount of martensite, decreasing from 100% at approximately $J_D = 3$ to 0 at approximately $J_D = 6$. (percent martensite

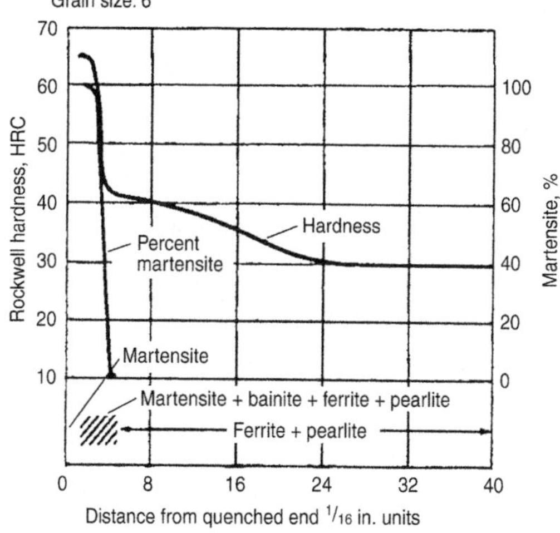

Fig. 9.13 Jominy hardness profile of 1080 steel. Source: Ref 9.1

is given on the right axis and HRC on the left axis.) In the region where $J_D = 3$ to 6, the bar contains a mixture of martensite, bainite, and pearlite, but beyond $J_D = 6$, the bar is all pearlite. The continued drop in hardness in the pearlite is due to the increased spacing of the plates of the pearlite at the lower cooling rates, which stabilizes at distances beyond $J_D = 24$.

Figure 9.1 illustrated the concept of hardenability by showing the hardness as a function of radius for a 5160 versus a 1060 steel. The Jominy data for these two steels also illustrate

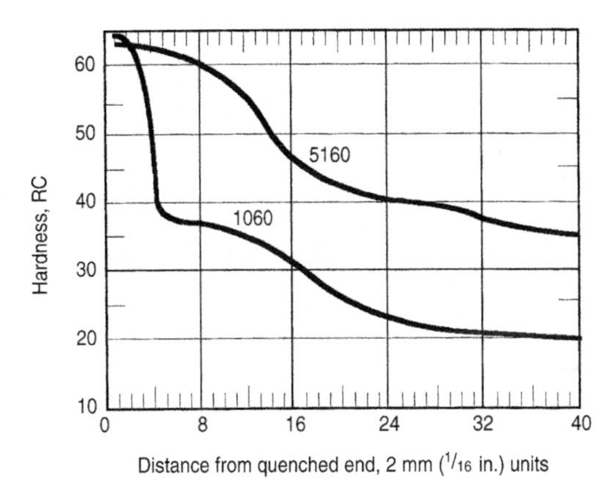

Fig. 9.14 Jominy hardness profile for 5160 and 1060 steels. Source: Ref 9.1, 9.5

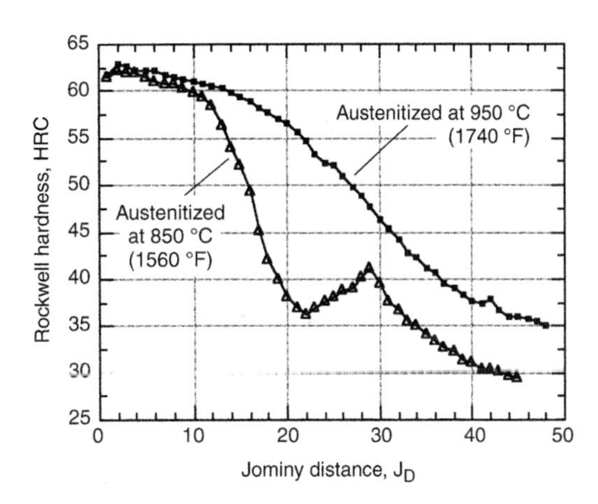

Fig. 9.15 Jominy hardness profile for 5160 steel austenitized at 850 and 950 °C (1560 and 1740 °F)

the superior hardenability of 5160, as shown in Figure 9.14, which superimposes the data for the two steels found in Ref 9.1 and 9.5. The increased depth to which the 5160 steel hardens is clearly seen by comparing the Jominy data. In the 1060 steel, the depth of 100% martensite extends to a J_D value of only approximately 1 to 2, whereas in the 5160, it extends to a J_D value of approximately 6 to 7. In the 1060 steel, all of the martensite is gone at a J_D of approximately 6. The significantly greater hardness of 5160 over 1060 at a $J_D = 40$ shows that some martensite is present in the 5160 even at this large distance from the quenched end.

The Jominy test is able to detect changes in hardenability produced by small changes in the heat treating operation. Figure 9.15 presents data from Jominy tests on 5160 steel where the steel was austenitized at two different temperatures for the same time of 1 h. The higher-temperature austenitization produced a dramatic increase in hardenability. This could be caused by either an increase in grain size at the higher temperature or a change in the composition of the austenite at the higher temperature. The grain size of both steels was found to be the same within the scatter of the measurement, at approximately ASTM No. 10. The carbides in 5160 will contain significant amounts of chromium. Apparently, the 850 °C (1560 °F) austenitization for 1 h is not adequate to dissolve all of the carbides and fully homogenize the chromium in the austenite grains of the steel. This is another example of how the slow diffusion of the alloying elements contained in low-alloy AISI steels affect heat treating.

The Jominy data for a given steel provide quantitative information about the cooling rate needed to produce a given amount of martensite, such as 50 or 99+%. As discussed in Chapter 12, "Quenching," the cooling rate within the steel depends on the quenching method. For example, oil quenching produces lower cooling rates than water quenching. The cooling rates corresponding to the different types of quenchants have been analyzed, and Jominy data can be used to determine the martensite content at the center and various radii of bars for the different quenchants. If the percent martensite is known, it is possible to estimate the hardness versus the radius for various quenchants. The analysis is fairly complicated and is not presented here. However, the results of these analyses are presented graphically in Ref 9.1 for the special cases of mild water quenches and mild oil quenches. Also, information such as that at the top of Fig. 9.16 is provided for many steels. The figure provides equivalent Jominy positions for three locations along the radius (center, ¾ out, and surface) for mild oil and mild water quenches. For example, for a 25 mm (1 in.) diameter bar subjected to a mild oil quench, the equivalent Jominy positions are shown by the downward-pointing arrows at $J_D = 2.2$, 4, and 6.2 for surface, ¾ radius, and center positions, respectively. By knowing the equivalent Jominy positions, the Jominy data for the steel can be used to determine the corresponding hardness predicted for that position. For example, for a 25 mm (1 in.) steel bar and a mild quench, the Jominy data given by the lower curve predict the HRC hardness would be 60 at its surface, 59 at its ¾ radius, and 55 at its center. Similarly, if the Jominy data followed the upper curve, the hardnesses would be 65, 65, and 64 at its surface, ¾ radius, and center, respectively.

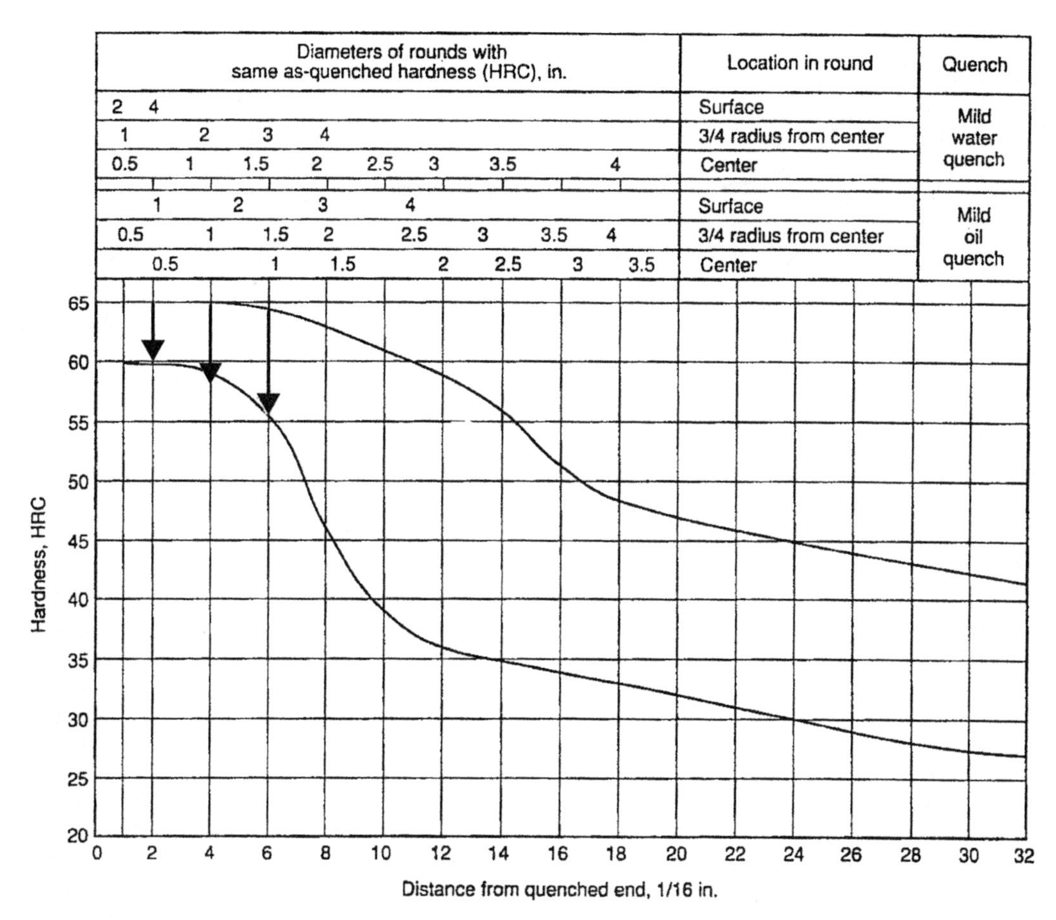

Fig. 9.16 Hardness profile illustrating hardenability band for 5160H and equivalent Jominy positions for mild water and oil quenches. Source: Ref 9.1

Table 9.2 Oil-quenched hardness (HRC) of 1060 and 5160 steels

Bar diameter		Surface		½ radius		Center	
mm	in.	1060	5160	1060	5160	1060	5160
13	1/2	59	63	37	62	35	63
25	1	25	62	32	62	30	60
50	2	30	53	27	46	25	43
100	4	29	40	26	32	24	29

Source: Ref 9.2

Reference 9.1 also provides simple straightforward data on hardness at different radii for oil-quenched samples in table form. The results given for 1060 and 5160 are presented in Table 9.2 as an illustration. The table gives the HRC values for the three locations—center, ½ radius, and surface—for the two steels after an oil quench of four different bar diameters. The data show that an oil quench will through harden 5160 bars of 25 mm (1 in.) diameter or lower but will not through harden a 1060 steel of 13 mm (½ in.) diameter.

Hardenability Bands. The hardenability is strongly controlled by the chemical compositions of the low-alloy steels. The composition of the steel is set at the steel mill that produced the steel, and there is always going to be some variation of the actual composition from the standard target values given in Table 6.3. Also, there can be grain size variation in the as-received steels. Composition and grain size variations will shift the Jominy data to the right or left, as discussed previously. To minimize such variations, steel companies supply steels that are guaranteed to produce a Jominy curve that falls inside a so-called hardenability band. Such steels are designated with an "H" at the end of their AISI name. For example, Fig. 9.16 represents the hardenability band for 5160H, and the steel company will guarantee that the Jominy data for their 5160H steel will lie somewhere between the two curves shown in the figure. The hardenability bands are presented in many references in addition to Ref 9.1. A particularly good reference is the SAE Handbook (Ref 9.6), which also contains other useful information on hardenability and the general heat treatment of steels.

Summary of the Major Ideas in Chapter 9

1. Quenched steels harden because martensite forms from the austenite. The hardness of the martensite depends on the carbon content; the higher the %C (up to approximately 0.8%), the harder the steel (Fig. 4.13).
2. Adding alloying elements, such as molybdenum, chromium, nickel, and manganese, does not increase the hardness of the martensite. However, it does increase the *hardenability* (measure of the depth to which martensite can be formed) of the steel.
3. If any of the three austenite quench products (ferrite, pearlite, or bainite) form before martensite forms, then martensite cannot form, and the hardness of the steel is decreased. The alloying elements inhibit the formation of all three quench products. Hence, they allow martensite to form at slower cooling rates.
4. Cooling rates are always slower at the center of quenched bars. Because alloying elements allow martensite to form at slower cooling rates, they increase the depth below the surface of a quenched bar where 100% martensite forms. This is what is meant when it is said that alloying elements increase hardenability. Figure 9.1 illustrates the effect of adding 0.8% Cr on the hardenability of a steel containing 0.6% C.
5. The extent to which an alloying element improves hardenability is measured by how much it reduces formation of the three quench products: ferrite, pearlite, or bainite. Isothermal transformation (IT) diagrams provide curves giving the time lapse after a quench needed to start formation of ferrite (the F_s curve), pearlite (the P_s curve), and bainite (the B_s curve) versus the temperature of the quench. The longer the times, the more the improvement in hardenability.
6. The start-time curves are shifted to longer times (increased hardenability) by: (1) increasing the %C in hypoeutectoid steels, (2) increasing the percent of the alloying elements added, and (3) increasing the grain size of the austenite. Because fine grain size is critical for improved toughness, it is important to use alloying elements in fine-grained steels to restore the hardenability lost due to the finer grains. The experiment in Fig. 9.6 illustrates the lack of hardenability in a fine-grained plain carbon 1086 steel versus a fine-grained alloy 5160 steel.
7. The IT curves require that the steel be quenched so fast that, even at its center, the temperature drops to the isothermal treatment level before any of the three quench products form in the austenite. Thus, IT diagrams are only accurate for thin samples. A second type of cooling diagram, the continuous cooling (CT) diagram, applies to steels cooled at slower rates, such as occurs near the center of steel samples without thin cross sections.
8. Similar to the IT curves, the CT curves also show the start curves for ferrite, pearlite, and bainite. Additionally, they show cooling curve lines for steels cooled at increasing cooling rates. By measuring the cooling rate of a steel and then finding the corresponding cooling rate on the diagram, the start times for the onset of formation of the three quench products—ferrite, pearlite, or bainite—can be estimated for that steel.
9. Most published IT diagrams, such as those in Fig. 9.3, and many published CT diagrams present the start curves without labeling which quench product is forming. For IT diagrams of plain carbon steels, one may make a good estimate of the temperature regions of the curves that are specifying the start time of ferrite, pearlite, or bainite from the information in Fig. 4.23. As an example, the unlabeled IT diagram of 1080 steel in Fig. 9.3 has been given the appropriate P_s and B_s labels in Fig. 9.7.
10. The IT diagrams identify the times when the quench products have finished forming, that is, the times at which all of the original austenite has transformed into the quench products of ferrite, pearlite, bainite, or combinations thereof. These lines are often labeled with a subscript "f," as, for example, the pearlite finish, P_f, and the bainite finish, B_f, in Fig. 9.7. In addition, it is common for IT diagrams to include a dashed line showing when 50% of the austenite has transformed into one or more of the three quench products.

11. The CT diagrams show a series of dashed lines that identify the extent of the austenite transformation. For example, the dashed lines in Fig. 9.9 for 5140 steel show when 1, 10, 50, 90, and 99% of the austenite is gone.
12. An alternate form of CT diagram describing the transformation progress only at the center of bars of known diameters was developed in Britain (Ref 9.4). An example is shown in Fig. 9.11.
13. The Jominy end quench is a test that measures the hardness at 1.6 mm (1/16 in.) intervals from the quenched end of a controlled heat-transfer experiment. Martensite forms at the quenched end. Progressing away from the end, the three quench products—bainite, pearlite, and ferrite—form as the cooling rate reduces. The order in which they form depends on the composition of the steel. The test provides a qualitative measure of the hardenability of a steel. High-hardenability steels will be 100% martensite at larger distances from the quenched end and will therefore remain harder to larger distances below the quenched surface. Figure 9.14 illustrates this effect by showing the Jominy hardness data for a 1060 and a 5160 steel. The 5160 steel is a 1060 steel with the addition of 0.8% Cr.
14. The Jominy end quench experiment can reveal useful information about the effect of austenitizing conditions on hardenability. For example, Fig. 9.15 illustrates the dramatic effect of increasing the austenitizing temperature from 850 to 950 °C (1560 to 1740 °F) in 5160 steel, an effect resulting from the slow homogenization of chromium into the austenite after the carbides, with their high chromium content, have dissolved. The slow homogenization is due to the low diffusion coefficient of chromium in austenite.
15. It is possible to use the Jominy end quench data to make theoretical predictions of the hardness of bars at various radial depths below the surface for different types of quenches, such as a water quench (a fast quench rate) or an oil quench (a slower quench rate). Results of such calculations are available in graphical form for many steels in Ref 9.1, such as that shown for 5160 steel in Fig. 9.16. By plotting the Jominy data for a 5160 bar on this graph and drawing lines down to a given hardness from the locations of its bar diameter found at the chart top, the chart predicts three radial locations (center, surface, and 3/4 out from center) where this hardness will occur for either a mild oil quench or a mild water quench. An example application would be to determine the maximum bar diameter that can be through hardened (100% martensite) to the bar center.
16. If the American Iron and Steel Institute (AISI) code label for a steel has an "H" at the end, such as the 5160H of Fig. 9.16, it means that the steel company guarantees that the composition and grain size of the steel is such that the Jominy data for the steel will fall within the published H-band (hardenability band) width for the steel. The two curves in Fig. 9.16 define the H-band width for 5160H. Graphs of the H-band widths for various AISI steels are available in Ref 9.1, 9.6, and many other sources.

REFERENCES

9.1. *Heat Treater's Guide: Practices and Procedures for Irons and Steels*, 2nd ed., ASM International, 1995
9.2. G. Vander Voort, *Atlas of Time-Temperature Diagrams for Irons and Steels*, ASM International, 1991
9.3. R.A. Grange, Estimating Critical Ranges in Heat Treatment of Steels, *Met. Prog.*, Vol 79, April 1961, p 73–75
9.4. M. Atkins, *Atlas of Continuous Cooling Transformation Diagrams for Engineering Steels*, British Steel Corp., Sheffield, 1977. See also p 297–452 of Ref 9.2 and p 309 of Ref 9.1
9.5. *Atlas of Isothermal and Continuous Cooling Diagrams*, American Society for Metals, 1977
9.6. *Materials,* Vol 1, *SAE Handbook*, Society of Automotive Engineers

CHAPTER 10

Tempering

THE IMPORTANCE OF TEMPERING can be illustrated through a simple experiment. The experiment is done with two short lengths of drill rod having a diameter of 3 mm (1/8 in.). Drill rod is either 1095 steel or W1 tool steel, both of which have essentially the same composition. One-inch long sections at the midpoints of the two rods are simultaneously heated with a propane torch to an orange color for approximately 20 s and immediately quenched in a beaker of water. The temperature produces austenite over a short length of the rods, and the quench converts it to martensite + retained austenite. If the first rod is placed in a vice with the discolored heated section approximately 13 mm (1/2 in.) from the grips and struck firmly with a hammer, the piece will break across the center of the heated section in a brittle manner, just like a piece of glass. If the discolored heated section of the second rod is polished shiny with a fine emery or sandpaper and gently heated with the torch until it turns a deep blue color, the rod can be bent in the vice or with a set of pliers to the shape shown in Fig. 10.1. This is the tempering step. Martensite has formed in the heated zone, and, by looking carefully at the everely bent rod, the boundaries between the martensite and nonmartensite regions can be seen, as shown by the arrows in Fig. 10.1. The nonmartensite region has the as-received spheroidized structure of Fig. 4.24. This structure is soft, and the metal flow within it on bending produces a surface distortion that terminates at the hard martensite zone and reveals the transition boundary.

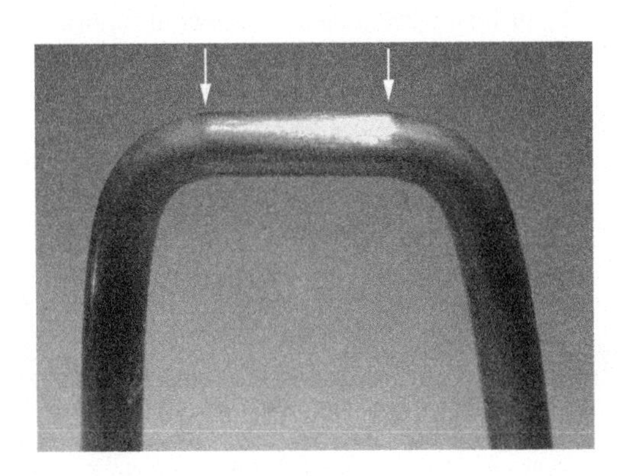

Fig. 10.1 Quenched and tempered 3 mm (1/8 in.) diameter drill rod after bending

If the tempering temperature is hot enough, the rod will not break through the martensite region, even when this region is bent to over 90° by hammering in the vice. If it is too hot, the martensite region will become softer than desired. On heating the polished steel rod, an oxide layer forms on its surface, and the thickness of the oxide layer produces various colors caused by the interference effect of light rays bouncing off the bottom and top of the oxide layer. The oxide thickness is controlled by the temperature of the steel, and a series of different colors occurs at the relatively low tempering temperatures. The temperatures corresponding to the various colors are known as temper colors, and Table 10.1 presents the correlation between colors and temperatures. The color-temperature designation has been used for millennia by blacksmiths when tempering steels.

The experiment shows that in the as-quenched condition, the steel is far too brittle to be useful

Table 10.1 Temper colors

	Temperature	
Color	°C	°F
Pale yellow	220	430
Golden yellow	240	470
Brown	255	490
Purple	280	530
Bright blue	290	550
Dark blue	315	600

for applications other than those that require extreme hardness with no bending, such as files. In practice, virtually all quenched steels are tempered. Figure 10.2 shows qualitatively how the quenching and tempering operation changes the stress-strain characteristics of the steel. The tempering operation sacrifices the high strength of the steel to gain improvements in ductility and toughness. Figure 10.3 presents actual data for an oil-quenched 4340 steel that has been tempered at increasing temperatures. Both the tensile and yield strengths fall as the tempering temperature is increased, but the percent elongation, or ductility, increases. Reference 10.1 shows similar curves for several other steels, demonstrating the same trend.

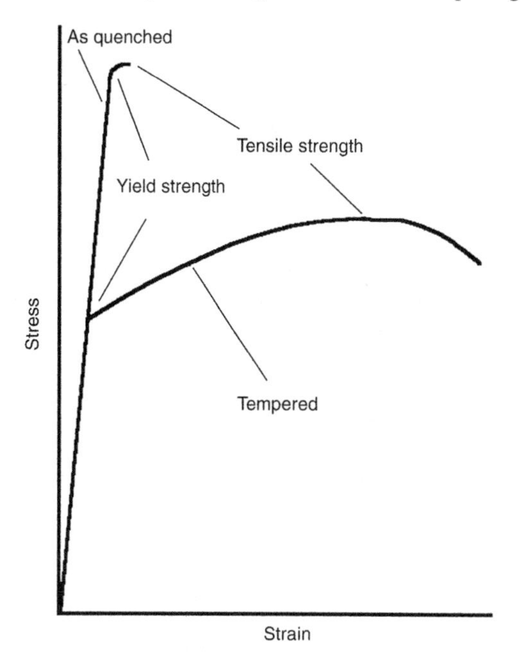

Fig. 10.2 Effect of tempering on stress-strain relationship

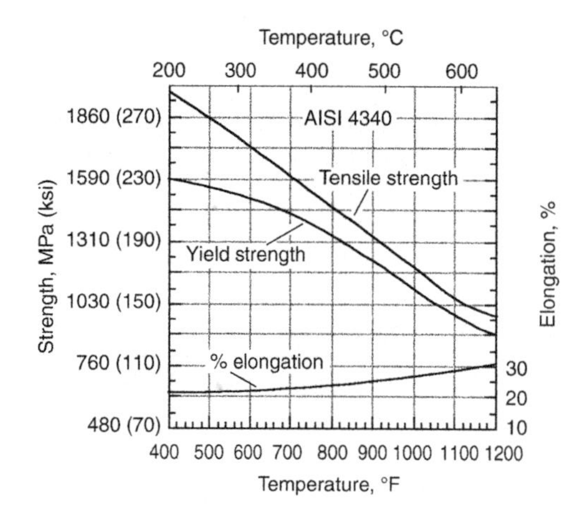

Fig. 10.3 Change in mechanical properties of 4340 steel versus tempering temperature. Source: Ref 10.1

Several processes occur in the steel during tempering that result in the loss of strength and the gain in ductility and toughness. First, the high degree of volume strain in the steel produced by the formation of the higher volume per atom martensite phase is relieved. Then, a series of internal structure changes occur that is generally partitioned into the three stages of tempering:

- *Stage 1:* Very small carbides are formed in the martensite—so small (approximately 10 nm) they can only be seen in an electron microscope. These first-formed carbides are metastable carbides (that is, they do not appear on the equilibrium phase diagram). In hypoeutectoid steels, epsilon carbide ($Fe_{2.4}C$) forms, and in hypereutectoid steels, Hagg ($Fe_{2.2}C$) and eta (Fe_2C) carbides form.
- *Stage 2:* Any retained austenite decomposes into carbides and ferrite. The percent retained austenite only becomes significant in high-carbon steels.
- *Stage 3:* The metastable carbides are replaced with small particles of cementite, the stable carbide of steels.* This stage occurs at the highest tempering temperatures.

As explained in Chapter 4, "The Various Microstructures of Room-Temperature Steel," toughness is a better measure of the ability of a steel to avoid failure than is ductility. Toughness is evaluated with impact tests, such as the Charpy test. The Izod test is an impact test similar to the Charpy test, and Fig. 10.4 presents a summary of scatter bands of data from impact tests on steels with two different levels of %C. These curves illustrate the following two important characteristics of heat treated steels: tempered martensite embrittlement and the effect of %C on toughness.

Tempered Martensite Embrittlement. Note that the impact energy drops significantly in the tempering range of 260 to 340 °C (500 to 650 °F). It is customary to call this loss of toughness tempered martensite embrittlement (TME).

*Strictly speaking, cementite should not be called a stable phase. In steel held in the 600 °C (1110 °F) range for extended times, the cementite will sometimes be replaced by graphite. This shows that graphite is the stable phase in the iron-carbon system, and the iron-carbon phase diagram in Chapter 3 is really not an equilibrium diagram. However, because graphitization of steels is rare, this point is mainly of academic interest.

(An old name for it is 500 °F, embrittlement.) Because of this problem, steels intended for use at high strength levels are not tempered above approximately 205 °C (400 °F). For applications requiring high toughness and less strength, tempering is done at temperatures above approximately 370 °C (700 °F). Notice that at temperatures of 540 to 650 °C (1000 to 1200 °F), excellent toughness is obtained. Of course, the hardness goes very far down, and one may think that air-cooled pearlitic steels would be cheaper for applications in this range of hardness. However, tempered martensite structures are generally tougher than pearlitic structures of the same hardness.

Temper embrittlement (TE) is a second type of embrittlement that occurs on tempering at high temperatures. Note the similarity of names, and avoid confusing the two types of embrittlement. (An older name for TE is temper brittleness.) Temper embrittlement occurs when tempering in the high-temperature range of approximately 600 °C (1100 °F). It is not a significant problem, because it can be avoided easily by quenching from the tempering temperature. As illustrated in Fig. 10.4, TE is not detected by simple plots of impact energy versus temperature but by more extensive impact testing that measures variation of the ductile-brittle transition temperature (DPTT) with the tempering temperature.

It is often possible to tell if a steel has failed because of one of these embrittlement problems by examining the fracture surface. A grain-boundary fracture mode is characteristic of both types of embrittlement. However, in steels with extremely low levels of phosphorus and sulfur, tempered martensite embrittlement can display a cleavage surface (see Chapter 5, "Mechanical Properties").

Effect of %C on Toughness. Figure 10.4 illustrates that increasing the %C from 0.4 to 0.5% produces a significant loss of fracture energy. Fracture energy values of less than 14 to 20 J (10 to 15 ft · lbf) are dangerously low for protection from brittle failures. For example, of all the steels used in an automobile, the maximum %C is only 0.4 in all parts except the springs, where %C is 0.6 (often 5160 in the leaf springs and 9260 in the coil springs), and the bearings, where %C is 1 (often 52100). This preference for low-carbon steels is directly related to the loss of toughness as %C increases.

Figure 4.13 shows that not much strength is really gained by increasing the %C above approximately 0.4. The Rockwell C hardness level at 0.4% C is 57 and corresponds to a tensile strength of 2210 MPa (320 ksi), which is adequate for most applications even after the strength reduction on tempering. Springs require a higher strength level because they are subject to

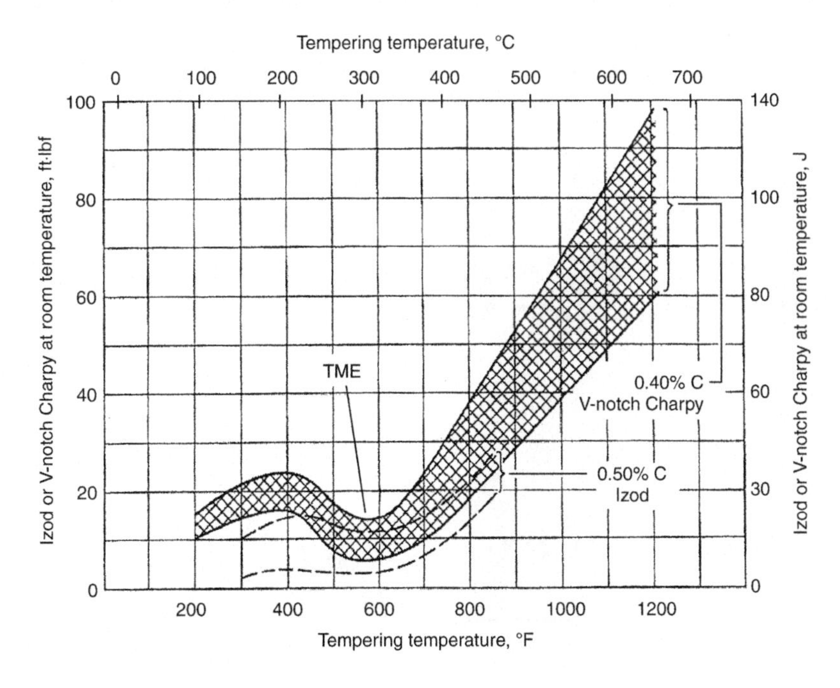

Fig. 10.4 Dependence of notched impact energy on tempering temperature for 0.4 and 0.5% C steels. TME, tempered martensite embrittlement. Source: Ref 10.2

fatigue failure, and high fatigue life is promoted by high strength.

Probably the main reason for using high-carbon steels is improved wear resistance. The wear resistance of a 0.8 to 0.9% C steel tempered to the same hardness as a 0.5 to 0.6% C steel is improved significantly. The improved wear resistance results from the increased volume percent of carbides in the tempered higher-carbon steels. The farm equipment industry uses high-carbon steels, such as 1077 to 1086, for the many implements that experience wear when dragged through dirt. High carbon levels are also an advantage for increasing wear resistance of knives made by bladesmiths. As in most applications, the highest combination of hardness and toughness is desirable, and a continuous battle of compromise occurs because the two properties vary inversely, as illustrated in Fig. 10.3.

The presence of carbides within the martensite matrix of bearing steels, such as 52100, produces increased wear resistance, because the carbides are even harder than the martensite. However, the presence of large carbides is detrimental to toughness. It is important in such steels to keep the carbides as small as possible and to realize that, even then, toughness will be inherently low; therefore, the steels should be used in applications not subject to tensile stresses, which is the case for most bearing applications. Another inherent problem contributing to the loss of toughness in high-carbon martensite is the formation of cracks in the plate martensite formed on quenching, may be seen in Fig. 4.15(b). An enlargement of a region showing quench cracks (Q.C.) is shown in Fig. 10.5. Such cracks often occur where martensite plates impinge on each other, leading to a severe loss in toughness. Lath martensite is an inherently better form of martensite than plate martensite when toughness is important.

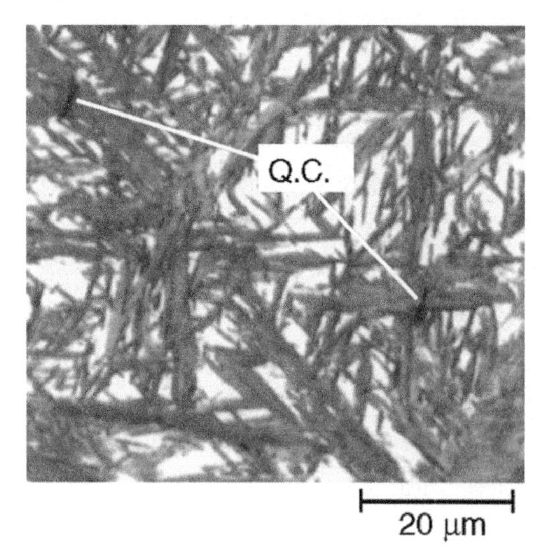

Fig. 10.5 Impingement of martensite plates leading to quench cracks (QC)

Effects of Alloying Elements

In addition to improving hardenability, the most important reason for adding alloying elements to plain carbon steels is to improve their tempering response. The effect of alloying elements in general is illustrated in Fig. 10.6 and 10.7. The upper curve of Fig. 10.6 presents the same data for the hardness of fresh martensite that was presented in Fig. 4.13. Note that this curve applies to both plain carbon and alloy steels. Tempering causes the hardness to drop, and the lower two curves show the hardness versus %C after tempering at the same temperature for the plain carbon and the alloy steel. The important difference is that the alloyed steel has a higher tempered hardness. The increased hardness is shown by the vertical arrowed line labeled ΔH, where the symbol Δ means change, and H means hardness.

The lower curve in Fig. 10.7 shows how the hardness of a 1040 steel drops with tempering temperature, and the upper curve shows how the addition of 0.24% Mo increases the hardness. Again, a higher hardness is seen for the alloy steel by an amount indicated as ΔH. The increment in hardness depends on the alloying element added, as well as the tempering

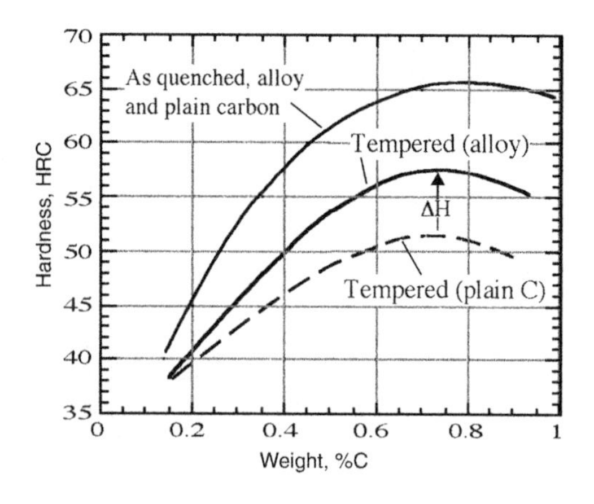

Fig. 10.6 As-quenched and tempered hardnesses of plain carbon and alloy steel versus %C. ΔH represents the incremental increase in hardness between plain carbon steels and alloy steels with the same %C.

temperature. As illustrated in Fig. 10.8, the four main alloying elements in the AISI steels, as well as the small amount of silicon present in them, produce a ΔH increment. The magnitude of the ΔH increment for each element varies with the tempering temperature, and Fig. 10.8 illustrates the general trends observed.

The main reason for this increment is that very small carbides contribute to hardness, but the contribution drops off as the size of the carbides increases. As explained in Chapter 8, "Control of Grain Size by Heat Treatment and Forging," the reactive alloying elements segregate into the cementite component of steel.

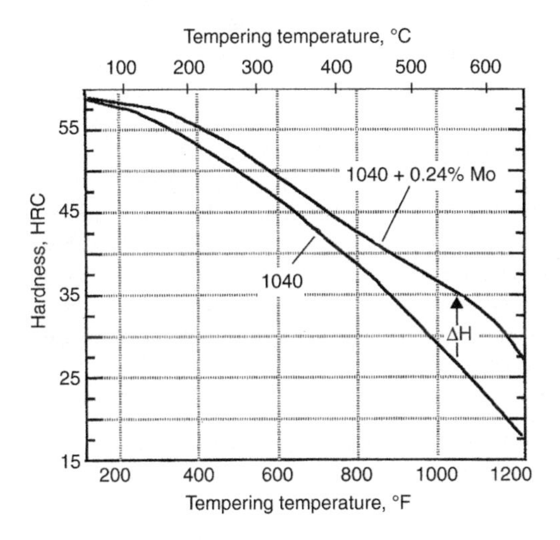

Fig. 10.7 Influence of a molybdenum alloy addition on the dependence of hardness on tempering temperature. The incremental hardness increase caused by alloying is represented by ΔH. Source: Ref 10.3

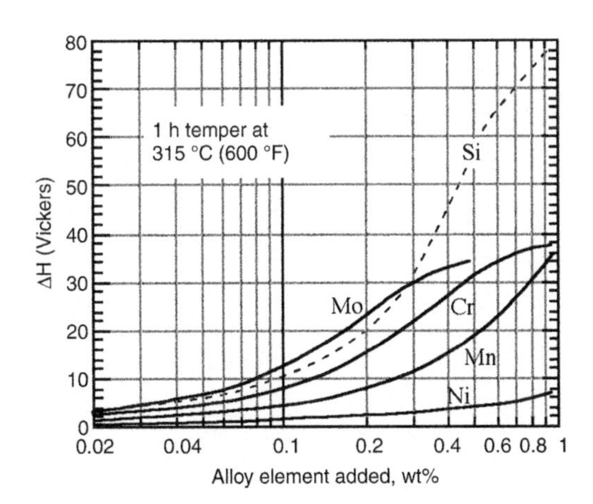

Fig. 10.8 Incremental increase in hardness on tempering for silicon and the four major alloying elements. Source: Ref 10.4

This changes the composition of the carbides from Fe_3C to $(Fe + X)_3C$ (often written as M_3C), where X refers to reactive alloying elements such as chromium and molybdenum. At higher tempering temperatures, the average carbide size increases as larger carbides replace smaller ones. However, this coarsening process requires diffusion, and the alloying element, X, cannot diffuse nearly as fast as carbon. Therefore, in the alloy carbides formed in low-alloy AISI steels, both $M_{2.4}C$ and M_3C, the carbides remain smaller to higher tempering temperatures and thereby give increased hardness after tempering

To illustrate the beneficial effects produced in tempered alloy steels, compare the two steels containing 0.6% C shown in Table. 10.2. The steels were quenched and tempered for 1 h at the four temperatures shown in Fig. 10.9. For a given reduction in area (RA) (ductility), the low-alloy AISI steel is harder (stronger). For example, at the ductility level of 42% RA, shown by the vertical dashed line, alloy 5160 has a hardness of HRC = 43 compared to HRC = 34 for 1060 (equivalent tensile strengths are 1393 and 1048 MPa, or 202 and 152 ksi, respectively). Another way to think about it is that for a given level of hardness, the alloy steel has improved ductility. At HRC = 32 (horizontal dashed line),

Table 10.2 Chemical composition of 1060 and 5160 steels

Steel	%C	%Mn	Alloying element
1060	0.6	0.75	None
5160	0.6	0.88	0.8% Cr

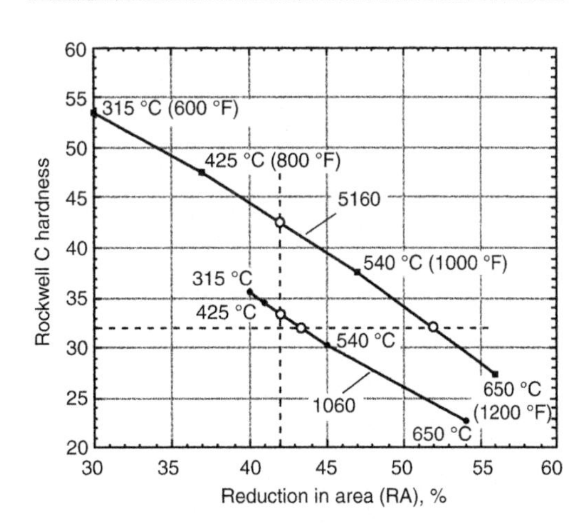

Fig. 10.9 Rockwell C hardness decreases with increasing reduction in area (ductility) for 1060 and 5160 steels after tempering for 1 h at temperatures shown. The dashed vertical line compares alloying effects at constant ductility. The dashed horizontal line compares alloying effects at constant hardness. Source: Ref 10.5

the percent RA of the 5160 is 52 compared to 43 for the 1060 steel.

The point was made in Chapter 8 that the reactive alloying elements tend to form carbides, which leads to two important results: (1) In steels containing cementite carbides, the reactive elements will segregate to the cementite, changing its composition from Fe_3C to M_3C, where M = Fe + alloying element; and (2) cementite is an iron carbide and is called a native carbide of iron. Similarly, all of the reactive alloying elements form native carbides. These carbides have a different crystal structure, a different melting point, and a different hardness than the cementite carbide, M_3C. For example, $Cr_{23}C_6$ and Mo_2C are native carbides of chromium and molybdenum. The hardnesses and the melting points of the native carbides of the reactive metals are higher than for the cementite carbide. The higher melting temperature indicates a reduced mobility of the atoms, which means that if these carbides form on tempering, they will coarsen less and give improved strength (hardness) at higher tempering temperatures. If adequate amounts of the reactive alloying elements, such as molybdenum or chromium, are added to a steel, their native carbides will form in significant amounts on tempering and give significant increases in tempered strength, as illustrated in Fig. 10.10 for molybdenum. As shown, the strength on tempering actually increases at higher tempering temperatures when the native molybdenum carbide forms. This strength increase is sometimes called the fourth stage of tempering or the secondary hardening peak. Such high-temperature strength is very important in tool steels.

In all of the discussion to this point, no mention has been made of the time of the hold at the tempering temperature. When tempering at a given temperature, the hardness decreases with increasing time. However, the drop in hardness with time is fairly slow and is often not shown on tempering curves. There have been some extensive studies on the effect of alloying elements, tempering temperature, and tempering time on the hardness produced on tempering (Ref 10.6, 10.7). These studies developed a correlating parameter that makes it possible to calculate a tempering curve for any AISI steel. Using the method described, the tempering curves for 1045 steel shown in Fig. 10.11 were determined. The curves show the effect of both time and temperature on hardness. Note that, compared to temperature variations, the time variation from 1 to 4 h is negligible. Reference 10.2 has a few plots of actual data on time and temperature variations that lead to this same conclusion. However, the tempering curves for most of the steels do not mention the time of tempering, which is assumed to be 1 h.

Summary of the Major Ideas in Chapter 10

1. Most quenched steels are tempered because the toughness of as-quenched steels is generally very poor. Tempering (heating to a low temperature for approximately 1 h)

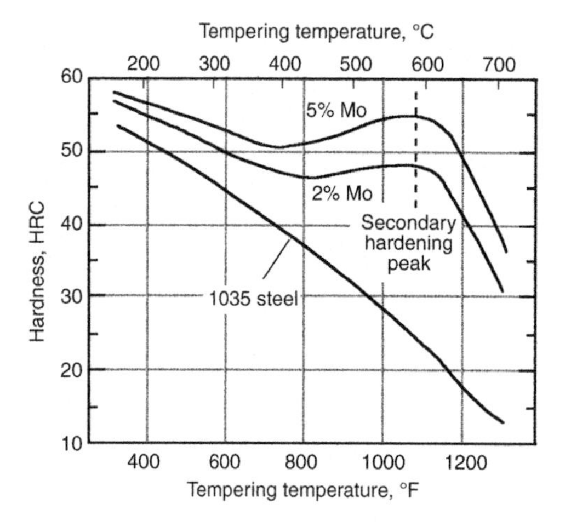

Fig. 10.10 Secondary hardening peak appears with large additions of molybdenum to a 1035 steel. Source: Ref 10.3

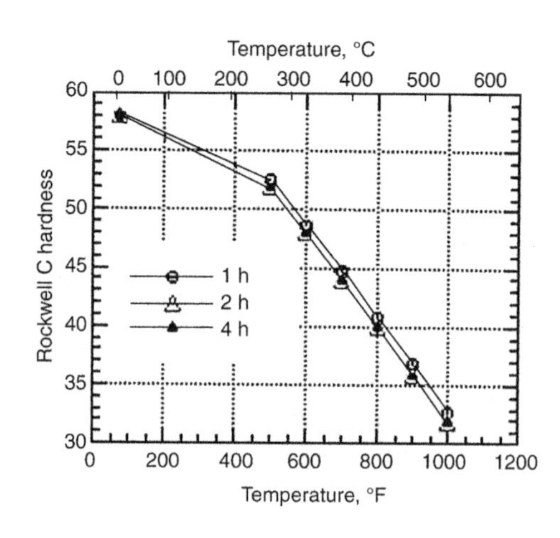

Fig. 10.11 Predicted Rockwell C hardness versus tempering temperature for 1045 steel

significantly increases toughness and ductility, at the sacrifice of both yield strength and tensile strength. Figures 10.2 and 10.3 show the general trend.

2. During tempering, the martensite begins to decompose into ferrite and carbides. At the lowest tempering temperatures, very fine metastable carbides (usually epsilon carbide, $Fe_{2.4}C$) form (called stage 1 tempering), and at higher temperatures they are replaced by the usual carbide, cementite (Fe_3C, or stage 3). At intermediate temperatures, if retained austenite is present, it will decompose (stage 2).
3. Toughness is best measured by notched impact tests such as the Charpy or Izod tests. Such tests, as shown in Fig. 10.4, show that toughness after tempering drops to dangerously low levels as the %C rises above approximately 0.4. Because of this fact, combined with the fact that 0.4% C is adequate to produce very strong quenched and tempered steel, the %C level of most steels used in industrial and commercial machinery, such as trucks and automobiles, is limited to 0.4.
4. Higher-carbon steels find widespread use because of their significant improvement in wear resistance. This wear resistance is due to the presence of carbides, which have hardnesses superior to martensite. The carbides promote brittleness, and this effect is minimized by keeping the carbide size small. Plain carbon steels, such as 1080 and 1086, have improved wear resistance due to the very fine carbides produced in stage 1 of the tempering process. Such steels find wide use in cutting and plowing applications. Bearing steels, such as 52100, as well as tool steels are heat treated to produce arrays of carbide particles of fine sizes. Although fine, approximately 1 μm (4×10^{-5} in.), the size range is still much larger than the carbides produced in stage 1 tempering.
5. Two embrittlement mechanisms can occur on tempering steels. The first, tempered martensite embrittlement (TME), occurs when tempering in the range of 260 to 340 °C (500 to 650 °F). This temperature range is to be avoided for tempering. The second, temper embrittlement (TE), occurs for tempering in the range of 540 to 650 °C (1000 to 1200 °F). It is easily avoided by quenching from the tempering temperature.
6. The strength of quenched steels decreases on tempering. The fine carbide arrays that form in stage 1 tempering produce a strengthening effect that tends to reduce the inherent rate at which strength drops during tempering. As tempering temperature rises, the fine carbides grow in size (coarsen) and reduce their strength contribution. The addition of the reactive alloying elements that prefer to form carbides, such as chromium or molybdenum, makes the carbides coarsen more slowly and produces higher strengths (hardnesses) after tempering, as shown in Fig. 10.7. The coarsening rates are reduced because the alloying elements have low diffusion rates in austenite and are therefore difficult to remove from the carbides, and the carbides remain smaller at higher temperatures.
7. A major advantage of the low-alloy AISI steels versus plain carbon steels is this improved tempering response. The improvement shows up as an increase in hardness (or strength) at the same level of ductility. The addition of 0.8% Cr converts a 1060 steel to the low-alloy 5160 steel. As shown in Fig. 10.9, after tempering to a ductility of 42% RA (reduction in area), the hardness of 5160 is HRC=43 compared to 34 for the 1060 steel (equivalent tensile strengths of 1393 and 1048 MPa, or 202 and 153 ksi, respectively). Alternatively, after tempering to an HRC of 32, the ductility of 5160 is superior to 1060 by 52 versus 43% RA. The small chromium addition produces this enhancement by reducing the coarsening rate of the small carbides formed in the stage 1 tempering.
8. The carbides that form in the low-alloy AISI steels of Table 6.3 are epsilon and cementite carbides that incorporate the alloying element and change the carbide from $Fe_{2.4}C$ to $M_{2.4}C$ or Fe_3C to M_3C, where $M = Fe + X$ and $X = Cr$ and/or Mo. In high-alloy steels, such as tool steels, carbide-forming elements such as molybdenum are added at higher levels and, on tempering, produce carbides native to that element and different from $M_{2.4}C$ or M_3C. These native carbides, such as $Cr_{23}C_6$ or Mo_2C, resist coarsening to higher tempering temperatures and often can produce strengths even higher than the as-quenched strengths at tempering temperatures just above a red heat, approximately 500 °C (930 °F) (Fig. 10.10).
9. Strength and hardness decrease during tempering with increases in either the tempering temperature or tempering time. For the low-alloy AISI steels, the reduction is influenced

much more strongly by tempering temperature. As shown in Fig. 10.11, increasing the tempering time for a 1045 steel from 1 to 4 h drops the hardness by only approximately 1 HRC unit compared to approximately 5 units for a 55 °C (100 °F) increase in tempering temperature. For this reason, many of the useful tempering graphs found in Ref 10.8 present tempered hardness versus tempering temperature with no mention of tempering time. In such cases, it is safe to assume the tempering time was 1 h.

REFERENCES

10.1. *Modern Steels and Their Properties, Handbook 2757*, Bethlehem Steel Corp., 1972

10.2. A. Grossmann and E.C. Bain, *Principles of Heat Treatment*, American Society for Metals, 1964

10.3. E.C. Bain and H.W. Paxton, *Alloying Elements in Steel*, American Society for Metals, 1966, p 198, 200

10.4. R.A. Grange et al., *Metall. Trans. A*, Vol 8, 1977, p 1780

10.5. ASM Databook, *Met. Prog.*, Vol 112, June 1977

10.6. J.H. Hollomon and L.D. Jaffee, *Trans. Metall. Soc. AIME*, Vol 162, 1945, p 223

10.7. R.A. Grange and R.W. Baughman, *ASM Trans.*, Vol 48, 1956, p 165

10.8. *Heat Treater's Guide: Practices and Procedures for Irons and Steels*, 2nd ed., ASM International, 1995

CHAPTER 11

Austenitization

THE FIRST STEP in the hardening of steel is heating the steel hot enough to form austenite, from which martensite can form on quenching. For some steels, it is desired or required that carbides be present during this austenitization step, and this is called two-phase austenitization. Because the presence of the carbides produces some significant changes, single-phase austenitization and two-phase austenitization are treated separately in the following discussion.

Single-Phase Austenitization

Plain carbon and low-alloy AISI steels are generally austenitized at temperatures that produce single-phase austenite. This means the steel must be heated above the A_3 or the A_{cm} temperature. The temperature range for the plain carbon steels is shown in Fig. 4.23, and Ref 11.1 presents specific recommended temperatures for all the individual steels.

Room-temperature steel will generally consist of ferrite and cementite. In hypoeutectoid steels, the cementite is usually present as pearlite, and in hypereutectoid steels, it will be present either as pearlite or a mixture of pearlite and cementite particles. In plain carbon steels, ferrite is essentially pure iron (Fe+0.02% or less C), and in hypoeutectoid steels, basically all of the carbon is contained in the pearlite. Therefore, homogenization of the carbon requires heating the steel at a hot enough temperature for a long enough time for the carbon to diffuse from the pearlite regions to the center of the largest ferrite regions. Because the alloying elements are also present at different compositions in the room-temperature cementite and ferrite, the same rule applies to homogenization of the alloying elements during austenitization. By simply heating to higher temperatures, this homogenization process can be accelerated. However, it is important not to heat so hot that excessive grain growth occurs in the austenite, because large grains reduce toughness. The austenitization temperature and time should be selected to control two factors: homogenization of carbon and alloying elements, and minimization of austenite grain growth.

Homogenization with respect to carbon and the alloying elements occurs by the process of diffusion, as discussed in Chapter 7. Consider first the homogenization of carbon. The homogenization of a ferrite-cementite mixture into austenite occurs by a two-step process that is first described for austenitization of pearlite. Figure 11.(a) shows a ferrite plate (α) in pearlite between two cementite (Cm) plates. In the first step, austenite will form along the α/Cm boundaries on heating to A_{c_1}, as shown in Fig. 11.1(b), and grow into both the ferrite and cementite plates. Eventually, the growing austenite consumes all of the ferrite and cementite, thus completing the first step. At this point, the carbon composition is not homogeneous in the austenite; it will be highest at the center of the old cementite plates and lowest at the center of the old ferrite plate. In the second step, the carbon composition in the newly formed austenite homogenizes by the process of diffusion. In the first step, diffusion also occurs in ferrite, but analysis shows that the first step, as well as the second step, is controlled by the diffusion of carbon only in the austenite. To a good approximation then, the time it will take any ferrite-cementite mixture to homogenize can be determined by calculating how long it will take the carbon atoms to diffuse the required distance in austenite. In Fig. 11.1, the required distance will be half the spacing

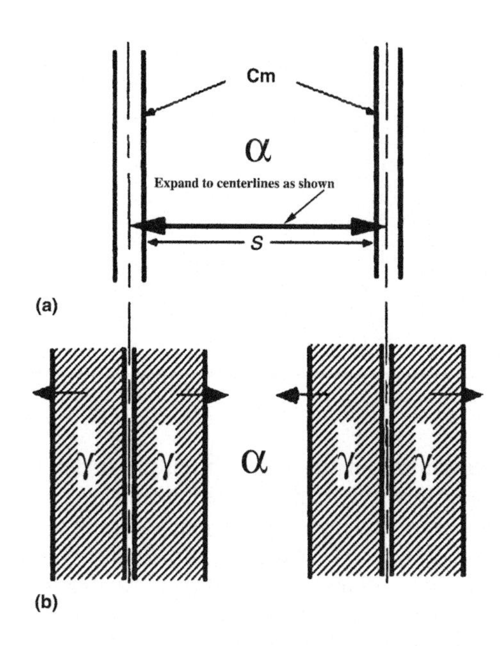

Fig. 11.1 (a) Pearlite with plate spacing = S. (b) Austenite (γ) formation at temperatures of A_{c_1} and above

between plates, $S/2$, because the carbon diffuses from both left and right into the ferrite plate shown.

If the diffusion coefficient, D, of carbon in austenite is known, Eq 7.1 can be used to calculate approximately how long it will take a carbon atom to diffuse some arbitrary distance, d. The values of D for carbon diffusion have been well measured and are given to a good approximation for most steels as $D = 0.12 \times \exp[-16000/(T + 273)]$ cm^2/s, where T is temperature in °C. With this information, it is possible to construct diagrams such as Fig. 11.2 that show the times required to move carbon atoms distances ranging from $d = 0.5$ to 500 μm as temperature increases. For example, the line labeled 5 indicates that at 800 °C (1470 °F), it will take 1 s for a carbon atom to diffuse 5 μm (2×10^{-4} in.). At 1000 °C (1830 °F), it will take only 0.1 s for a carbon atom to diffuse the same 5 μm distance. The pearlite spacing, S, is generally less than 1 μm (4×10^{-5} in.), giving a diffusion distance for homogenization of $d = S/2$ = less than 0.5 μm (2×10^{-5} in.). At the recommended temperature range for austenitization of steels of 820 to 840 °C (1500 to 1550 °F) (Ref 11.1), Fig. 11.2 shows that the pearlite should form homogeneous austenite in less than 10 milliseconds (0.01 s).

Now, consider a typical ferrite-pearlite steel, as shown in Fig. 11.3, that is to be austenitized. Essentially all the carbon in this steel is contained in the pearlite regions, marked P. The previous discussion shows that these pearlite regions will form homogeneous austenite in a manner of milliseconds after reaching the A_{c_1} temperature. The freshly formed austenite in the old pearlite regions will have the carbon composition of the old pearlite, approximately 0.77% C. To complete homogenization of the steel, this carbon must be redistributed to the center of the surrounding ferrite grains, a maximum distance of z in Fig. 11.3, or approximately ½ the ferrite grain size. As before, the homogenization occurs in two steps: (1) The austenite formed in the old P regions expands into the ferrite grains by migration of the austenite-ferrite boundaries. (2) This newly formed austenite will contain less carbon at the center of the old ferrite grain and more carbon at the edge.

The second step is homogenization of this carbon composition gradient by diffusion. As with the pearlite, the total homogenization process is controlled by the diffusion coefficient of carbon in austenite, and the time also may be approximated with Fig. 11.2. Suppose the ferrite grain size is relatively large, perhaps 100 μm (4×10^{-3} in.). This gives a diffusion distance, d, of 50 μm (2×10^{-2} in.), and, at typical austenitization temperatures, homogenization is predicted to be complete in approximately 1 min. At a more typical ferrite grain size of approximately 20 μm (8×10^{-4} in.), homogenization is predicted to require less than 10 s.

Hypereutectoid steels are supplied from the steel mill in the spheroidized form, with particles of cementite in a ferrite matrix. When these steels are austenitized, homogenization occurs by the formation of spherical regions of austenite that form around the carbides and grow into the surrounding ferrite. Again, the process occurs in two stages: first, the conversion of the surrounding ferrite to austenite as these spherical regions expand, and second, by the homogenization of the freshly formed austenite with carbon diffusing to the center of the ferrite regions from the positions of the dissolved cementite particles. Homogenization times again can be approximated with Fig. 11.2. Because the distance between the spheroidized cementite particles in hypereutectoid steels is generally less than approximately 5 μm (2×10^{-4} in.), the carbon homogenization times are predicted to be small, on the order of a matter of seconds after the austenitization temperature is reached.

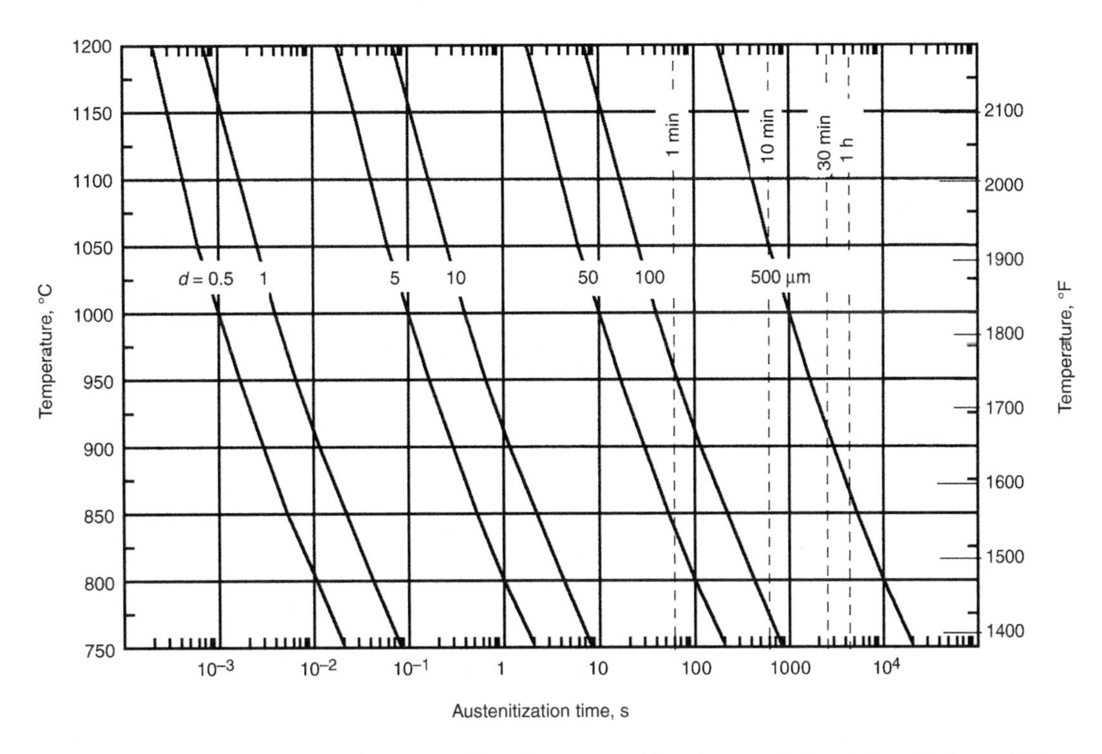

Fig. 11.2 Temperature dependence of time required for carbon atoms to diffuse distances, *d*, of 0.5, 1, 5, 10, 50, 100, and 500 μm in austenite

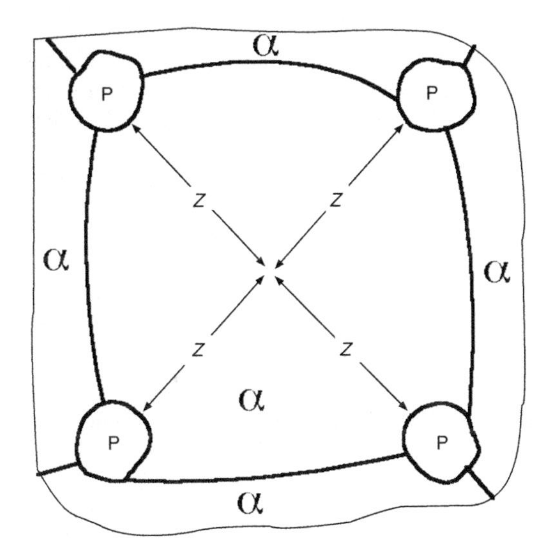

Fig. 11.3 Ferrite (α) grain with pearlite (*P*) at its grain corners. The maximum distance through which the carbon must be redistributed for complete homogenization is represented by *z*

It is important to bear in mind that these calculations give the time of homogenization after the steel has achieved the recommended austenitization temperature. After placing a piece of steel in a furnace set at the desired austenitization temperature, it will take some time for the center of the steel to reach this temperature. A general rule of thumb is that the time it takes for the steel to heat through is approximately equal to 1 h per inch of thickness. When working with propane-fired furnaces having considerable gas flow, as is often done by bladesmiths, the times to achieve the austenitization temperature can be significantly reduced. Similarly, austenitizing in salt pots will reduce times to heat to the austenitization temperatures, because the liquid salt transfers heat more rapidly than gases. Three techniques that produce extremely fast heating, rates are flame heating, induction heating, and laser heating. As shown in Fig. 11.2, even with these ultrafast heating rates, homogenization of austenite with respect to %C is not a big problem as long as the microstructure of the steel is fine grained. However, homogenization of austenite in alloy steels is more complicated.

As pointed out in Chapter 7, "Diffusion," the alloying elements have diffusion coefficients in austenite that are thousands of times lower than for carbon. Thus, homogenization of the alloying elements during austenitization is much slower than homogenization of carbon. For example, consider the alloying element chromium, whose diffusion distances are given in Fig. 11.4, as

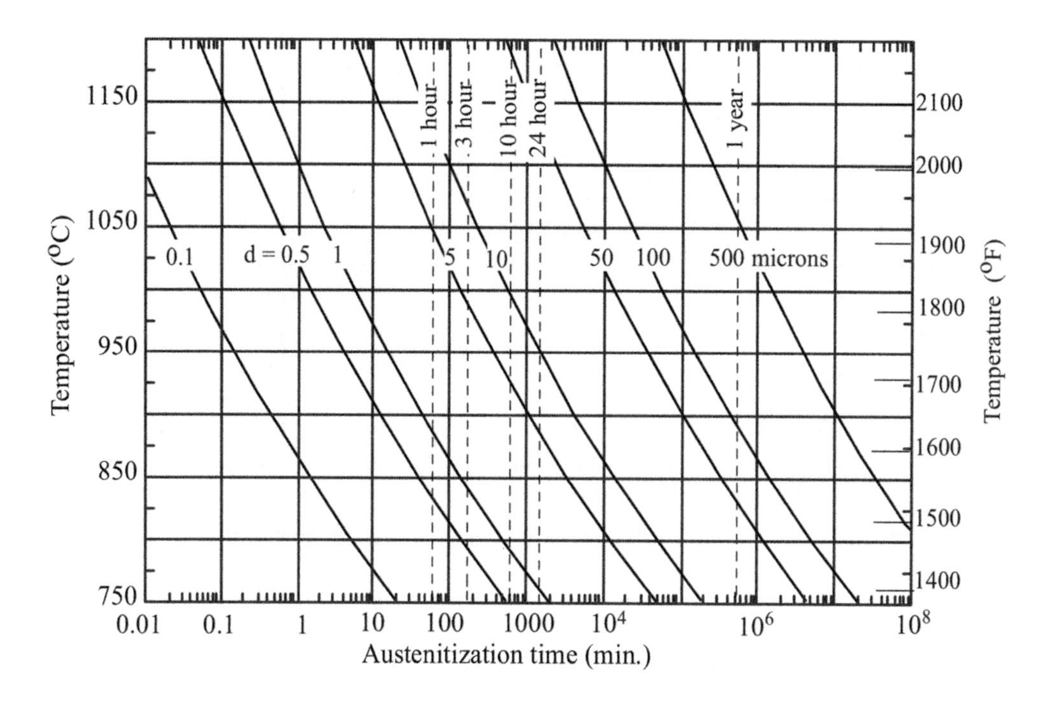

Fig. 11.4 Temperature dependence of time required for chromium atoms to diffuse distances, *d*, of 0.1, 0.5, 1, 5, 10, 50, 100, and 500 μm in austenite

calculated from its measured diffusion coefficient in austenite using Eq 7.1. Diffusion times for chromium are much longer than for carbon; for example, at 850 °C (1560 °F), a diffusion distance of 1 μm (4×10^{-5} in.) requires 5 h for chromium, compared to only 0.02 s for carbon. Air-cooled pearlite usually will have a plate spacing, *S*, of less than 0.2 μm (8×10^{-6} in.), giving a homogenization diffusion distance of 0.1 μm (4×10^{-6} in.). Figure 11.4 estimates that at the normal austenitization temperature for 5160 steel (830 °C, or 1530 °F) recommended in Ref 11.1, the pearlite should homogenize in approximately 5 min. However, if the chromium-containing carbides of a chromium steel are more widely separated than in pearlite, as in a spheroidized steel, homogenization times can be signi cantly longer. For example, if a spheroidized 5160 or 52100 steel with carbides spaced at 1 μm (4×10^{-5} in.)is austenitized at 840 °C (1540 °F), Fig. 11.4 shows that homogenization will take on the order of 60 min. This effect may explain the dramatic improvement of hardenability of a 5160 steel that is achieved by increasing the austenitization temperature from 850 to 950 °C (1560 to 1740 °F) (Fig. 9.15).

Another example of slow homogenization of alloying elements was pointed out in Chapter 10, "Tempering." In alloys containing carbides, the rst step of homogenization is to dissolve the carbides into the surrounding austenite that forms on them. The rate of this step is dramatically reduced by the low diffusion coefficient of carbide-forming alloying elements. During stage 1 tempering, $M_{2.4}C$ epsilon carbides remain smaller than the $Fe_{2.4}C$ epsilon carbides of plain carbon steels, resulting in higher tempered strengths at a given tempering temperature, as shown for a molybdenum steel alloy in Fig. 10.7 and a chromium steel alloy in Fig. 10.9.

Austenite Grain Growth. The principles of grain growth were discussed in Chapter 8, "Control of Grain Size by Heat Treatment and Forging." As illustrated in Fig. 8.3, austenite grain size increases rapidly as austenitization temperature is increased. The figure also shows that, in general, the austenite grain size increases more rapidly with increasing temperature than increasing time.

At normal heating rates, the new austenite grains will nucleate on the boundaries between cementite and ferrite, with their subsequent growth rate controlled by carbon diffusion in the freshly formed austenite. When heated very rapidly to temperatures higher than 910 °C (1670 °F), however, new austenite grains can nucleate directly on old ferrite-ferrite boundaries with the same composition as the ferrite, so that no carbon diffusion is required. To understand why rapid heating rates and high temperatures are required, remember that the ferrite component of room-temperature steel is nearly pure iron. Figure 3.5 shows that the ferrite component will contain approximately 0.02% C, and if heated above the A_c temperature of approximately 910 °C (1670 °F), it can transform directly into austenite with no composition change required. However, it must be heated rapidly or it will be transformed into austenite from austenite regions that nucleate first at the cementite-ferrite boundaries and then grow into the remaining ferrite. As discussed in Chapter 8, the first-formed austenite grains will have small diameters and grow rapidly as temperature increases. As shown, the small grains can effectively be retained by using the trick of rapid cyclic heat treating. It is also shown in Chapter 8 that alloying additions reduce austenite grain growth by two mechanisms: the formation of small particles (Fig. 8.9, 8.10) and grain-boundary segregation (Fig. 8.12).

Two-Phase Austenitization

Steels with superior wear properties normally contain carbides in a tempered martensite matrix. This means that the carbides must be present during the austenitization step in order for them to be present after the austenite transforms to martensite on quenching. Common examples include bearing steels, such as 52100, tool steels, and heat treated cast irons, such as ductile cast iron. Many tool steels and most cast irons cannot be heat treated to single-phase austenite. The reason can be understood by examining the full iron-carbon phase diagram shown in Fig. 11.5. (The iron-carbon phase diagrams presented in Chapter 3, "Steel and the Iron-Carbon Phase Diagrams," are just small portions of the full diagram and do not show any of the liquid region that occurs at very high temperatures.)

In the initial discussion of phase diagrams in Chapter 2, "Solutions and Phase Diagrams," the salt-water system was chosen to illustrate the major ideas. Figure 2.2 illustrates that the freezing temperature of water-salt solutions decreases as the percent salt in the liquid increases, but it decreases only to a certain point, called the eutectic point, which occurs at the eutectic temperature of −49.8 °C (−57.6 °F) and the eutectic composition of 30% salt content.

Figure 11.5 shows that the same thing happens in steel, with the freezing temperature falling to a eutectic temperature of 1148 °C (2098 °F) at a composition of 4.3 wt% C in the liquid steel. Most cast irons have carbon compositions above 2.1 and below 4.3% C and can be modeled

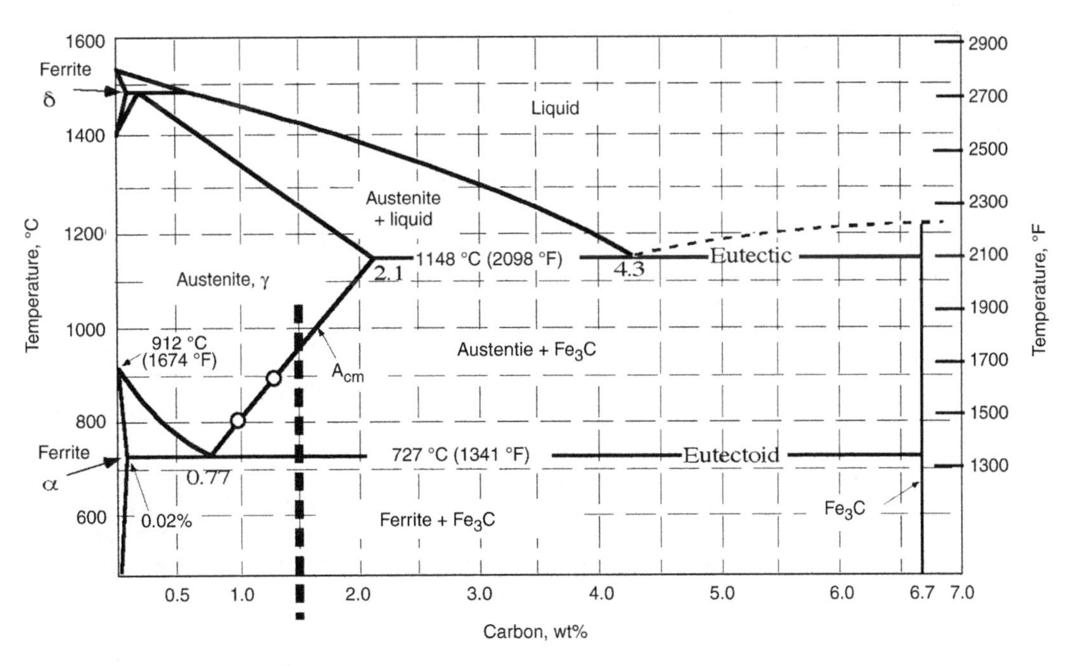

Fig. 11.5 Full iron-carbon phase diagram

with the pure iron-carbon diagram of Fig. 11.5.* It shows that an iron-carbon alloy in the cast iron composition range of 2.1 to 4.3% C will begin to melt and become a mixture of liquid steel and cementite at temperatures above 1148 °C (2098 °F). At temperatures below 1148 °C (2098 °F), these alloys will always contain some cementite (Fe_3C) mixed in with the austenite. The diagram shows that for iron-carbon compositions above 2.1%, there is no way to form 100% austenite. Hence, most cast irons cannot be heat treated to form single-phase austenite; they will begin to melt before the cementite can fully dissolve into the austenite. The same is true of many tool steels, so their heat treatment always involves two-phase austenitization.

Such is not the case for steels that have carbon compositions under 2.1%. Consider the alloy of Fe + 1.5% C shown at the vertical dashed line in Fig. 11.5. If this alloy is austenitized at temperatures above its A_{cm} value of approximately 960 °C (1760 °F), it will become single-phase austenite. Now, consider what must happen if this steel is austenitized at either of the two lower temperatures of 800 or 900 °C (1470 or 1650 °F). In both cases, the steel will contain cementite in the austenite and will be an example of two-phase austenitization.

There is a major difference in single-phase and two-phase austenitization. To understand this difference, it may be helpful to review the discussion of the 1095 alloy whose microstructure is shown at the bottom of Fig. 3.5. After heating to 760 °C (1400 °F), the alloy is located in the two-phase austenite + cementite (γ + Cm) region of the diagram at a position shown with the circle. According to the phase diagram, at 760 °C (1400 °F), any alloy having a composition between points *O* and *P* will be a two-phase mixture of austenite and cementite. Furthermore, it gives the compositions of these two phases.

The austenite composition is given by the intersection of the horizontal line at 760 °C (1400 °F) with the A_{cm} line, point *O*. This composition is less than the 0.95% C in the alloy, because not all of the carbon is contained in the austenite; some is lost to the cementite carbides. Now, return to consideration of the 1.5% C alloy shown with the vertical dashed line in Fig. 11.5. According to this expanded phase diagram, the austenite composition that forms at 800 or 900 °C (1470 or 1650 °F) is given by the open circles located at the A_{cm} line for these temperatures. Hence, the %C in the austenite part of the steel will change with temperature, being approximately 1% C at 800 °C (1470 °F) and 1.3% C at 900 °C (1650 °F). *Therefore, the %C in austenite for two-phase austenitization is controlled by the austenitization temperature.* This is a very important result that must be clearly understood when heat treating such alloys as 52100 bearing steel, as is now discussed.

Figure 4.18 shows the amount of retained austenite that is obtained in a quenched steel as the %C of the quenched austenite increases. In hypoeutectoid steels (%C less than 0.77), the amount of retained austenite is generally small enough to be of no consequence. However, in hypereutectoid steels, the amount of retained austenite is large enough to have significant effects on both strength and tempering response. The effect on hardness (strength) is shown by the data in Fig. 4.13, where the drop in hardness when %C is above 0.8 is due to the retained austenite. Because two-phase austenitization can occur in hypereutectoid steels, the austenitization temperature will have an effect on the amount of retained austenite and therefore on the as-quenched hardness of these steels.

Such an effect is illustrated nicely by the data for an 8695 steel, shown in Fig. 11.6. (The 8620 steel is often used for carburizing, and 8695 is a typical carburized composition for this steel.) Figure 11.6(a) shows the seemingly strange result that the as-quenched hardness of the steel drops as the austenitizing temperature increases from 760 to 870 °C (1400 to 1600 °F). The A_1 and A_{cm} temperatures are shown in Fig. 11.6(a), and two-phase austenitization must be occurring between these two temperatures. This means that the %C in the austenite is increasing as the temperature rises from A_1 to A_{cm}. The increase in %C produces an increase in percent retained austenite, as shown in Fig. 11.6(b), which, in turn, accounts for the drop in as-quenched hardness. When austenitizing above the A_{cm} temperature, the austenite has a constant %C composition equal to that of the overall alloy (0.95% C here), and the hardness becomes constant, because the amount of retained austenite is no longer changing with austenitization temperature.

Bearing steels such as 52100 are deliberately austenitized below A_{cm} in order to produce arrays of cementite carbides in the final martensite for

* Actual cast irons also contain significant amounts of silicon, and its effect on their phase diagram is discussed in Chapter 16, "Cast Irons."

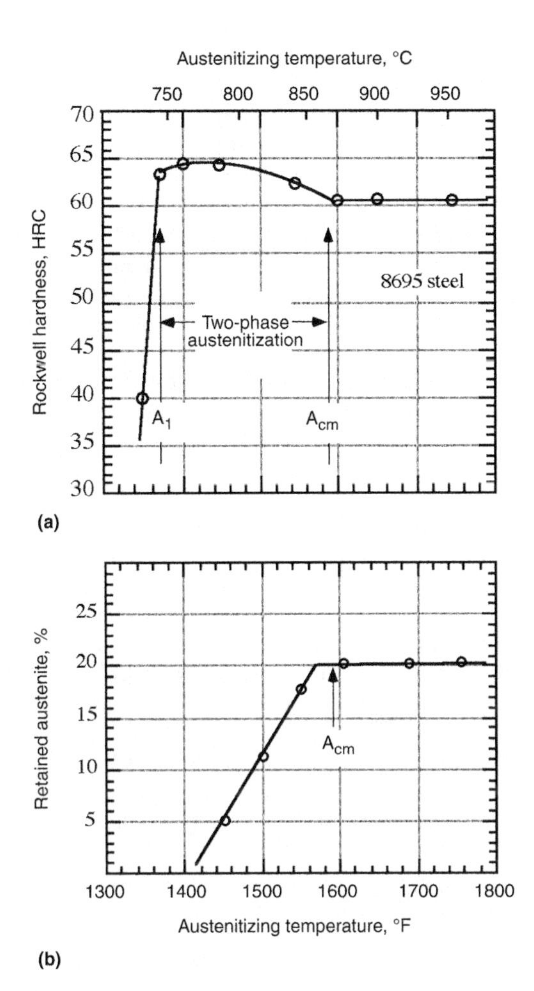

Fig. 11.6 Effect of austenitizing temperature on (a) the as-quenched hardness of 8695 steel and (b) the amount of retained austenite at room temperature. Source: Ref 11.2. Copyright: American Metal Market

increased wear resistance. The recommended austenitizing temperature for this steel (Ref 11.1) is 845 °C (1555 °F), and Fig. 6.4 is an approximate phase diagram for 52100, showing that, at this austenitizing temperature, the %C in the austenite will be approximately 0.78. Thus, only a modest amount of retained austenite is expected in the quenched product. In addition, the martensite should consist of a mixture of lath and plate forms. If, however, this steel were austenitized at approximately 950 °C (1740 °F), it would become single phase and produce the more brittle plate martensite with more retained austenite. Figure 4.18 predicts that the amount of retained austenite would increase from approximately 8 to 15%.

Data are available on 52100 steel (Ref 11.3) that nicely illustrate the effect of the chromium

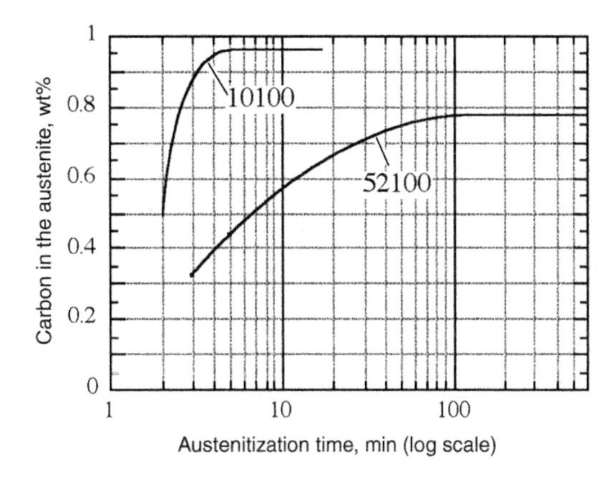

Fig. 11.7 Weight percent carbon increase in austenite at 845 °C (1550 °F). Adapted from Ref. 11.3

alloying element on both the phase diagram and the rate of austenitization. For austenitization at the recommended 845 °C (1555 °F), Fig. 11.7 presents data showing the measured %C in the austenite as a function of austenitization time for both 52100 and the corresponding plain carbon steel of the same %C value, 10100. The chromium addition in the 52100 steel does two things. First, it reduces the rate at which the cementite carbides dissolve. In the plain carbon 10100 steel, the carbides dissolve after approximately 5 min, while it takes approximately 100 min for the carbides to dissolve to their final size in the 52100 steel. Second, the alloy addition raises the A_{cm} temperature, so that austenitization is occurring in the two-phase region for 52100 (Fig. 6.4). In the 52100 steel, the %C in the austenite has been reduced to the value on the A_{cm} line (0.78% C),* while for the 10100 steel, single-phase austenitization occurs, and the %C is the same as in the starting alloy. (Note: Apparently this 10100 steel contained only approximately 0.95% C.)

Summary of the Major Ideas in Chapter 11

1. At room-temperature, most steels consist of a mixture of ferrite and cementite, with the cementite present either as the fine plates in pearlite or as discrete particles. Both carbon atoms and alloying element atoms have different compositions in ferrite and cementite.

* Strictly speaking, a ternary isothermal diagram is needed to evaluate A_{cm} for steel alloys, but Fig. 6.4 is a reasonable approximation.

For example, cementite contains 6.7% C compared to only 0.02% C in ferrite, a huge difference. For carbide-forming alloying elements such as chromium and molybdenum, the differences in %Cr and %Mo in cementite versus ferrite are less than for carbon but can be significant. Hence, when steels are heated to form austenite, the fresh austenite inherits a very nonuniform distribution of carbon and alloying element atoms. It is desirable to make the distribution uniform in the austenite prior to quenching. Therefore, an important process that occurs in the austenitization step of heat treating is the homogenization of the freshly formed austenite with respect to %C and percent alloying elements.

2. Two important processes occur during austenitization of steels: (1) homogenization of the composition of carbon and alloying elements in the freshly formed austenite, and (2) grain growth of the new austenite grains. Increasing the austenitization temperature will speed up homogenization, but it will also speed up grain growth. Because small grains are required for improved toughness, the austenitization temperature is chosen to produce a compromise between these two processes. Reference 11.1 presents recommended austenitization temperatures for the standard hardening of virtually all commercial steels.
3. Homogenization is controlled by the rate of diffusion of carbon and the alloying elements in austenite. These atoms must diffuse from the center of the cementite regions to the center of the ferrite regions during homogenization. The homogenization time is dependent on two variables: the diffusion coefficient of the atoms, D, and the distance the atoms must diffuse, d. Both variables have strong effects. Increasing temperature increases D rapidly and decreases the homogenization times. Increasing d, the distance from the center of the cementite regions to the center of the largest ferrite regions, increases the homogenization times rapidly. Figure 11.2 shows how the required austenitization time to achieve homogenization of carbon varies as temperature and diffusion distance is changed.
4. When austenitizing a steel containing pearlite and ferrite, the homogenization occurs in two stages. First, the pearlite converts to a homogeneous austenite with the same composition as the pearlite, 0.77% C. Second, the ferrite surrounding the old pearlite regions converts to austenite and homogenizes on holding at temperature. The first stage occurs in seconds after the austenitization temperature is reached, because the plate spacing of pearlite is virtually always very small (less than 1 μm, or 4×10^{-5} in.) and the diffusion distance is very small. The second stage is larger because the diffusion distance is now set by the ferrite grain size (Fig. 11.3), which can have diffusion distances 10 to 100 times larger than in pearlite. However, even with large ferrite grain size, homogenization times for the second stage are still low for carbon, on the order of minutes.
5. Because diffusion coefficients for the alloying elements are on the order of thousands of times smaller than for carbon, the homogenization times for alloying elements can be very large at the normal austenitization temperatures. Data are presented for chromium alloys in Fig. 11.4, showing that air-cooled pearlite (which has a fine spacing) will homogenize in minutes, but homogenization to the center of the ferrite grains in the second stage can easily take many hours at normal austenitization temperatures. Raising the austenitization temperature can speed up homogenization and produce dramatic increases in hardenability in alloy steels, as shown for 5160 steel in Fig. 9.15. However, possible grain growth at the higher temperatures is a concern.
6. Grain growth must be minimized during austenitization. Grain growth was discussed in Chapter 8, "Control of Grain Size by Heat Treatment and Forging." Grain growth increases with increases in both austenitizing temperature and austenitizing time. The temperature effect is much stronger (Fig. 8.3). Grain growth is decreased by the addition of alloying elements, which allows alloy steels to be austenitized at higher temperatures than plain carbon steels. Ultrafine austenite grain sizes can be achieved by cyclic heat treatments, as demonstrated by the experiment discussed in Chapter 8 (Table 8.2).
7. For most steels, austenitization is carried out at temperatures that produce single-phase austenitization, and the general temperature ranges are shown in Fig. 4.23. For high-carbon wear-resistant steels, two-phase

austenitization is employed. In this case, cementite carbides are present in the austenite, so that quenching gives arrays of cementite particles in a martensite matrix. The carbides enhance wear resistance because they are harder than the martensite. An important example of two-phase austenitization is the common bearing steel, 52100, and Fig. 6.5 shows the typical carbide arrays that are produced in this steel.

8. Two-phase austenitization is tricky, because the carbon content of the austenite is controlled by the austenitization temperature. Because austenitization is occurring below the A_{cm} temperature of the steel, the %C in the austenite is given by the location of the A_{cm} line for the temperature of choice, as illustrated for austenitization of a 1.5% C steel in Fig. 11.5. In this example, raising the austenitization temperature from 800 to 900 °C (1470 to 1650 °F) increases the %C in the austenite from 1.0 to 1.3%.
9. Two-phase austenitization occurs for high-carbon steels, tool steels, and cast irons. As seen in Fig. 4.18, at higher carbon levels, significant amounts of retained austenite can occur on quenching, which results in reduced hardness and strength. As the two-phase austenitization temperature is raised, the %C in the austenite will increase, and this can lead to reductions of as-quenched hardness due to increases in percent retained austenite. Figure 11.6 illustrates this effect for 8695 steel (a carburized 8620 steel). Austenitizing this steel above its A_{cm} temperature produces single-phase austenitization, and as-quenched hardness no longer depends on austenitization temperature, because the %C in the austenite is now limited to the overall composition of 0.95% C.

REFERENCES

11.1. *Heat Treater's Guide: Practices and Procedures for Iron and Steels,* 2nd ed., ASM International, 1995

11.2. A.R. Troiano and R.F. Hehemann, Effect of Austenitizing Conditions on Medium Alloy Steels, *Iron Age,* Vol 174 (No. 21), 1954, p 151–153

11.3. J.M. Beswick, The Effect of Chromium in High Carbon Bearing Steels, *Metall. Trans. A,* Vol. 18, 1987, p 1897–1906

CHAPTER 12

Quenching

THE QUENCHING PROCESS has played an important role in the production of hardened steel implements since ancient blacksmiths first discovered the tricks of hardening carburized iron blades, an event that probably occurred in approximately 1200 B.C. (Ref 12.1). When reading historical accounts of methods for hardening steel, one cannot help but be impressed by the great emphasis placed on the quenching media, as well as the great variety of quenchants that were recommended. Writing in 1923, Roberts-Austen (Ref 12.2) gives a short history of quenchants and states: "The belief in the efficacy of curious nostrums and solutions for hardening steel could hardly have been firmer in the third century B.C. than in the sixteenth of our era." He quotes a recipe from a 1531 book as: "Take snayles, and first drewn water of a red die, of which water being taken in the two first monthes of harvest when it raynes, boil it with the snails, then heate your iron red hot and quench it thererin, and it shall be hard as steele." C.S. Smith has published a book titled *Sources for the History of the Science of Steel* (Ref 12.3) that illustrates the importance placed on the quenchant. A 1532 pamphlet, called *On Steel and Iron*, gives several recipes for hardening steel, one of which is: "Take the stems and leaves of vervain, crush them, and press the juice through a cloth. Pour the juice into a glass vessel and lay it aside. When you wish to harden a piece of iron, add an equal amount of a man's urine and some of the juice obtained from little worms known as cockchafer grubs." A discussion by Wertime (Ref 12.4) shows that there was a widespread belief that special waters were important: "The ever reliable Pliny informs that Roman steelworkers carefully distinguished the waters of various rivers for quenching, a tradition that persisted in Western Europe for two millennia and was extant in Asia. In England the waters of the river Derwent, in France of the river Fure were highly regarded."

All of these stories reflect the fact that hardening steel was largely an art in early times, plus, as pointed out by C.S. Smith, the importance of the tempering step was often not understood. By slowing down the quench rate, some of the exotic recipes probably provided a degree of tempering. Also, recent progress in evaluating the quenching power of water has shown that it is sensitive to various impurity additions, which may well account for some of the early variations that were used.

Special Quenching Techniques

By inserting a thermocouple into a piece of steel, one can measure the temperature versus time during the quench operation. A plot of this temperature versus time is called a cooling curve. If the cooling curve plot is superimposed on a continuous transformation (CT) diagram for the steel, a reasonable estimate can be made of what austenite transformation products (pearlite, ferrite, bainite, or martensite) will form during the quench. Except for thin steel pieces, such as knife blades, the surface of the steel will cool faster than the center region. Figure 12.1 presents hypothetical cooling curves at the surface and center of a rapidly quenched 5140 steel superimposed on its CT diagram, which was previously discussed as Fig. 9.9(a). Because both the surface and the center have dropped below the M_{90} line without ever crossing the CT lines for ferrite start (F_s), pearlite start (P_s), or bainite start (B_s), the bar is expected to be 100% martensite to its center. (Note: It is common to use the M_{90} temperature as a reasonable approximation of the martensite finish temperature, M_f.)

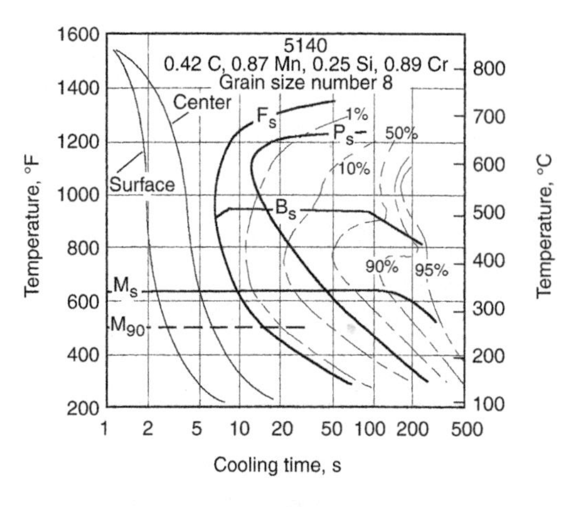

Fig. 12.1 Cooling curve at the surface and center of a rapidly quenched bar of 5140 steel superimposed on the continuous transformation curve

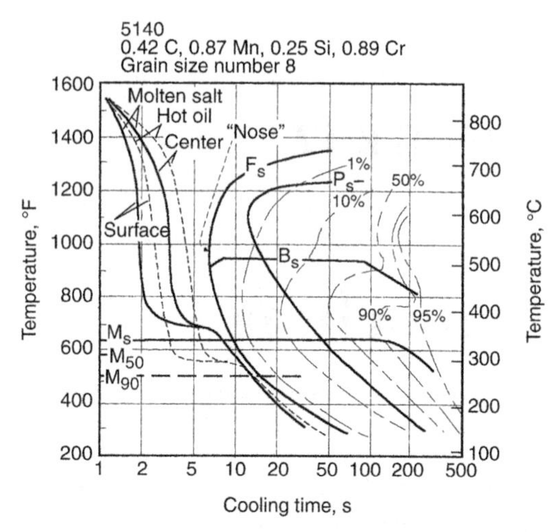

Fig. 12.2 Cooling curve at the surface and center of a martempered bar of 5140 steel superimposed on the continuous transformation curve. Solid curves are molten salt quench baths; dashed curves are hot oil quench baths

Martempering. A major problem with rapid quenching of steel is the formation of both distortion and quench cracks in the piece. Experiments have shown that these problems arise from two main causes: (1) a nonuniform temperature distribution in the piece as its temperature falls through the M_s-M_f range, and (2) too high a cooling rate as the temperature falls through the M_s-M_f temperature range. Quench crack formation leads to embrittlement, and susceptibility to this problem depends strongly on the %C in the steel and the cooling rate through the M_s-M_f range. Kern (Ref 12.5) claims that water quenching of steels with %C greater than 0.38 will produce quench cracking unless the piece has a simple shape, such as round with no holes. He recommends the use of a slower oil quench for the higher-%C steels, which is the usual industrial practice. Distortion on quenching is due mainly to the first cause, large temperature gradients (Ref 12.6). For the example in Fig. 12.1, extreme temperature gradients would be expected, because when the surface reaches M_{90}, the center is at approximately 730 °C (1350 °F), which is 470 °C (850 °F) hotter than M_{90}. Both of these problems can be significantly reduced by employing a special quenching technique first suggested in the 1940s (Ref 12.7) that is usually called martempering but sometimes called marquenching.

Martempering is generally done by quenching into a molten salt bath at a temperature just above the M_s temperature. The piece is held in the salt just long enough to equalize the temperature throughout, and then it is air cooled. Figure 12.2 shows possible cooling curves plotted on the CT curve for the 5140 steel. The process produces both a uniform temperature in the piece and a reduced cooling rate as it cools through the critical M_s-M_f temperature range, where the martensite forms from the austenite. A problem with martempering is that it requires the steel to have a fairly high hardenability, which means that the nose of the CT curve must be far enough to the right (high enough cooling times) to allow the center of the piece to cool fast enough to miss it. Except for very thin pieces (such as knife blades), martempering is difficult to use properly with plain carbon and some low-alloy AISI steels.

Reference 12.8 (p 104) offers an excellent discussion of martempering and gives recommended salt bath temperatures and compositions as well as maximum cross sections for some plain carbon and several low-alloy AISI steels. Martempering can also be done using hot oils, so-called martempering oils, which can be used at higher temperatures than ordinary quenching oils. The dashed cooling curves on Fig. 12.2 correspond to a martempering oil. Experiments (Ref 12.6) show that even though the hot oil temperature is below the M_s temperature, significant reduction of distortion is still obtained. Reference (12.8) provides a good discussion of the advantages and disadvantages of salt versus oil for martempering.

Austempering. The microstructure now called bainite was first discovered with an isothermal quenching experiment in the early 1930s by Bain and Davenport at U.S. Steel. As shown by

Fig. 4.23, bainite has two forms, called upper and lower bainite. It forms by isothermal holding at temperatures from just above to just below the M_s temperature. To form steels with 100% bainite, it is necessary to quench the steel to temperatures above M_s fast enough to avoid the formation of pearlite or ferrite and hold it at temperature until all the austenite transforms to bainite. For plain carbon steels, this means that the workpiece must be cooled extremely rapidly, as illustrated for a 1080 steel in Fig. 12.3 (which uses the isothermal transformation (IT) diagram in Fig. 9.3). In this example, the center of the workpiece must cool below approximately 540 °C (1000 °F) in only 1 s, which illustrates that only quite thin samples of 1080 steel can be successfully austempered. As illustrated in Fig. 9.4, alloy additions shift the pearlite start to longer times, so low-alloy AISI steels can be austempered in larger section sizes. A problem with austempering the alloys, however, is that the bainite finish curves are shifted to longer times, which, in some cases, leads to prohibitively long hold times to complete decomposition of austenite into bainite.

Shortly after the discovery that fully bainitic steels could be produced by isothermal quenching, researchers at U.S. Steel found that the ductility and toughness of such steels were superior to those of quenched and tempered martensites at the same hardness level. Their early work evaluated the hardness of steels transformed completely at isothermal temperatures and is summarized in Fig. 12.4. The hardness of fresh martensite at room temperature is found in the upper left of the figure, and the values are what one expects from the data presented in Fig. 4.13. The curves show a significant hardness increase as the bainite is formed at decreasing temperatures, which is a reflection of the fact that lower bainite is stronger than upper bainite. The M_s temperatures for the steels are marked by the short vertical lines on each curve. The hardest fully bainite structures formed at the M_s temperatures (vertical bars) are not as hard as fresh martensite in each steel. To better illustrate, Table 12.1 lists a comparison of these hardnesses. The curves of Fig. 12.4 also illustrate that the hardness of fully pearlitic steels depends on the %C of the steel. At higher %C, the cementite plates of the pearlite increase in relative thickness and make a larger contribution to the total hardness.

To illustrate the advantages of austempered plain carbon steels, the bar graph in Fig. 12.5 presents the early U.S. Steel data comparing the ductility, measured as percent reduction in area (%RA), and the impact strengths of austempered steels (open bars) versus quenched and tempered steels (solid bars) at the various hardness levels.

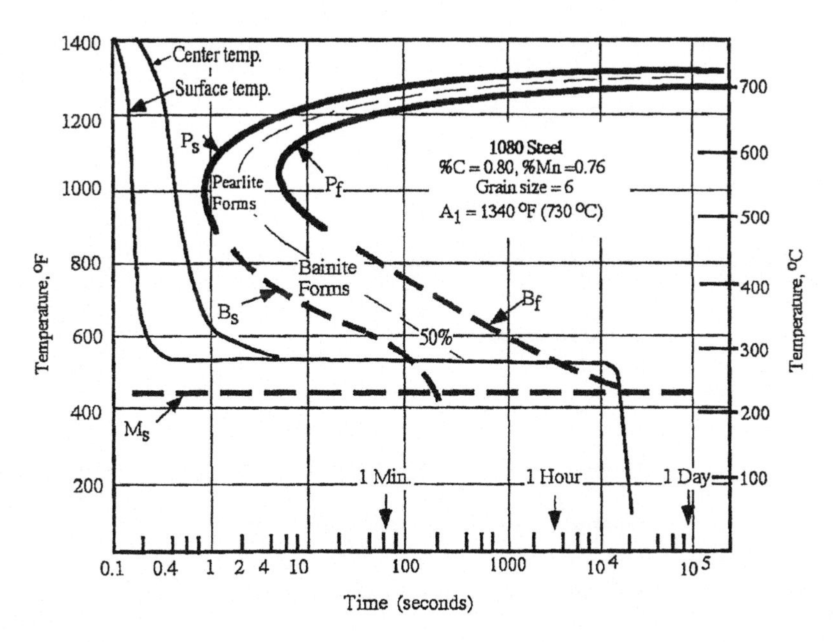

Fig. 12.3 Cooling curve at the surface and center of an austempered bar of 1080 steel superimposed on the continuous transformation curve

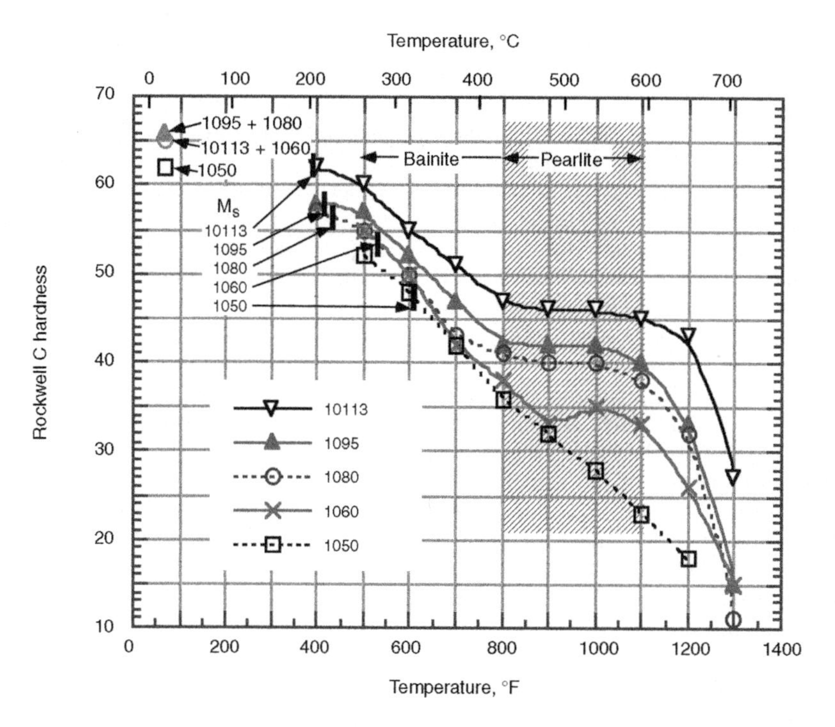

Fig. 12.4 Hardness of several plain carbon steels after isothermal quenching. All steels are 0.80 to 0.90% Mn except 1095 and 10113, which are 0.30% Mn. Source: Ref 12.9

Table 12.1 Comparison of fresh martensite hardness to maximum bainite hardness

Steel	Martensite, R_c	Max bainite, R_c
10113	65	62
1095	66	58
1080	66	57
1060	65	53
1050	62	48

The austempered steels have significantly better ductility and impact strength in the hardness range of HRC = 45 to 50. This enhancement in mechanical properties extends to the higher hardness levels possible with austempering, as illustrated dramatically by Fig. 12.6. The austempered rods on the left could be bent to nearly 180° without breaking, while the quenched and

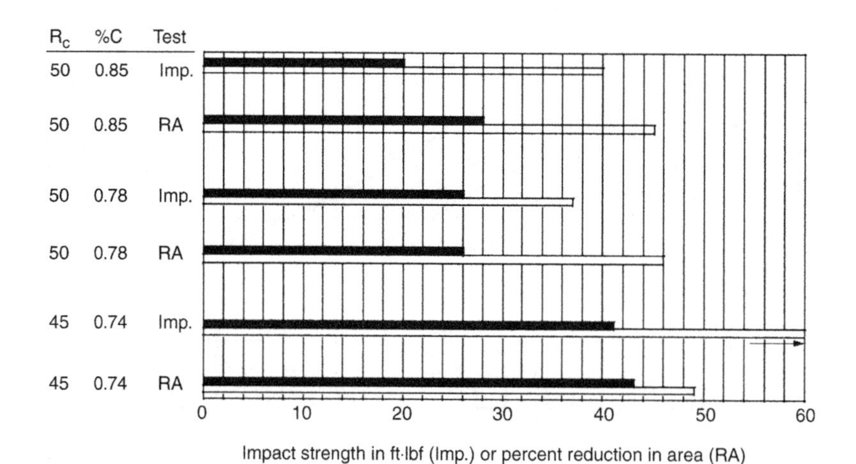

Fig. 12.5 Impact strength and percent reduction in area for austempered steels (open bars) compared to quenched and tempered steels (solid bars) with carbon levels of 0.74, 0.78, and 0.85%. Impact test on unnotched Charpy 7 mm (0.28 in.) rod. Source: Ref 12.10

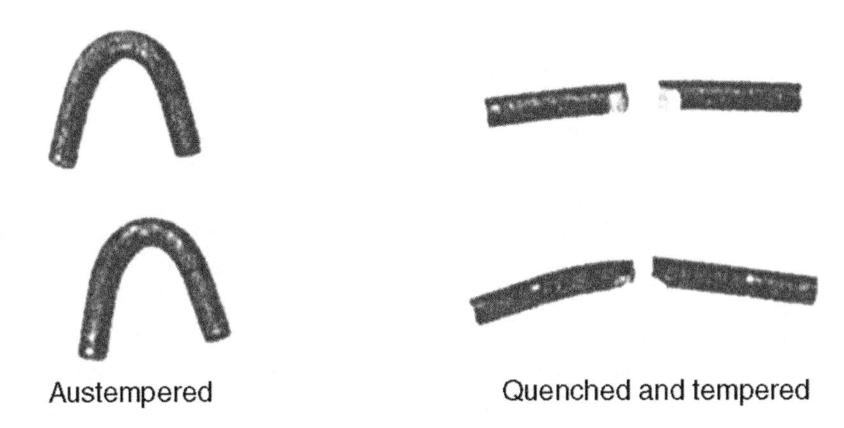

Fig. 12.6 Rods 3 mm (0.1 in.) in diameter of 1% C and 0.4% Mn produced at HRC = 58 and bent under equivalent conditions. Source: Ref 12.11

tempered rods at the same hardness level of HRC = 58 fractured with little prior bending strain.

Early industrial work with austempering of plain carbon steels established guidelines for the largest round bar diameters that could be successfully austempered, and these data are presented in Table 12.2. Plain carbon steels are available with low and high levels of manganese, and the table illustrates that the hardenability improvement of manganese allows for larger cross sections. The final column was added by using heat-transfer calculations to estimate sheet thicknesses that have the same cooling rate at their centers for an oil quench as occurs for a round bar.

Austempering should be an attractive method of hardening steels for knives, because the thin geometry lends itself, especially near the cutting edge, to fast cooling rates. The final column of Table 12.2 provides a guideline for maximum blade thicknesses of plain carbon steels that should be successfully hardened with austempering. In addition to the improvement in ductility and toughness provided by austempering, it also has the same advantage as martempering for reducing distortion. There is evidence that bainitic structures often show superior wear to that of quenched and tempered steels at the same level of hardness (Ref 12.12).

Austempering does not always produce an improvement in toughness over the quenched and tempered condition. A review of the literature shows that austempering improves toughness only at the higher hardness levels. The crossover hardness appears to be at approximately HRC = 40. Below this value, the quenched and tempered condition has a toughness as good or better than the austempered condition (Ref 12.13, 12.14).

Variations on Conventional Austempering. Austempering heat treatments can be used to produce mixtures of bainite and martensite. There is some industrial experience with such mixed structures that indicates they also have excellent ductility and toughness compared to quenched and tempered steels at the same hardness level (Ref 12.15). As an aid to understanding these unconventional austempering treatments, the IT diagram at temperatures near the M_s for a 1075 steel is shown in Fig. 12.7, which may also be found in Ref 12.9. The diagram shows the times required to transform the

Table 12.2 Maximum section size for austempering [12.10]

			Max diameter		Max sheet thickness	
%C	%Mn	Max HRC	mm	in.	mm	in.
0.95–1.05	0.30–0.50	57–60	3.76	0.148	3.00	0.118
0.95–1.05	0.60–0.90	57–60	4.75	0.187	3.81	0.150
0.80–0.90	0.30–0.50	55–58	3.96	0.156	3.18	0.125
0.80–0.90	0.60–0.90	55–58	5.54	0.218	4.42	0.174
0.60–0.70	0.60–0.90	53–56	4.75	0.187	3.81	0.150
0.60–0.70	0.90–1.20	53–56	7.14	0.281	5.72	0.225
0.60–0.70	1.60–2.00	53–56	15.9	0.625	12.7	0.50
1.0 + 0.4–0.6 Cr	0.40–0.60	57–60	7.92	0.312	6.35	0.250

Source: Ref 12.10

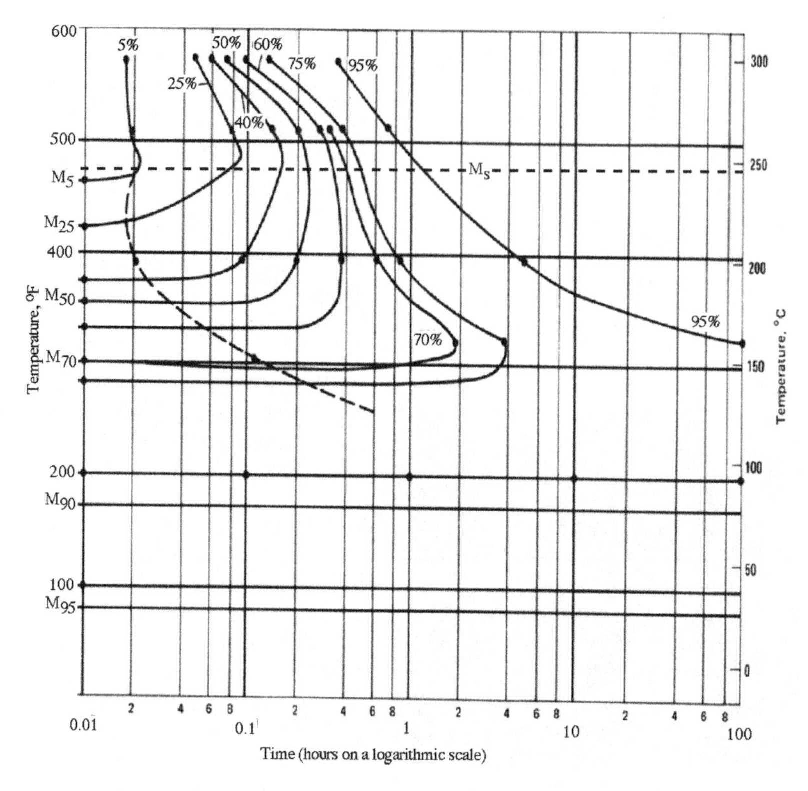

Fig. 12.7 Isothermal transformation diagram for a 1075 steel (0.75% C, 0.50% Mn) showing the transformation times to lower bainite just above M_s and the times for combined transformation to martensite and bainite just below M_s. Source: Ref 12.16

austenite by the percentage values given. Interpretation above the M_s temperature is straightforward. For example, by heat treating at 260 °C (500 °F), the austenite transforms to lower bainite at a rate of 5% transformed in approximately 0.02 h (72 s) to 95% transformed in 0.8 h (48 min).

Below the M_s temperature, the interpretation is a little more complicated. Consider a steel held at the M_{25} temperature of approximately 220 °C (430 °F). After a few milliseconds, this steel would contain 25% martensite (because that is what the M_{25} temperature designates). The downward-curving dashed line indicates when the remaining 75% austenite begins to transform to lower bainite, which, in this case, is after approximately 0.02 h (72 s). If a horizontal line is drawn at the M_{25} temperature, it crosses the 50% line at approximately 0.25 h (15 min). At this point, the austenite is 50% transformed, so the steel will be 25% martensite (which has been tempered for 15 min), 25% lower bainite, and 50% retained austenite. The 95% line is crossed at approximately 3 h, so the steel would be 25% tempered martensite, 70% lower bainite, and 5% retained austenite. Therefore, this diagram provides information on the combined formation of martensite and lower bainite at temperatures below the M_s temperature. The diagram indicates that below the M_{75} temperature, lower bainite will not form in the retained austenite even on extended holding times.

With this information as background, two types of modified austempering heat treatments are described, both of which produce steels with mixed bainite-martensite structures. Table 12.3 presents the steps in the austempering process following the austenitization step, which is, of course, the beginning step of the process. Conventional austempering (CA) is a two-step process, as shown in the table, and it produces steels that are 100% bainite. The term ΔT is a conventional scientific nomenclature specifying a range of temperatures. Using as an example the steel in Fig. 12.7 that is austempered at 260 °C (500 °F), the value of ΔT would be 15 °C (27 °F), because M_s for this steel is 245 °C (473 °F), with $\Delta T = 260 - 245$ °C. For CA, the final hardness is controlled by the two methods shown below the first box in Table 12.3.

Table 12.3 Three variations of austempering

Conventional austempering (CA)	
Step 1	**Step 2**
Quench to (M_s+ΔT) and hold to 100% bainite.	Air or oil quench (piece will be 100% fresh bainite).

Hardness control:
(a) Temper to desired HRC.
(b) Adjust hardness with ΔT. Higher values give softer bainite, often upper bainite.
Small values give hardest bainite, lower bainite.

Modified austemper No. 1 (MA1)		
Step 1	**Step 2**	**Step 3**
Quench to (M_s $-\Delta T_1$) and hold short time.	Up-quench to (M_s+small ΔT_2) and hold until all of the retained austenite is transformed to bainite.	Air or oil quench. Piece will contain tempered martensite+fresh lower bainite.

Hardness control:
(a) Increase ΔT_1 of step 1 to increase %M/%B ratio and increase HRC.
(b) Increase ΔT_2 of step 2 to obtain softer bainite and drop HRC.
(c) Temper the fresh bainite at temperatures above its formation temperature to drop HRC.

Modified austemper No. 2 (MA2)	
Step 1	**Step 2**
Quench to ($M_s$$\pm\Delta T$) and hold time needed to give desired %B. (Note: Is martempering if %B = 0 and ΔT small)	Air or oil quench. Piece will contain martensite+fresh bainite. For positive ΔT values, martensite is fresh; for negative values, it will be partially tempered. Type of bainite (lower vs. upper) and hardness set by ΔT

Hardness control:
(a) Temper to desired HRC.
(b) Reduce hold time in step 1 to increase %M/%B ratio and increase HRC prior to tempering.
(c) Change ΔT of step 1 to control hardness of bainite component. Increase ΔT for lower hardness, and decrease ΔT (even to negative values) to increase hardness.

In the first modified austemper (MA1) process in Table 12.3, the mix of martensite and bainite is controlled in step 1 by controlling how far the bath temperature is set below M_s, which controls the percent martensite that ends up in the sample. Because the sample is held at temperature only a short time, little to no bainite will form in this step. The bainite is formed in step 2 after the up-quench (M_s + small ΔT_2). Note that the martensite in the steel will be in the tempered condition, having been tempered at the temperature of step 2. Three methods of controlling final hardness are listed below the middle box in Table 12.3.

An example design of an MA1 experiment to produce 50% martensite/50% lower bainite in 5140 steel can be made using the CT diagram in Fig. 12.2. After austenitizing, the sample would need to be quenched to a temperature of 300 °C (580 °F) (M_{50} temperature), with the center dropping below the nose temperature in less than approximately 6 s. After the center cools to 300 °C (580 °F), the steel should be up-quenched to approximately 340 °C (650 °F) and held for at least 500 s before air or oil quenching. The sample should consist of 50% fresh lower bainite and 50% martensite that has been tempered at 340 °C (650 °F) for 500 s.

In the second modified austemper (MA2) process in Table 12.3, the mix of martensite and bainite is controlled in step 1 by controlling the percent bainite that will end up in the sample. Note that, in this case, the bath temperature may be set above M_s (positive values of ΔT) or below M_s (negative values of ΔT). For negative values of ΔT, the martensite formed in step 1 will be tempered, but martensite formed in step 2 is fresh.

Again, the CT diagram in Fig. 12.2 can be used to illustrate how one would design an MA2 process to produce 50% lower bainite in 5140 steel. After austenitizing, the sample would be quenched in salt to 340 °C (650 °F). By drawing a line across Fig. 12.2 at 340 °C (650 °F), it can be seen that the 50% transformation line is crossed at approximately 60 s. Therefore, the sample should be held at 340 °C (650 °F) for 1 min and then oil quenched to room temperature. It should now consist of 50% fresh bainite and 50% fresh martensite.

These variations in austempering techniques offer the bladesmith a rich variety of possibilities for producing knives of varying toughness and hardness. As pointed out previously, 100% bainite will not be as hard as full martensite. By going to mixed structures, hardnesses can be achieved above the limits for 100% bainite given in Table 12.2. There is industrial evidence that such steels have excellent toughness (Ref 12.17).

An interesting possible steel for austempering of knives is 52100 steel. The small carbides in it that produce such good wear properties for its primary application as a bearing steel also make it an attractive material for knives. It is sometimes austempered to minimize distortion in bearing races. There is one study in the literature where mixed martensite-bainites were produced in 52100 through the MA2 process, and, in this case, the ΔT was set at 0 (Ref 12.18). Figure 12.8 presents the hardness and impact energies of 52100 as a function of the holding time in step 1. At the shortest times, the steel was 100% martensite, and at the longest times, it was 100% bainite. While the bainite improves toughness, it is not possible to say that the toughness would be higher than the 100% martensite after it was tempered to the same hardness level. However, the known benefits of bainite on toughness would lead one to think the austempered samples would show improved toughness over tempered martensite at the same levels of hardness.

Very little if any systematic studies have been done on the effect of mixed structures on toughness of austempered versus quenched and tempered steels of equal hardness. There is one study showing the effect of percent bainite on ductility in three plain carbon austempered steels (Ref 12.17), summarized in Fig. 12.9. This study produced the mixed bainite-martensite structures of various hardnesses using method MA1(a, b). At hardnesses of HRC = 50 and 56, increased bainite improves ductility, and more so as %C increases from 0.65 to 0.95. There are two studies on 4340 alloy steels showing that mixed, austempered structures have superior room-temperature toughness for percent bainite levels up to approximately 30% (Ref 12.19, 12.20).

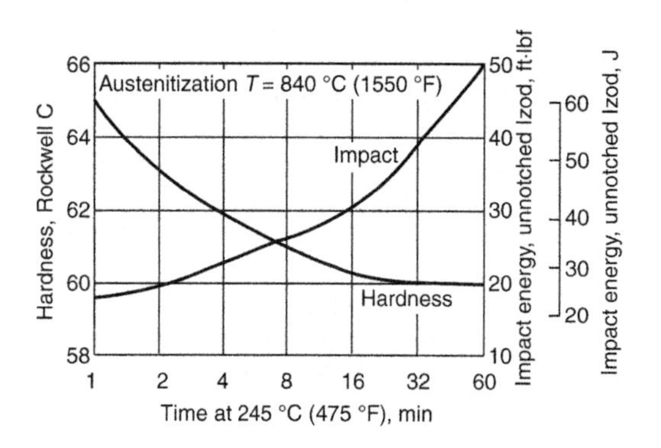

Fig. 12.8 Hardness and impact strength of 52100 steel versus hold time at M_s prior to quenching. The percent bainite goes from 0 to 100% as time goes from 1 to 60 min. Copyright: American Metal Market

In a study on a steel with composition of 0.9C, 1.6Mn, and 0.55Cr (which is not too different from 52100), the toughness of the austempered and the quenched and tempered condition at specific hardnesses was compared and is shown in Fig. 12.10 (Ref 12.21). Toughness was measured with the fracture load in a bend test. The M_s for this steel was 195 °C (380 °F), and the mixed bainite-martensite structures were produced by the MA2 method using both positive and negative ΔT values. In the first case, the samples were austempered at 300 °C (570 °F) ($\Delta T = +105$ °C, or +190 °F) to yield 10% bainite, and in the second case, the austempering temperature of 180 °C (360 °F)($\Delta T = -15$ °C, or −27 °F) yielded 13% bainite. In both methods, final

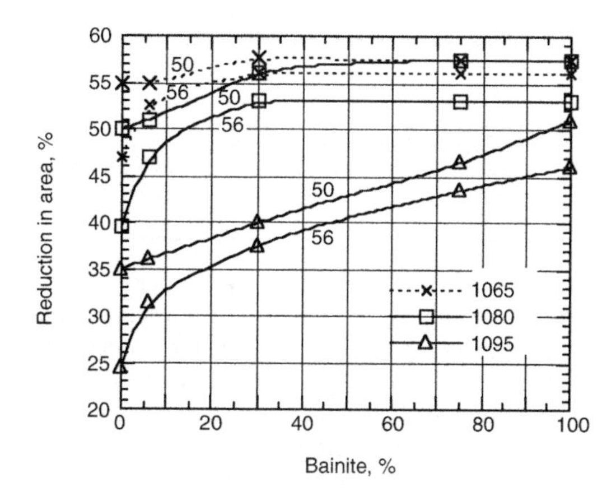

Fig. 12.9 Effect of adding bainite to martensite on ductility at constant hardness levels in plain carbon steels. Source: Ref 12.17

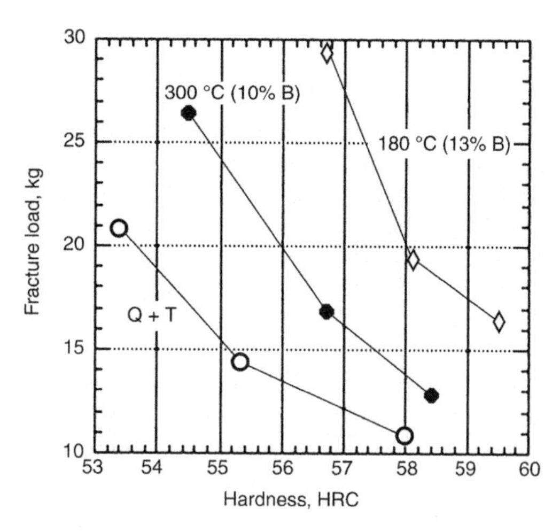

Fig. 12.10 Comparison of bend test fracture strengths for a quenched and tempered (Q & T) steel and two mixed martensite-bainite structures in a 0.9% C steel

hardness was controlled by tempering; the three points on each curve of Fig. 12.10 are for tempering at 200, 250, and 300 °C (390, 480, and 570 °F). According to Fig. 12.10, the negative ΔT method appears to produce a better toughness at a given hardness level.

Perhaps the most widespread use of austempering is in the foundry industry, where ductile cast iron is processed in such a way as to produce austempered ductile iron (ADI). The performance of ADI gears rivals that of conventional carburized gears, and production is cheaper and more energy efficient. The wear resistance of bainitic ADI is superior to quenched and tempered steels. Improved wear resistance may not carry over to the bainite formed in conventional steels, because there is evidence that the improved wear resistance is related to a high degree of retained austenite in the ADI bainites, resulting from the high level of silicon in the cast irons. See Chapter 16, "Cast Irons," for more discussion of ADI.

Characterization of Quench Bath Cooling Performance

There was a considerable amount of research done on water and oil quenching at approximately the time period when Jominy developed his test in the 1930s and 1940s. The cooling intensity was characterized by a heat-transfer parameter called *H*, which has units of 1/inches. Larger values of *H* correspond to higher quench intensities, as summarized in Table 12.4. The table provides a qualitative measure of the relative increase in quenching rate as the quench medium is varied from air to oil to water to brine (a salt solution) and as agitation of the quench piece is increased. Heat-transfer rates vary enormously during a quenching operation, and the use of a single value of *H* to characterize it is an oversimplification. However, like most simplifications, it is quite useful for some practical purposes.

Quenching a piece of hot steel in water or oil produces copious amounts of vapor around the piece, often called the vapor blanket. In water, the blanket is steam, and in oil, it is vaporized oil. The presence of this vapor phase around the hot steel leads to the very complex mode of heat transfer during quenching. There was considerable research on this problem in the later decades of the 20th century, and a method developed in England for characterizing quench fluids has been adopted as international standard ISO 9950.

The test uses an Inconel 600 alloy cylinder (basically the same nickel-base alloy used for the heating elements of a household electric stove), 12.5 mm (0.5 in.) in diameter by 60 mm (2 in.) long. A metal-clad type K thermocouple (see Appendix A, "Temperature Measurement") is fitted into a hole along the center, and the thermocouple is monitored during the quench. The output produces a cooling curve such as that shown for an oil bath in Fig. 12.11. The heat transfer during the quench is partitioned into three stages, conventionally called *A*, *B*, and *C*. During the *A* stage, heat transfer is relatively slow, because the heat must pass through the vapor blanket that initially surrounds the immersed sample. Notice that the *B* stage begins with a rapid increase in the rate at which the temperature drops. When the oil (or water) begins to penetrate through the vapor blanket, it contacts the hot steel and immediately boils. The heat required to boil the liquid is removed from the steel, and this mode of heat transfer is extremely efficient. It is called nucleate boiling heat transfer, thus the name of stage *B* shown in Fig. 12.11. When the boiling stops, heat is transferred directly to the liquid touching the steel, causing the liquid temperature to increase, which, in turn, decreases its density. As this less dense liquid rises, it is replaced by colder liquid that comes

Table 12.4 Heat-transfer parameter (H) values for various quenches

	H, severity of quench, in.$^{-1}$			
Movement of piece	Air	Oil	Water	Brine
None	0.02	0.3	1.0	2.2
Moderate	...	0.4–0.6	1.5–3.0	...
Violent	...	0.6–0.8	3.0–6.0	7.5

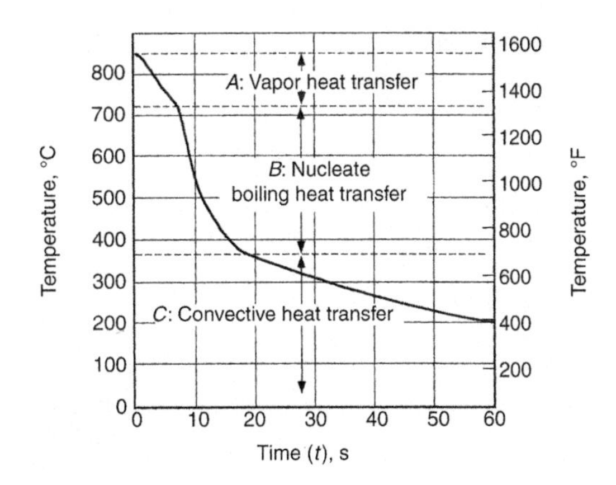

Fig. 12.11 Cooling curve measured in the ISO 9950 test using an oil quench bath

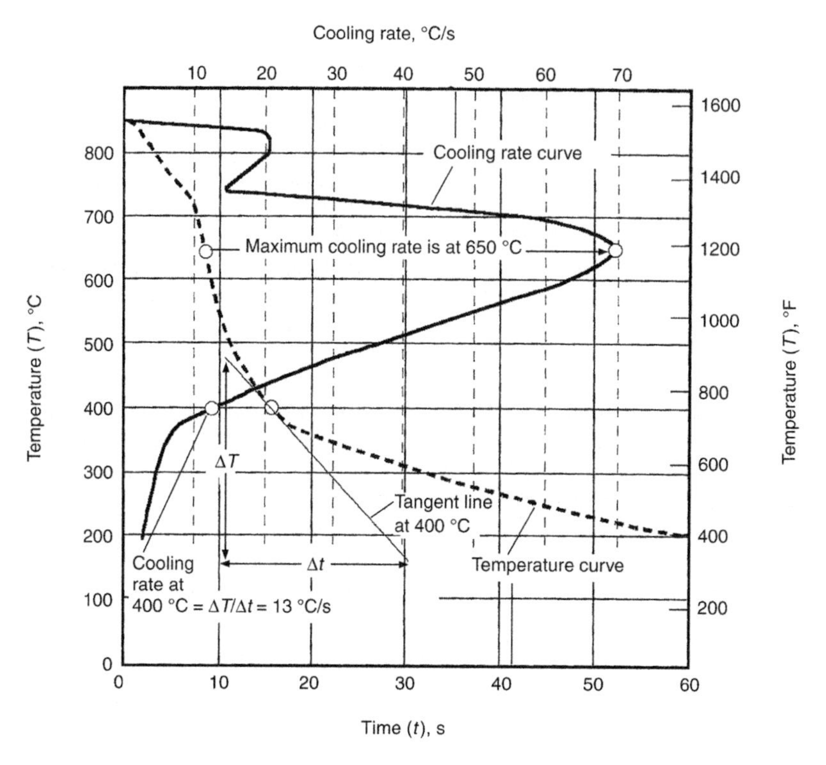

Fig. 12.12 Construction used to convert the cooling curve of Fig. 12.11 to a cooling rate curve

in contact with the steel piece. The motion of the quenching liquid is called convection, also the name of stage *C* shown in Fig. 12.11.

When evaluating quenching intensity, the primary interest is how fast the temperature is falling, so the most useful parameter is the cooling rate, which has units of °C/s (°F/s). The quenching power of a quenchant is commonly characterized with a plot of cooling rate versus temperature of the Inconel rod, and Fig. 12.12 shows how such curves are related to the simple cooling curve of Fig. 12.11. To obtain the cooling rate at 400 °C (750 °F), simply construct a line tangent to the cooling curve at the temperature, as shown. Then, moving from the bottom of this line to the top, measure the rise, ΔT, and the run, Δt. The ratio of the rise divided by the run, $\Delta T/\Delta t$ (called the slope in geometry classes), has units of °C/s and is the cooling rate when the Inconel center is at 400 °C (750 °F).

With computer software, it is a simple matter to determine the cooling rate at each temperature, and the solid line is the corresponding cooling rate curve, with the cooling rates in °C/s given along the top of the diagram. Notice that the maximum cooling rate is 70 °C/s, and it occurs at approximately 650 °C (1200 °F) for this oil. There is an excellent discussion of water, oil, and polymer quenchants in Ref 12.8 that shows how these cooling rate curves are widely used to characterize the quenching power of the various commercially available quenching oils and polymer quenchants. Figure 12.13 is a set of cooling rate curves for several common quenching fluids.

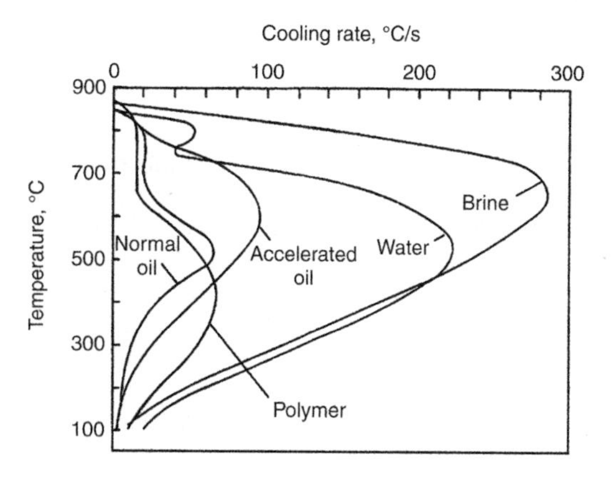

Fig. 12.13 Typical cooling rate curves for several types of quenchants at 30 °C (85 °F) and not agitated. Source: Ref 12.22

Notice that the maximum cooling rate occurs at different temperatures and varies from a high of 285 °C/s (510 °F/s) for salt water to a low of 65 °C/s (120 °F/s) for a typical oil.

For a given quenchant, the speed of the quenching process will depend on the temperature of the bath as well as any agitation used on the workpiece during the quenching operation. The quench rate of water can be decreased significantly by heating the water. The cooling rate curves given in Ref 12.8 show that the maximum cooling rate for water is decreased from 225 °C/s (400 °F/s) for a bath at 20 °C (68 °F) to only 90 °C/s (160 °F/s) for a bath at 80 °C (176 °F), a decrease of 60%.

The cooling rate data given in Ref 12.8 for hot oils are less sensitive than water to bath temperature and vary in a more complicated manner. As the bath temperature is changed from 200 to 150 to 50 °C (390 to 300 to 120 °F), a maximum cooling rate of 83 °C/s (150 °F/s) occurs at the intermediate bath temperature of 150 °C (300 °F) and drops to 80 °C/s (145 °F/s) at the higher temperature of 200 °C (390 °F). At the lowest bath temperature of 50 °C (120 °F), the cooling rate is 75 °C/s (135 °F/s).

Oil Quenchants

There are three categories of oil quenchants: normal, accelerated, and marquench (Ref 12.23). The accelerated oils contain additives that produce increased quench rates, as illustrated by the cooling rate curves in Fig. 12.13. The martempering oils are formulated for use at higher temperatures and are available in accelerated and nonaccelerated formulations. An oil bath must not be heated too close to its flash point temperature, because oils can ignite if heated too high. Reference 12.8 (p 80) presents a table with flash point temperatures for the three types of oils. Conventional and accelerated oils have flash points of approximately 175 °C (350 °F) compared to 300 °C (570 °F) for marquenching oils. The table mentions the GM quenchmeter, which refers to a test developed by General Motors and used in the past to evaluate quenching rates. As described in more detail in Ref 12.23, the test measures the time for the quenchpiece temperature to fall to the magnetic transition temperature (the Curie temperature) of a nickel ball, 354 °C (670 °F). The test number has units of seconds, and smaller values mean faster quench rates.

Figure 12.14 shows cooling rates for two different oils held at 40 °C (105 °F) and illustrates the usefulness of cooling rate curves. With the GM quenchmeter test, the cooling rates for both oils were essentially the same, 18.1 s for oil *A* and 17.6 s for oil *N* (Ref 12.24). However, the cooling rate curves illustrate that there are significant differences in the two oils. Oil *N* has a faster cooling rate down to temperatures of approximately 550 °C (1020 °F), which means that it would be preferred for cooling fast enough to miss the ferrite and pearlite noses of the CT diagrams on low-hardenability steels. The temperature axis shows the M_s and M_{90} temperatures for a 1050 steel. Notice that the cooling rates at temperatures between M_s and M_{90} are significantly lower for oil *N*. This means that oil *N* would be preferred for avoiding distortion and quench cracking in parts subject to those problems. Also, notice that this occurs even though oil *N* has a faster maximum cooling rate, which, by itself, would indicate a higher tendency to cause quench cracking and distortion.

The effect of agitation of the quenching oil can be evaluated with these cooling rate curves, and Fig. 12.15 presents the results of one study (Ref 12.22). In this case, the oil was agitated with a two-blade impeller, and the agitation speeds in revolutions per minute (rpm) are given on the graph. The agitation simply shifts the curves to higher cooling rates.

Polymer Quenchants

These quenchants consist of water with a polymer dissolved in the water at some specific concentration level. There are several types of

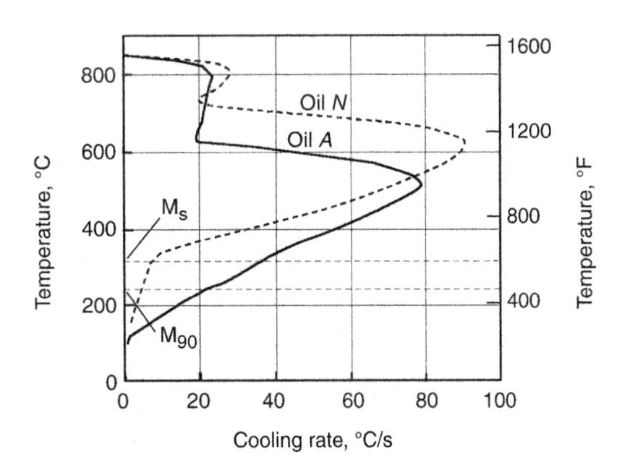

Fig. 12.14 Cooling rate curves for two different oils. Source: Ref 12.24

polymers that have been developed for quenching purposes, and the most popular is called polyalkalene glycol (PAG) (Ref 12.8). The quench rates of polymer quenchants are intermediate to those of water and oils. As illustrated in Fig. 12.16, the quenching rate of a PAG quenchant is progressively reduced as the concentration of the polymer in the water increases from 10 to 20 to 30%. All quench baths were at 40 °C (105 °F). A problem that sometimes occurs with water quenching is that the cooling can be nonuniform, leading to soft regions in the steel.

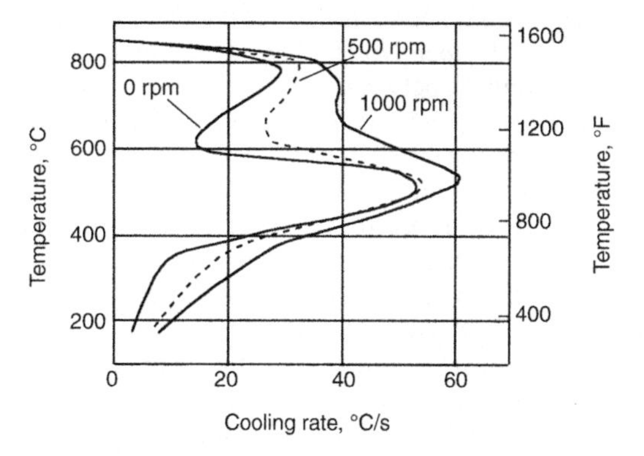

Fig. 12.15 Cooling rate curves for conventional oil at 40 °C (105 °F), showing effects of agitation. Source: Ref 12.22

The addition of just a small amount of the polymer causes more uniform contact between the quenchant and steel surface during the initial cooling stages. A more uniform quench is achieved with only a small reduction in quench rate. Also, the polymer quenchants produce much cleaner surfaces than can be obtained with oil quenchants. Notice that the polymer quench rate in the usual M_s-M_f range increases the chances of quench cracking versus oils.

There are problems with both oil and polymer quenchants related to aging of the bath. Immersion of very hot steel can lead to oxidation of the liquid components, and the continual removal of the steel pieces can produce a selective loss of liquid components. This latter effect is called *drag out*. There is an excellent discussion of the problems of quench bath maintenance in Ref 12.23. Interestingly, cooling rate curves taken on oil baths over periods of months show that cooling rate increases with oxidation. Also, with oil baths, water buildup in the bath can change the quenching rate significantly. Because the polymer quench rate is strongly dependent on polymer concentration in the water solution, maintaining constant concentration is difficult. A method for monitoring polymer concentration with the refractive index is described in Ref 12.23.

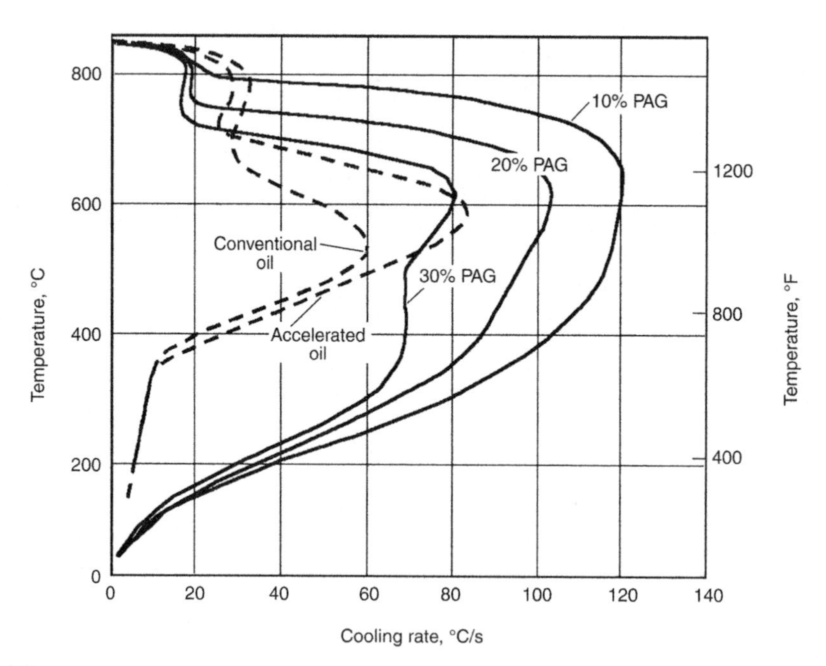

Fig. 12.16 Cooling rate curves for polyalkalene glycol (PAG) polymer quenchant compared to oil quenchants. Source: Ref 12.25

Salt Bath Quenchants

Salt baths composed of mixtures of molten salts can be used for both the quenching and the austenitizing steps of heat treating. The high-temperature salt baths used for austenitizing are usually mixtures of either potassium and sodium nitrates (KNO_3-$NaNO_3$) or potassium-lithium-sodium chlorides (KCl-LiCl-NaCl). The lower-temperature salts used for austempering and martempering are mixtures of nitrate and nitrite salts of potassium and sodium. The salts are supplied from the manufacturer with recommended temperature ranges of heat treatment. The advantages of austenitizing in molten salts include very rapid heating, uniform temperature distribution, and avoidance of surface oxidation and decarburization. (See Ref 12.23, p 309, for a detailed discussion.)

Salt bath quenching is used when the goal is to quench very rapidly to elevated temperatures for such processes as austempering and martempering. Tool steels are sometimes quenched in molten salts to the 500 to 600 °C (930 to 1110 °F) range prior to additional cooling to reduce scaling (oxidation) and distortion and to minimize cracking. This high-temperature quenching is sometimes called interrupted quenching. When hot steel is immersed rapidly in molten salts, the vapor blanket does not form as it does with water, polymer, and oil quenchants. This results in a faster cooling rate initially. However, because the salt is at an elevated temperature, the cooling rate at longer times will be slower than for the other quenchants that are held at lower temperatures.

This is illustrated in Fig. 12.17 for a stainless steel probe quenched into 220 °C (430 °F) salt.

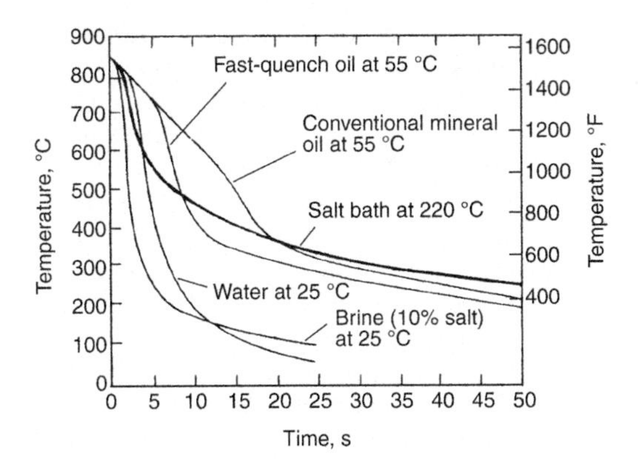

Fig. 12.17 Temperature of a stainless steel probe quenched in various quench baths. Source: Ref 12.26

The figure shows that the salt cools down more quickly to temperatures of 600 °C (1110 °F) than does a water bath at 25 °C (75 °F). Also, the salt bath cools more quickly to 490 °C (915 °F) than does a fast-quench oil bath at 55 °C (130 °F). For martempering operations of low-hardenability steels, the initial cooling rate is required to be extremely rapid. For such applications, it has been shown that the quench rate of salts can be increased by the controlled addition of small amounts of water to the bath (Ref 12.23, p 313). In addition to Ref 12.23, reviews of salt bath quenching are given in Ref 12.26 and 12.27.

Molten metals are an ideal liquid bath for rapid quenching because of the very high thermal conductivity of liquid metals and the lack of vapor blanket formation. Molten lead does not react with steel, and it melts at 327 °C (620 °F). It can be alloyed with bismuth, which also does not react with steel, giving a eutectic molten bath temperature of 125 °C (260 °F). These baths are normally covered with powdered graphite to avoid surface oxidation, but they must be used with great caution so that the metallic vapors are not inhaled, which would be extremely hazardous to heat treaters.

Summary of the Major Ideas in Chapter 12

1. Heat transfer in the quenching process is complicated by the formation of a vapor blanket around the immersed steel piece, because heat transfer through a gas is relatively slow. For a given quenchant, two main factors increase cooling rate: increased agitation of the liquid and reduction in the temperature of the bath.
2. A relatively new technique has been accepted (ISO 9950) for characterizing the cooling rate of the various liquid quenchants used industrially. Figure 12.13 shows how these cooling rate curves illustrate the relative quenching rates of common liquid quenchants.
3. To through harden a piece of steel, the center must cool fast enough to decrease its temperature below the nose of the continuous transformation (CT) diagram for that steel, as illustrated for a 5140 steel in Fig. 12.1. For low-hardenability steels, a quench that is very fast is desirable. As illustrated in Fig. 12.13, brine and water

quenches produce the fastest cooling. However, for steels with %C above approximately 0.4, these quenches will lead to quench cracking and distortion.

4. Quench cracking and distortion are caused primarily by large temperature gradients in the steel piece as martensite is forming. Hence, a critical factor in avoiding these problems is to have a low cooling rate in the M_s-M_f temperature range. A desirable characteristic of the quenchant is that it cools the piece very rapidly from the austenitizing temperature to just above the M_s temperature and then slowly to room temperature. For steels with %C above 0.4, oil quenches are often the best compromise.
5. Martempering is a technique that minimizes the problems of quench cracking and distortion. As illustrated in Fig. 12.2, the quench is interrupted at temperatures just above the M_s temperature, allowing the temperature gradients in the workpiece to flatten out prior to the formation of martensite. Martempering is generally done by quenching into hot salt baths.
6. Oil quenchants are available in three formulations: normal oils, accelerated oils, and martempering oils. Special additives are used to produce the latter two types. As illustrated in Fig. 12.13, the accelerated oils have a significantly higher maximum cooling rate but still have a relatively low cooling rate in the M_s-M_f range. The martempering oils can be used at higher temperatures and are effective for martempering some steels, even though the temperature equalization is done just below the M_s temperature, as shown in Fig. 12.2.
7. Polymer quenchants produce cooling rates intermediate to those of oils and water, as illustrated in Fig. 12.16. These baths are solutions of polymers in water, and the higher the concentration of polymer in the water, the more the quench rate is reduced. Cleanup following quenching is much easier than for oil baths, but maintenance of constant polymer concentration over extended use can be a problem.
8. Molten salt bath quenchants are useful over specific temperature ranges, which are dictated by the type of salts. They offer very rapid initial quench rates, but because they are used hot, quench rates at longer times are reduced. In addition to quenching, these baths can also be used for austenitizing and annealing and provide protection from oxidation.
9. Austempering is a quenching technique designed to produce bainite in steels. It is usually done by quenching into hot salt baths and holding for times adequate to produce all bainite. Such times may be found on the CT curves specific to steels of interest (Fig. 12.3).
10. Bainitic steels display superior toughness to quenched and tempered steels at hardness levels of approximately HRC = 40 and above and are therefore an excellent choice for knives (Fig. 12.5, 12.9, and 12.10). As illustrated in Fig. 12.4, hardness levels in the high-50s range may be obtained in bainitic steels with higher carbon levels.
11. Table 12.3 outlines two modified forms of austempering that produce a mixture of martensite and bainite. These methods give increased hardness over fully bainitic steels, and there is limited experimental evidence (Fig. 12.10) indicating that toughness can be superior to the quenched and tempered condition (i.e., full martensite) at the same hardness.

REFERENCES

12.1. R. Maddin, A History of Martensite: Some Thoughts on the Early Hardening of Iron, Chap. 2, *Martensite*, G.B. Olson and W.S. Owen, Ed., ASM International, 1992

12.2. W.C. Roberts-Austen, *An Introduction to the Study of Metallurgy*, Charles Griffin and Co., London, 1923, p 128–129

12.3. C.S. Smith, *Sources for the History of the Science of Steel 1532–1786*, MIT Press, 1968

12.4. T.A. Wertime, *The Coming of the Age of Steel,* University of Chicago Press, 1962, p 193

12.5. R.F. Kern, *Steel Selection,* Wiley-Interscience, 1979, p 132

12.6. J.M. Hampshire, User Experience of Hot Oil Quenching, *Heat Treat. of Met.*, 1984.1, p 15–20

12.7. B.F. Shepherd, Martempering, *Iron Age,* Vol 151, Jan 28,1943, p 50–52

12.8. *Heat Treater's Guide,* 2nd ed., ASM International, 1995 (Note: Expanded versions of the treatment on martempering and austempering may be found in

Heat Treating, Vol 4, *ASM Handbook,* ASM International, 1991, p 137–163)

12.9. *Atlas of Isothermal Transformation and Cooling Transformation Diagrams,* American Society for Metals, 1977

12.10. E.E. Legge, The Industrial Application of Austempering, *Met. Alloys,* Vol 10, Aug 1939, p 228–242

12.11. A. Grossman and E.C. Bain, *Principles of Heat Treatment,* American Society for Metals, 1964, p 178

12.12. K.H.Z. Gahr, *Microstructure and Wear of Materials,* Elsevier, 1987, p 277

12.13. P. Payson and W. Hodapp, *Met. Prog.,* Vol 35, 1939, p 358–362

12.14. D.L. Turner and F.J. Worzala, *Proc. Int. Symp. Test. and Failure Anal.* (Long Beach, CA), 1958, p 352–356

12.15. Q.D. Mehrkam, Austempering in Actual Practice, *Met. Prog.,* Vol 86, 1964, p 134–136

12.16. *Heat Treater's Guide,* 2nd ed., ASM International, 1995, p 203

12.17. H.J. Elmendorf, The Effect of Varying the Amount of Martensite upon the Isothermal Transformation of Austenite Remaining after Controlled Quenching, *Trans. ASM,* Vol 33, 1944, p 236–260

12.18. H.E. Boyer, Controlling Physical Properties by the Interrupted Quench, *Iron Age,* Vol 160, July 1944, p 49–54

12.19. R.F. Hehemann, V.J. Luhan, and A.R. Troiano, The Influence of Bainite on Mechanical Properties, *Trans. ASM,* Vol 49, 1957, p 409–426

12.20. Y. Tomita and K. Okabayashi, Mechanical Properties of 0.4%C-Ni-Cr-Mo High Strength Steel Having a Mixed Structure of Martensite and Bainite, *Metall. Trans.,* A, Vol 14, 1983, p 485–492

12.21. S.M. Sundaram and A.K. Mallik, Effect of Lower Bainite on Mechanical Properties after Tempering, *J. Iron Steel Ins. London,* Vol 205, 1967, p 876

12.22. D. Close, The Wolfson Engineering Group Quench Test—Scope and Interpretation of Results, *Heat Treat. Met.,* Vol 10, 1984.1, p 1–6

12.23. G.E. Totten, C.E. Bates, and N.A. Clinton, *Handbook of Quenchants and Quenching Technology,* ASM International, 1993

12.24. D.A. Guisbert and D.L. Moore, Correlating Quenchant Test Data, *Adv. Mater. Process.,* (No. 10), 1996, p 40Q–40V

12.25. R.T. von Bergen, Advances in Quenching Technology, *Heat Treat. Met.,* Vol 10, 1984.1 p 7–8

12.26. C. Skidmore, Salt Bath Quenching—A Review, *Heat Treat. Met.,* Vol 12, 1986.2, p 34–38

12.27. R.W. Foreman, Salt Bath Quenching, *Conf. Proc. Quenching and Distortion Control,* G.E. Totten, Ed., ASM International, 1992, p 87–94

CHAPTER 13

Stainless Steels

A BIG PROBLEM WITH STEELS is that they are easily corroded when exposed to damp conditions, particularly if a salty or acidic atmosphere is present. For example, a piece of steel left in a room in which acids are used will very quickly become highly rusted with the characteristic dirty-brown color of steel rust. (Chemically, rust is a hydrated form of iron oxide.) The two major types of stainless steels were both discovered in 1912 by accident because of their unusual resistance to corrosion (Ref 13.1). The austenitic stainless steel type was discovered in Germany by E. Maurer when he noticed that certain chromium-nickel steels did not rust after being left in a room with acid fumes. The ferritic stainless steel type was discovered in England by H. Brearley when he observed difficulty in trying to etch gun barrels made from a chromium steel. Brearley first used the name stainless steel for the ferritic type of stainless, and the name was later applied to the austenitic stainless.

The good corrosion resistance of stainless steels is the result of a very thin chromium-rich oxide film that forms on the surface of the steel. (It is so thin, approximately 2 nm, or 0.002 μm, that it is transparent.) The oxide film protects the underlying steel from reacting further with the environment. When the protective film is present, the steel is said to be *passivated* or in the passive state. In general, a stainless steel is passivated when in contact with an oxidizing aqueous solution but not when in contact with a reducing solution.

Figure 13.1 shows the corrosion rate of iron-chromium alloys as a function of the %Cr in the alloys when tested in an oxidizing acid, nitric acid, and a reducing acid, sulfuric acid. In nitric acid, there is a significant reduction in corrosion rate as the %Cr content increases to levels of 11 to 12, followed by a plateau (constant corrosion rate) to 15% Cr, and then a second strong reduction as the %Cr increases to almost 20. The steels are generally said to be passivated in nitric acid environments at chromium contents of 11 to 12% and above. It is therefore commonly stated *that for a steel to be considered stainless, it must contain a minimum of just under 12% Cr.* Figure 13.1 also illustrates that passivation depends on the type of aqueous liquid contacting the steel. Notice that in the reducing acid (sulfuric), the passivated film is actually lost with increasing %Cr.

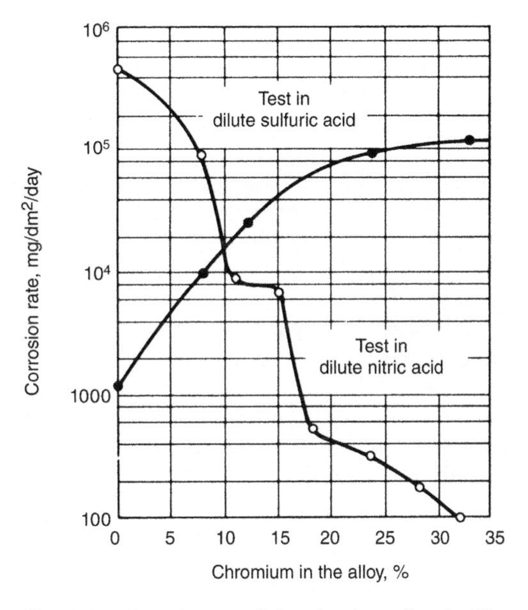

Fig. 13.1 Corrosion rate of chromium-iron alloys in dilute nitric and sulfuric acids. Source: Ref 13.2

Stainless steels exhibit improved resistance to oxidation at higher temperatures. Figure 13.2 illustrates that oxidation resistance (as measured by weight loss after 48 h in 1000 °C, or 1830 °F, air) also improves as %Cr is increased.

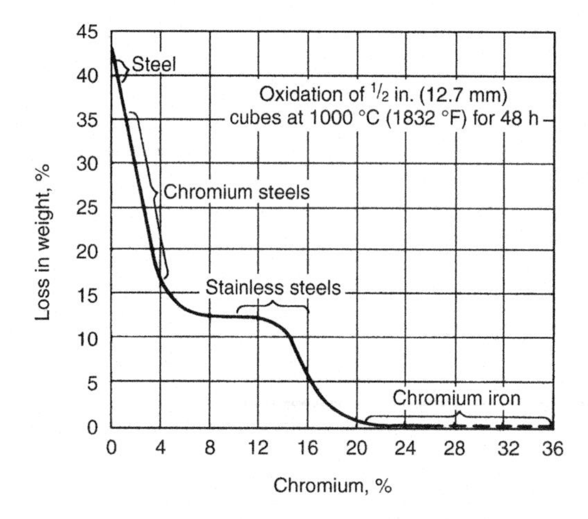

Fig. 13.2 Oxidation of chromium steels at 1000 °C (1830 °F). Source: Ref 13.3, p 461

Again, there are two composition ranges where the improvement is dramatic, and they are also separated by a plateau region, but the plateau now extends from 15% Cr to lower %Cr values, approximately 6.

In general, passivity is promoted by small additions of molybdenum or nickel to the stainless steel. Loss of passivity is promoted by the presence of chlorine ions, such as are present in salt water, and, as already discussed, by reducing conditions.

Ferritic Stainless Steels

As a first approximation, ferritic stainless steels may be considered to be alloys of chromium-iron with chromium compositions exceeding 12%. As with ordinary steels, a phase diagram is helpful for understanding the structure of these alloys, and Fig. 13.3 presents the phase diagram of Fe + Cr alloys with no carbon present. The diagram gives the melting point of pure iron as 1538 °C (2800 °F) and pure chromium as 1863 °C (3380 °F). The solid line extending between these two temperatures is called the liquidus, that is, the freezing temperature of all the alloy compositions. The symbol L above this line identifies the region where alloys are fully liquid. The dashed line between the two pure melting points is called the solidus, that is, the melting point of all alloy compositions. The large open region below the solidus is labeled with the Greek letter alpha (α), because all alloys within are ferrite (the body-centered cubic structure).

The looped region at the left center shows where austenite (γ) can exist on this temperature-composition map. In the region labeled sigma (σ) at the bottom center, a structure forms that is not body-centered cubic ferrite. It has a different crystal structure that turns out to be quite brittle, and consequently, this sigma phase is to be avoided. As may be suspected, its

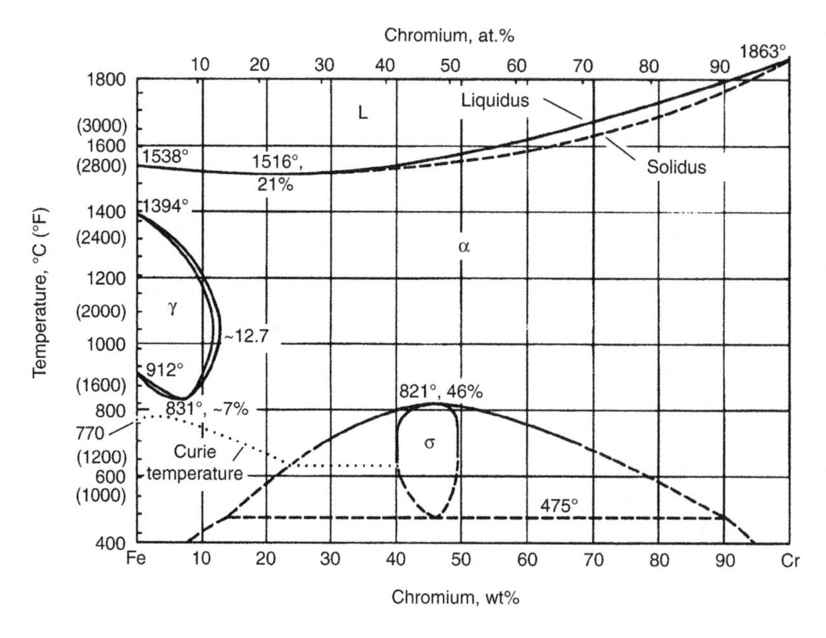

Fig. 13.3 Iron-chromium phase diagram. Source: Ref 13.4

formation is a problem in the heat treatment of stainless steels that have high-chromium compositions. Notice that the lines extending out from the top of the central sigma region are dashed. At the temperature/compositions of these dashed regions, the sigma phase will not form without extended holding at temperature. Hence, the diagram predicts that if an iron-chromium alloy with %Cr in the stainless range of 12 to 26 is hot forged and cooled to room temperature, the σ phase will not form unless very slow cooling rates are used.

In the real world, it is extremely difficult to make ferrous alloys with no carbon in them. The addition of even a very small amount of carbon has a large effect on the appearance of the chromium-iron phase diagram. When carbon is present, the phase diagram becomes a ternary diagram of Cr-Fe-C. Such diagrams were discussed briefly in Chapter 6, "The Low-Alloy AISI Steels." By drawing a vertical section through the Cr-Fe-C ternary diagram at a fixed %C, a diagram is obtained that approximates the pure chromium-iron phase diagram as the %C approaches zero. Figure 13.4(a) shows a vertical section through the Cr-Fe-C diagram at %C of only 0.05. Comparing to Fig. 13.3, it is obvious that the small carbon addition has two very pronounced effects on the diagram. (Note: The sigma, σ, phase lines have been removed from Fig. 13.4 to make comparison more simple.)

Two major changes occur. First, the size of the austenite (γ) region increases significantly. Notice that in the pure alloy (Fig. 13.3) at 1050 °C (1920 °F), 100% ferrite is obtained for compositions as low as 12.7% Cr. When small amounts of carbon are present (Fig. 13.4a), the austenite region is expanded, and 100% ferrite at 1050 °C is only obtained for compositions down to 20% Cr.

The second effect involves the formation of chromium carbides. The element chromium is a fairly strong carbide-forming element, and if iron is not present, it forms a whole series of pure chromium carbides with carbon, all of which have different crystal structures than the iron carbide, cementite (Fe_3C). (As explained in Chapter 10, "Tempering," in alloy steels, the cementite carbide is shown as M_3C, where M means Fe + *X*, with *X* referring to the added alloying element. In an Fe-Cr-C alloy, the *X* would be chromium.) The first two pure chromium carbides have been labeled K_1 and K_2 carbides by German-speaking authors. The chemical formulas for these carbides are $K_1 = M_{23}C_6$ and $K_2 = M_7C_3$, where M means Fe + Cr in Fe-Cr-C alloys such as stainless steel. The K_1 carbide is perhaps the most important carbide in stainless steels, because its formation often occurs along grain boundaries and promotes a type of localized corrosion called intergranular corrosion (Table 13.12). There is basically no change in the phase diagrams of Fig. 13.4 down to room

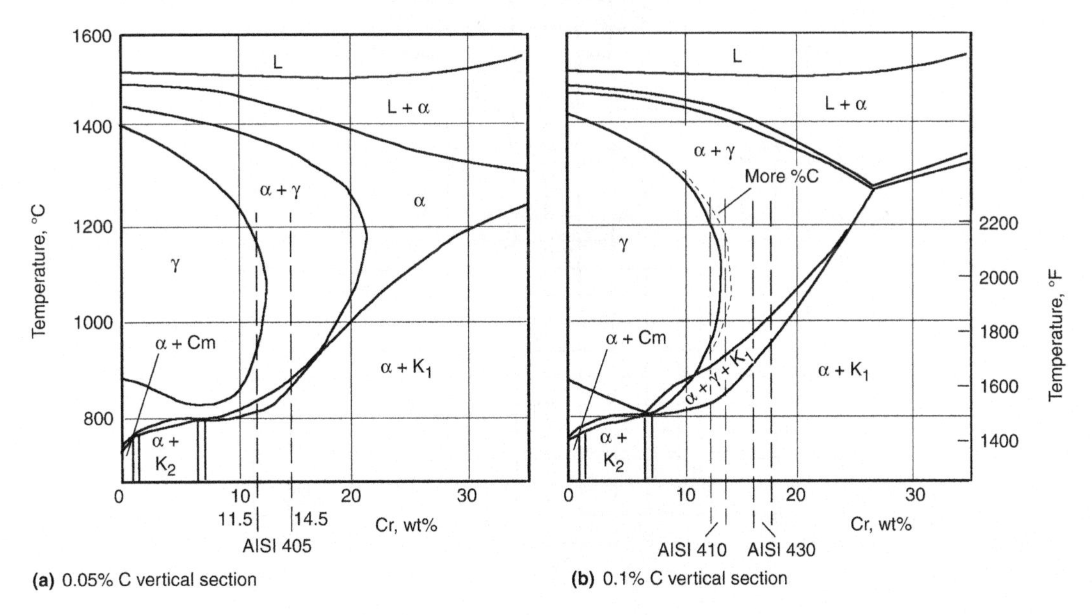

Fig. 13.4 Chromium-iron vertical sections of Cr-Fe-C ternary phase diagram at wt%C values of (a) 0.05 and (b) 0.1. Source: Ref 13.5, p 9-6. Copyright: 1958 Verlag Stahleisen GmbH, Düsseldorf, Germany

temperature, which means that the ferritic stainless steels (12% Cr or more) will consist of ferrite (α) and K_1 carbides at room temperature, even with %C values as low as 0.05.

There are three major AISI grades of ferritic stainless steels, and they are listed in Table 13.1. Notice that the three grades have %Cr compositions corresponding to three distinct levels of corrosion resistance shown in Fig. 13.1. Notice also that the maximum %C level increases as the %Cr increases. The range of chromium compositions for AISI 405 is shown on the 0.05% C vertical section of Fig. 13.4(a), and that of AISI 430 on the 0.1% C vertical section of Fig. 13.4(b).

AISI 430 is the most popular grade of the ferritic stainless steels. At common forging temperatures (approximately 1040 to 1150 °C, or 1900 to 2100 °F), this steel will contain a mixture of ferrite and austenite (α + γ). On cooling, there is the possibility that the austenite will transform to martensite. In Chapter 9, "Hardenability of Steel," it was shown that high concentrations of chromium promote high hardenability, so avoiding martensite in 430 may require very slow cooling rates. The cooling rates required to avoid martensite can be estimated from the isothermal transformation (IT) diagram for 430, which is shown in Fig. 13.5. Notice that the steel was austenitized at 1090 °C (2000 °F), and Fig. 13.4(b) predicts the steel should be a mixture of austenite and ferrite (α + γ) at this temperature. Figure 13.5 indicates that the austenitized steel contains 60% ferrite + 40% austenite. The diagram shows that the steel must be cooled below approximately 760 °C (1400 °F) in less than a second to avoid precipitation of the K_1 carbides. The nose of the transformation curve for the start of ferrite formation occurs at approximately 700 °C (1300 °F), with transformation starting at 60 s and not being completed until just under 2 h. If this stainless steel is to be annealed to the fully ferritic condition, it should be held at 760 °C (1400 °F) for at least 2 h. The diagram shows that the hardnesses of the fully ferritic structures vary from HRB = 78 for an 815 °C (1500 °F) anneal to HRB = 86 for a 590 °C (1100 °F) anneal. Quenching from the austenitizing temperature produced HRB = 101.

Ferritic stainless steels are subject to embrittlement problems when held at relatively low temperatures. The embrittlement occurs by two different mechanisms. One involves the formation of the sigma (σ) phase, shown in Fig. 13.3, and the other the formation of ferrite that has a

Table 13.1 The AISI ferritic stainless steels

AISI No.	%C	%Cr	Other(a)
405	0.08 max	11.5–14.5	0.1–0.3 Al
430	0.12 max	16–18	...
446	0.20 max	23–27	...

(a) Both manganese and silicon present at approximately 1%

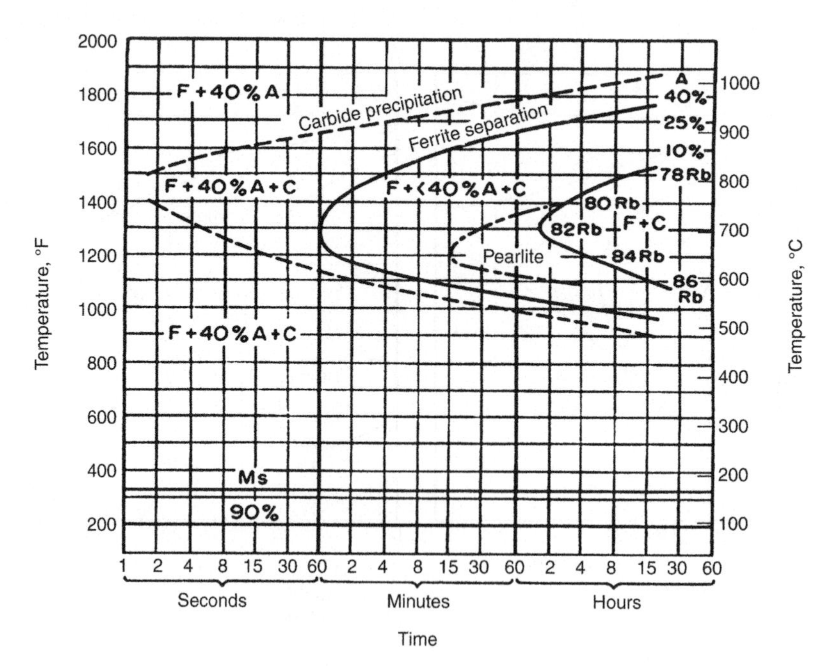

Fig. 13.5 Isothermal transformation diagram for a 430 stainless steel austenitized at 1090 °C (2000 °F) for 15 min. Source: Ref 13.6

very high percentage of chromium in it. This second mechanism is referred to as 475 °C embrittlement, because it occurs for steels held between 400 and 540 °C (750 and 1000 °F). Figure 13.6 presents a schematic IT diagram for the formation of these embrittlement mechanisms. Both problems are more severe as the %Cr in the stainless becomes higher, and these curves are representative of alloys containing approximately 25% Cr. (Higher %Cr shifts the curves to the left.) Because of these embrittlement problems, it is often recommended that ferritic stainless steels be rapidly cooled following an annealing heat treatment. See Ref 13.7 (p 759) for a good brief discussion of annealing practice for the AISI ferrite stainless steels.

There is a relatively new class of ferritic stainless steels, called the superferritics, that was developed in the mid-20th century. These steels have improved corrosion resistance, and Table 13.2 presents compositions of three of them. The improved corrosion resistance is due to the addition of molybdenum and sometimes nickel and to the very low levels of carbon and nitrogen. The low levels of carbon and nitrogen are produced by special processing techniques, and consequently, these steels are more costly than the regular AISI ferritic stainless steels. Reference 13.1 presents an excellent review of their corrosion resistance properties in comparison to other stainless steels as well as specialty nickel-base alloys such as Inconel and Hastelloy.

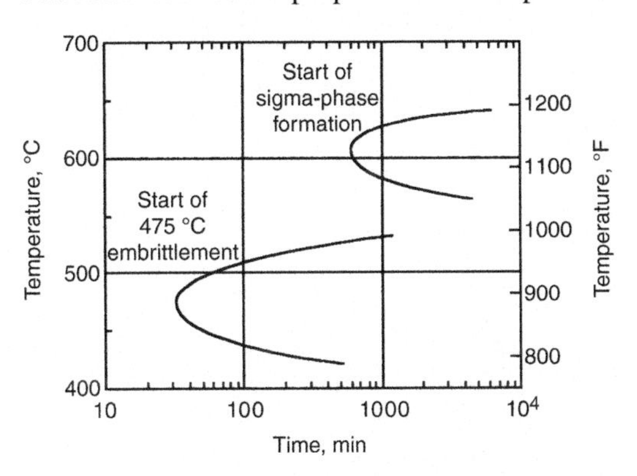

Fig. 13.6 Isothermal transformation diagram of start times for formation of sigma phase and 475 °C embrittlement in ferritic stainless steel

Table 13.2 Examples of superferritic stainless steels

Name	%Cr	%Mo	%Ni	Max %C	Max %N
18-2(a)	18	2	...	0.025	0.025
26-1(a)	26	1	0.1	0.005	0.015
29-4-2	29	4	2	0.01	0.02

(a) Low levels of titanium and/or columbium (niobium) added as scavenger elements

Martensitic Stainless Steels

As the name implies, the martensitic stainless steels can be heat treated to form martensite. All of the martensitic stainless steels classified by the AISI have been assigned numbers in the 400 series. Table 13.3 lists the compositions of five of the more popular AISI martensitic stainless steels.

Hardening heat treatments of these steels are basically the same as in plain carbon and low-alloy steels in that three steps are involved: austenitization, quenching, and tempering. In general, however, processing must be done more slowly because of two factors: (1) the carbides dissolve much more slowly into the austenite than the cementite carbide because of the presence of chromium in them, and (2) the thermal conductivity is much smaller. Factor 1 requires that austenitization times be longer to ensure the carbides have dissolved into the austenite. Factor 2 causes temperature gradients to be much higher, which can lead to cracking and warpage even on heating. Therefore, it is often recommended that complex parts be given an intermediate, low-temperature soak on the austenitization heat-up step to smooth out temperature variations and a martempering treatment on cooling.

Just as for plain carbon and alloy steels, the hardness of the martensite increases as the %C rises. Table 13.4 presents an approximate guideline for the as-quenched hardness dependence on %C in the AISI martensitic stainless steels.

Table 13.3 Some AISI martensitic stainless steels

AISI No.	%C	%Cr	Other(a)
410	0.15	11.5–13	...
431	0.20	15–17	1.25–2.5 Ni
440A	0.65–0.75	16–18	0.75 Mo
440B	0.75–0.95	16–18	0.75 Mo
440C	0.95–1.2	16–18	0.75 Mo

(a) All contain 1% Mn and 1% Si

Table 13.4 Hardness of as-quenched AISI martensitic stainless steels

Carbon range, wt%C	As-quenched hardness, R_c
0.06–0.14	38–49
0.2–0.4	44–54
0.65–1.2	56–61

Heat treatment of these steels is complicated by a number of factors that are discussed in the next section, which is devoted to optimizing composition and heat treatment for cutlery applications.

Some guidelines for the heat treatment of AISI martensitic stainless steels are presented in Ref 13.7. As an example, consider AISI 410. The composition range for this steel has been included on the 0.1% C vertical section of Fig. 13.4(b). The increased %C in 410 up to 0.15 causes the austenite (γ) region to expand to higher %Cr values, as is indicated by the dashed line labeled "more %C." The %C composition range for AISI 410 is shown on the diagram, and one sees that the 410 steels are expected to be fully austenite in a narrow temperature range of approximately 1000 to 1200 °C (1830 to 2190 °F). If the steel is austenitized slightly above or below this range, it will contain some ferrite that will reduce the hardness of the quenched steel. (Notice in Figure 11.5 that the ferrite formed in iron-carbon alloys at approximately 1450 °C, or 2640 °F, is designated delta, δ, ferrite and that formed below 912 °C, or 1670 °F, is called alpha, α, ferrite. So, δ-ferrite forms at high temperatures and the α-ferrite at low temperatures. However, both ferrites are the same body-centered cubic phase, and to avoid confusion on ternary diagrams, ferrite formed at high temperatures is sometimes called α-ferrite rather than δ-ferrite, as in Fig. 13.4.) As illustrated with micrographs in Ref 13.7, it is common to find stringers of delta ferrite in as-forged samples of this steel.

As with ferritie stainless steels, the high %Cr levels in martensitic steels produce very high hardenability conditions. Figure 13.7 is an IT diagram for a 410 steel austenitized at 980 °C (1800 °F). Notice that both the position of the transformation curves and the hardness of samples held to full transformation between 540 and 760 °C (1000 and 1400 °F) are similar to the IT diagram for 430 steels shown in Fig. 13.5. A major difference is that after austenitization, the 410 steel of Fig. 13.7 is 100% austenite, compared to only 40% for the 430 steel in Fig. 13.5. Even though IT diagrams show that martensitic stainless steels can be air cooled to form fully martensitic structures, it is generally recommended that samples be oil quenched to optimize corrosion resistance. The quenching minimizes formation of K_1 carbide precipitate formation on cooling, which, as explained in the next paragraph, tends to promote corrosion. (Note: The transformation-start curve for the K_1 carbide precipitation is shown in Fig. 13.5 but not in Fig. 13.7.)

The corrosion resistance of martensitic stainless steels is generally poorer than ferritic and austenitic stainless steels. This results from effects of the increased amount of carbon present in the martensitic stainless steels. As is explained in the next section, the higher carbon levels reduce the amount of chromium that will dissolve

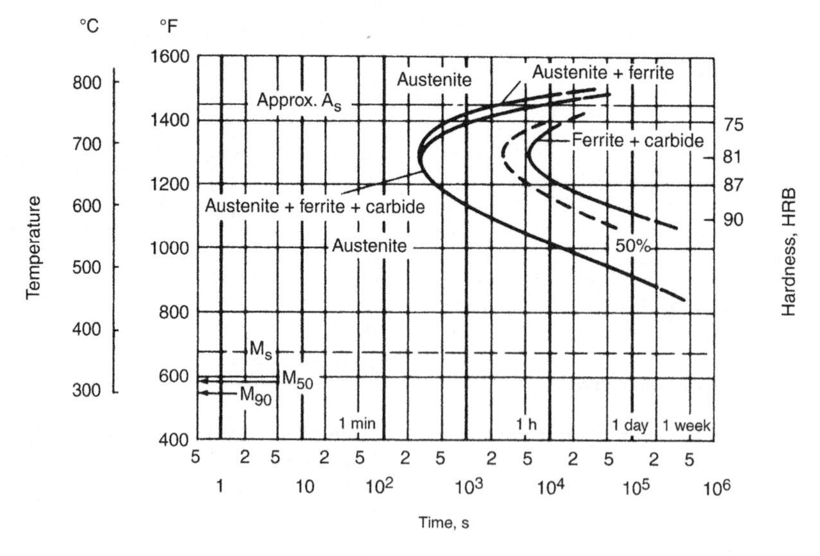

Fig. 13.7 Isothermal transformation diagram for a 410 stainless steel containing 0.11% C and 12.2% Cr austenitized at 980 °C (1800 °F). Source: Ref 13.7

into austenite. Also, the higher carbon level increases the possibility of forming the K_1 carbide as small precipitate particles on low-temperature heating. Recall that the K_1 carbide has the chemical formula of $(Cr,Fe)_{23}C_6$, which means that it contains significant amounts of chromium. When a small particle of K_1 forms in a matrix of either ferrite or austenite, the chromium atoms in the carbide are depleted from the matrix iron immediately surrounding the particle. If the depletion causes the %Cr to drop much below 12 in the surrounding matrix, the steels become susceptible to corrosion and are said to be *sensitized*. Because the carbides prefer to form along the grain boundaries of the matrix, the corrosion occurs along the grain boundaries, and therefore, this type of corrosion is called intergranular corrosion. Sensitization is said to occur when a low-temperature heat treatment, such as tempering, causes K_1 precipitate particle formation to the extent that intergranular corrosion may occur. Figure 13.8 presents corrosion data on a 410 stainless steel showing the loss of corrosion resistance produced by tempering. The steel became sensitized on tempering in the range of 400 to 700 °C (750 to 1290 °F).

The martensitic stainless steels generally are tempered to enhance toughness, just as are plain carbon and alloy steels. The variations of hardness on tempering the various AISI martensitic stainless steels are presented in Ref 13.7. Figure 13.9 presents a summary of data showing the tempered hardnesses of 12% Cr stainless steels as the %C in the steels varies from 0.055 to 0.14. The higher hardnesses in the range shown correspond to the higher %C values. Very little reduction in hardness occurs until the tempering temperature exceeds approximately 480 °C (900 °F). The large reduction of hardness above this temperature has been shown to be due to formation of particles of the K_1 carbide precipitate.

As with the plain carbon and alloy steels, it is generally necessary to temper the martensitic stainless steels to improve toughness. Reference 13.5 (page 20–21) presents Izod impact data on a 410 stainless steel after tempering to various temperatures. The data show that toughness is improved for tempering to temperatures up to nearly 260 °C (500 °F). Above 260 °C (500 °F), toughness (impact energy) drops to a minimum at 480 °C (900 °F) and then rises strongly at temperatures above approximately 540 °C (1000 °F). The peak loss of toughness at approximately 480 °C (900 °F) is due to the 475 °C embrittlement discussed previously for ferritic steels (Fig. 13.6). It is caused by the formation of carbide precipitates containing chromium, and consequently, it is more of a problem as the %Cr increases in these stainless steels.

Table 13.5 summarizes Izod impact data for the five AISI martensitic stainless steels in the

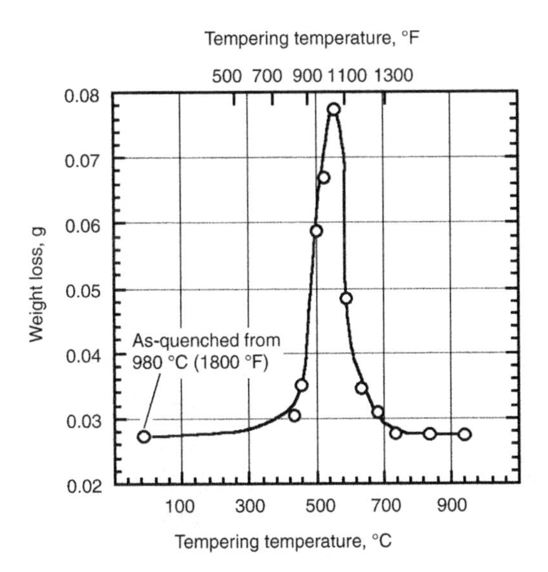

Fig. 13.8 Intergranular corrosion produced by sensitizing a 410 steel on tempering. Corrosion test: 14 days in 20% salt fog. Source: Ref 13.8. Copyright: NACE International, 1953

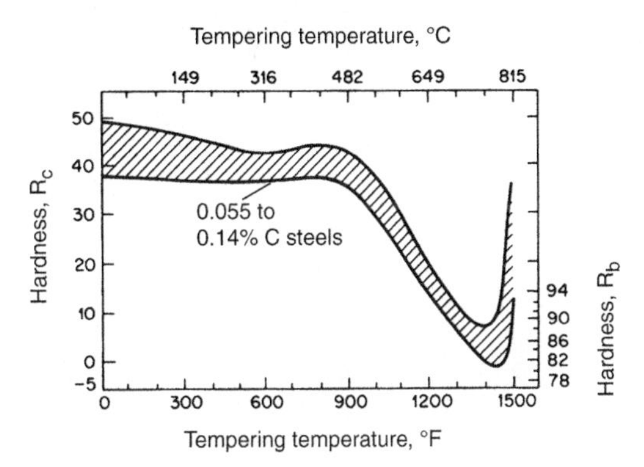

Fig. 13.9 Tempering response of 12% Cr stainless steels tempered for 2 h. Source: Ref 13.9

Table 13.5 Izod impact data for annealed martensitic stainless steels

AISI No.	Izod energy J	Izod energy ft · lbf
410	122	90
431	68	50
440A	20	15
440B	7	5
440C	7	5

Source: Ref 13.2

annealed condition. These data show that, even in the annealed condition, the 440 martensitic stainless steels suffer from poor toughness. This results, in large part, from the large carbide particles present at the high levels of carbon in these steels.

Optimizing Martensitic Stainless Steel for Cutlery Applications

To produce a knife with optimal properties, it is desirable to have high hardness levels and a fine array of carbides to enhance wear resistance and maintain sharpening ability, combined with adequate %Cr to produce corrosion resistance. There are two basic requirements:

1. The austenite phase that is quenched requires %C levels of approximately 0.6 or higher to produce Rockwell C hardnesses in the 63 to 64 range.
2. The austenite phase that is quenched requires %Cr levels of 12 or higher to ensure reasonable corrosion resistance.

To understand the various trade-offs involved between composition and austenitization temperature for optimal results, it is helpful to learn how to interpret isothermal sections of the Fe-Cr-C ternary phase diagram. In Chapter 3, "Steel and the Iron-Carbon Phase Diagram," it was shown that the iron-carbon binary Phase diagram is a map with temperature on the vertical axis and composition on the horizontal axis. The various regions on the phase diagram map the temperature-composition values where the various phases of the iron-carbon alloys may exist. Ternary phase diagrams are similar to binary diagrams in that the temperature is mapped along a vertical axis. However, composition can no longer be mapped (plotted) along a single line, and instead, it is mapped along a plane, called the composition plane, with temperature being the height above this plane. The composition plane for an Fe-Cr-C ternary phase diagram is shown in Fig. 13.10. The %C is plotted along the horizontal axis, and the %Cr along the vertical axis of the composition plane. Consider a 440C stainless. According to Table 13.3, the average composition for this Fe-Cr-C alloy is 17% Cr and 1.075% C. A black dot, labeled 440C, is plotted on the composition plane of Fig. 13.10, and this dot locates the composition of 440C on the ternary diagram.

With composition plotted along a plane and temperature plotted as distance above the

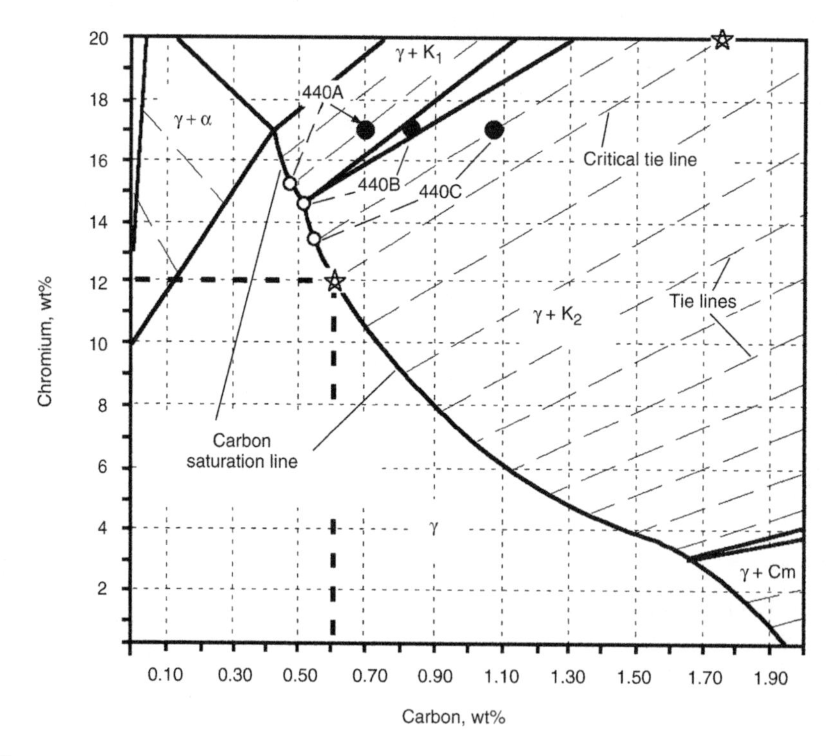

Fig. 13.10 Isothermal section of Fe-Cr-C ternary phase diagram at 1100 °C (2010 °F). Source: Ref 13.10

plane, the ternary phase diagram becomes a three-dimensional plot. These complex three-dimensional diagrams are used by making sectional cuts through them, either vertically or horizontally. The diagrams in Fig. 13.4 are vertical sections through the Fe-Cr-C ternary along a section that holds the %C fixed, at 0.05 in (a) and 0.1 in (b). Consequently, these vertical sections look similar to binary phase diagrams with temperature plotted vertically and %Cr along the horizontal axis.

Figure 13.10 is a horizontal sectional cut through the Fe-Cr-C ternary phase diagram. It is called an isothermal section because the cut is made at a specific temperature (which is what isothermal means). As indicated by the caption, Fig. 13.10 is the 1100 °C (2010 °F) isothermal section, and it maps the compositions (%Cr and %C) of the various phases of Fe-Cr-C alloys that can exist at 1100 °C (2010 °F).

Figure 13.11 helps explain Fig. 13.10. The large white area is labeled γ, which is austenite. Compositions of alloys in Fig. 13.11 that fall in the γ area will be 100% austenite when held at 1100 °C (2010 °F). The shaded areas of Fig. 13.11 show the composition regions in which a second phase is present within the austenite at 1100 °C (2010 °F). For example, the composition of alloy 440C falls inside the region labeled γ + K_2. This predicts that if 440C is heated to 1100 °C (2010 °F), it will consist of austenite with particles of K_2 carbide in it, and on quenching, the martensite formed from the austenite will have K_2 carbides dispersed within it. (Note: Figure 13.11 should be regarded only as a good approximation for heat treating commercial alloys. For example, experiments on a 14% Cr/7.4% C alloy, Ref 13.11, show that the carbides present at 1100 °C, or 2010 °F, are a mixture of K_1 and K_2, with K_1 being predominant.)

The heat treatment of martensitic stainless steels often involves two-phase austenitization, which was discussed in Chapter 11, "Austenitization." To help understand the complications that result from this effect for martensitic stainless steels, consider the four alloys A to D in Table 13.6. All of the alloys contain 13% Cr, slightly above the minimum required for stainless passivity. The compositions of the four alloys have been plotted in Fig. 13.11. Alloy A lies within the austenite (γ) region and will be 100% austenite at 1100 °C (2010 °F), with a composition of 13% Cr and 0.3% C. Notice that in Fig. 13.10, the right boundary of the austenite region is labeled "carbon saturation line." Alloy

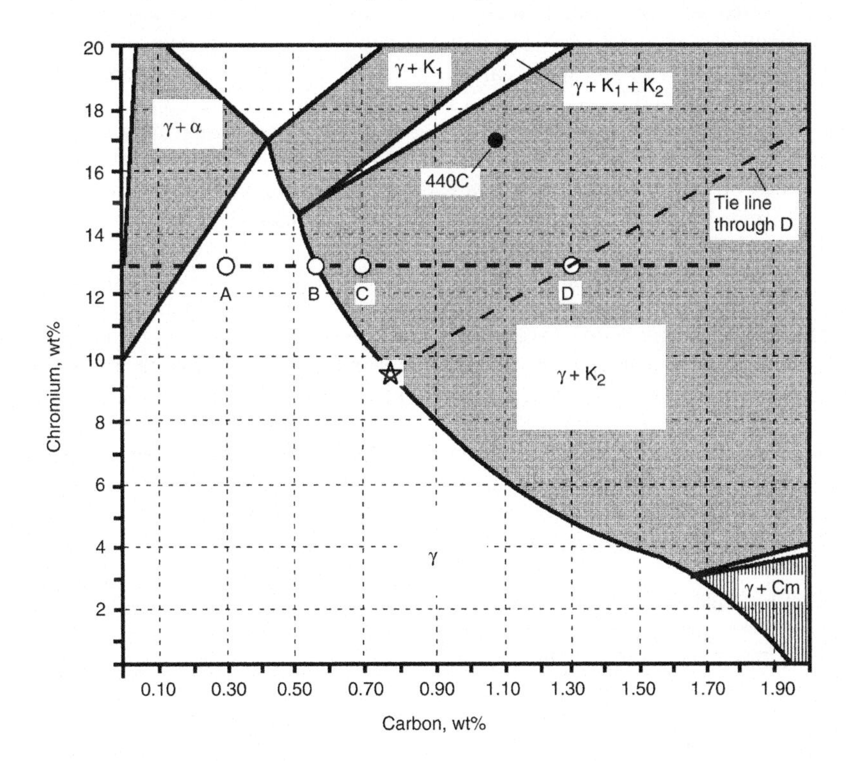

Fig. 13.11 The important regions on the 1100 °C (2010 °F) isothermal section of Fig. 13.10

B lies directly on this line, with a carbon level of 0.58%. The carbon saturation line is very important. It is like the A_{cm} line in the binary iron-carbon phase diagram. It marks the maximum amount of carbon that austenite can dissolve within itself before addition of more carbon results in the formation of carbide particles. Alloy C, at 0.7% C, is in the two-phase region, austenite + carbide ($\gamma + K_2$), and the alloy will consist of austenite with K_2 carbide particles dispersed within it. The diagram provides a measure of how much carbide lies within the austenite. The further the alloy composition lies to the right of the saturation line, the larger the volume percent of K_2 carbide it will contain. Hence, alloy D is expected to contain many more carbide particles than alloy C, because it lies further to the right from the saturation line.

The composition of the austenite formed at 1100 °C (2010 °F) in alloys A and B will be the same as the overall compositions given in Table 13.6. Such will not be the case for alloys C and D, which contain the K_2 carbides at 1100 °C. These carbides are rich in both chromium and carbon, and so the austenite is depleted in both elements relative to the overall composition. The composition of the austenite in alloys C and D can be determined from the isothermal phase diagrams. Notice the dashed slanted lines in Fig. 13.10 that are labeled "tie lines." The full isothermal diagram will contain many more such nearly parallel lines, and Fig. 13.11 shows the tie line that passes through alloy D. The austenite composition for alloy D is found at the intersection of this tie line and the carbon saturation line; it is labeled with a star. Even though this alloy contains 1.3% C and 13% Cr overall, the austenite formed in it at 1100 °C (2010 °F) will contain only approximately 0.78% C and 9.5% Cr. Hence, the martensite formed on quenching this alloy would have quite poor corrosion resistance, with a %Cr level well below the minimum 12 needed for good passivity.

With this background, it is possible to demonstrate shortcomings of the 440 stainless steels for knifemaking applications. The three alloys, 440A, 440B, and 440C, are shown in Fig. 13.10 with a solid arrow pointing at their average overall compositions and a dashed arrow pointing at the compositions of austenite that form in them at 1100 °C (2010 °F), which are tabulated in Table 13.7. In all three alloys, the %Cr in the austenite is reduced from the overall 17% Cr amount but is still above the 12% Cr needed for good passivity. Notice, however, that the %C in the austenite is lowered a small amount below the 0.6 value needed to achieve R_c values in the 63 to 64 range. In addition, the highest %C alloy, 440C, is found to contain large primary carbide particles produced in the solidification process (discussed further in Chapter 15, "Solidification"). Reference 13.7 (p 777) presents a series of micrographs of 440C that illustrate the relatively large primary carbide particles present in this stainless steel.

To illustrate what effect lowering the austenitizing temperature will have on austenite composition and volume fraction of carbides, the change in the position of the carbon saturation line is shown in Fig. 13.12 when the temperature is reduced from 1100 to 1000 °C (2010 to 1830 °F). The overall composition of the 440C on the superimposed diagrams remains fixed. However, one sees that the %C/%Cr in the austenite drops from 0.56% C/13.6% Cr to 0.39% C/12.1% Cr. Reference 13.7 recommends that 440C be austenitized at 1010 to 1065 °C (1850 to 1950 °F). Further, it recommends austenitizing at the upper end of the range for maximum corrosion resistance and strength. The reason for this recommendation is apparent in light of the previous discussion: higher austenitization temperatures will increase both %C and %Cr in the austenite.

To produce stainless steel blades that can be heat treated to as-quenched hardnesses in the range of R_c = 63 to 64, it is necessary to be able to produce austenite with %C values above approximately 0.6 and %Cr levels above approximately 12. This composition is shown by the star in Fig. 13.10, and it happens to fall on the carbon saturation line for heat treatment at 1100 °C (2010 °F). As shown for alloy D in Fig. 13.11, if

Table 13.6 The four alloys shown in Fig. 13.11

Alloy	%Cr	%C
A	13	0.3
B	13	0.58
C	13	0.7
D	13	1.3

Table 13.7 Overall composition versus that in 1100 °C (2010 °F) austenite

Alloy	Overall, %C/%Cr	In austenite, %C/%Cr
440A	0.70/17	0.48/15.1
440B	0.85/17	0.52/14.6
440C	1.07/17	0.56/13.6

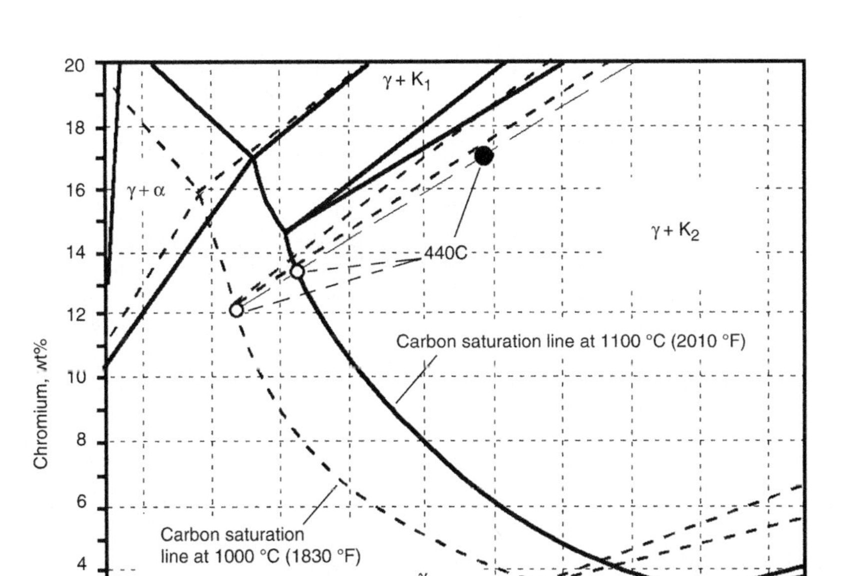

Fig. 13.12 Shift of lines of Fe-Cr-C isothermal section when temperature decreases from 1100 to 1000 °C (2010 to 1830 °F). Source: Ref 13.10

the %C is increased above 0.6 to 1.3 and chromium is held at 13%, the tie lines through the resulting composition predict that the resulting austenite will produce a high-hardness martensite (%C is ≈ 0.78), but its %Cr will fall to approximately 9.5, well below the 12% Cr required for good passivity (see the star composition in Fig. 13.11).

If both the %C and the %Cr are increased along the tie line marked with a star at both ends in Fig. 13.10, then the austenite composition will match the 0.6% C/12% Cr requirement. This tie line is called the critical tie line. In alloys with overall compositions along this line, the austenite will contain K_2 carbides with a volume fraction dependent on how far the overall composition lies to the right of the lower star on the carbon saturation line. The compositions of the 440 series lie well above the critical tie line. However, there are commercial stainless steels with compositions that lie close to the critical tie line, and two of them are shown in Table 13.8. (Note: The positions of the compositions of the steels in Table 13.8 lie slightly above the critical tie line. The positions of the carbon saturation line in Fig. 13.11 and 13.12 are, at best, only good approximations for real commercial alloys. In choosing heat treat temperatures for particular alloys, it is best to consult the technical data provided by the supplier.)

From the previous discussion, it should be clear that the austenitization temperature and times are important variables in the heat treat process. In addition, the quench temperature and quench rates are important, because these alloys are prone to formation of retained austenite. It was demonstrated in Chapter 10, "Tempering," that during two-phase austenitization, the austenitization temperature controls %C in the austenite, which can change the amount of retained austenite in the quenched samples. Even though the %C may increase in the austenite, which will result in harder martensite, the hardness of the quenched bar can decrease because the amount of retained austenite also increases with increased carbon in the austenite.

Table 13.8 Optimal steels for stainless knives

Steel	%C	%Cr	%Si	%Mn
Sandvik 12C27	0.60	13.5	0.4	0.4
Uddeholm AEB-L	0.65	12.8	0.4	0.65

In steels prone to retained austenite, such as hypereutectoid alloy steels, tool steels, and these stainless steels, there is an important and subtle effect that can occur, called *stabilization*, which is now briefly discussed.

The usual way to explain stabilization is with the curves shown in Fig. 13.13. The upper curve of Fig. 13.13 can be explained by referring to Fig. 4.16. The upper solid curve labeled "direct quench" shows the amount of martensite formed in a steel quenched directly to various temperatures between the martensite start temperature, M_s, and the martensite finish temperature, M_f. The curve shows that if this steel is quenched to room temperature, T_{room}, it will contain 40% retained austenite. The dashed curve shows the amount of martensite formed in this steel if it is quenched by a two-step process: (1) quench to room temperature, and (2) hold at room temperature for some time, followed by a quench into a cold quenchant. The dashed curve shows that the amount of martensite formed by the cold quench after the room-temperature delay is significantly less than would have occurred if the steel had been quenched to the cold temperature with no room-temperature delay. A direct quench to the cold temperature, T_c, produces no retained austenite for this steel, but after the room-temperature delay, 10% retained austenite remains. The room-temperature delay is said to have stabilized the austenite.

Steels are virtually always quenched to cold temperatures in a two-step process, because a direct quench would promote quench cracking, and, other than ice brine, there are severe problems with vapor blanket formation with cold quenchants (such as liquid nitrogen). In order to minimize retained austenite, it is important to minimize the room-temperature hold time prior to the cold quench. In addition, if the piece is heated slightly during the room-temperature hold, the stabilization effect is increased.

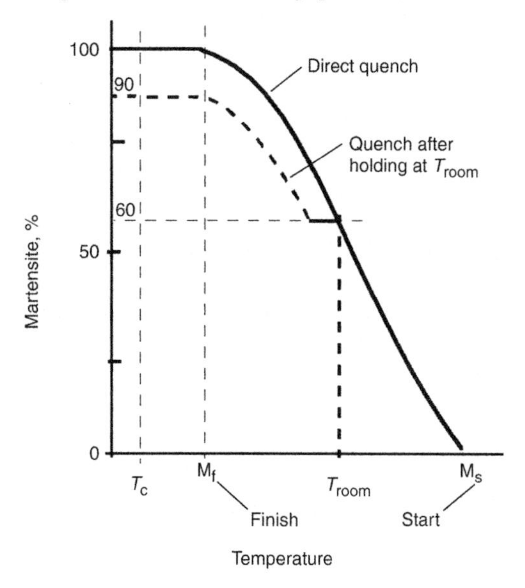

Fig. 13.13 Stabilization of austenite through a two-step quenching process

With highly alloyed steels such as stainless, there are some subtle effects that result from stabilization, which are shown in the study of Sandvik 12C27 in Ref 13.12. A series of heat treatments was carried out in which the amount of retained austenite was measured as a function of the austenitization hold time, with four different decreasing quenching rates: (1) water, (2) forced air plus water, (3) forced air plus oil, and (4) forced air alone. In all four cases, the amount of retained austenite increased steadily as the austenitization time increased from 4 to 10 min. At longer hold times, more carbides dissolve, increasing the %C and %Cr in the austenite and decreasing the M_s-M_f range. Perhaps more interestingly, at any given austenitization time, the amount of retained austenite increased with decreasing quench rate. For example, with an 8 min austenitization, it increased from 7.5 to 10 to 14.3 to 15.5% for four quench rates. This increase is due to stabilization as the austenite is cooled more slowly through the temperature range from approximately 300 °C (570 °F) to room temperature. For the cooling rates that include forced air, the steels were first cooled with forced air to 200 °C (390 °F) and then either water quenched (rate 2) or oil quenched (rate 3) to room temperature. These results show that the amount of retained austenite formed in this stainless steel is significantly influenced by the cooling rate through the low temperatures. This is a subtle stabilization effect. Thus, there are two reasons to cool these stainless steels by quenching rather than air cooling: (1) to reduce the amount of retained austenite, as illustrated here, and (2) to reduce K_1 carbide precipitation on cooling and improve corrosion resistance.

Example Heat Treatment Using AEB-L. To illustrate some of the important features involved in heat treating stainless steels, experiments carried out by the author on Uddeholm AEB-L strip of 0.7 mm (0.03 in.) thickness are presented. The composition of this alloy is shown on the combined 1000 and 1100 °C (1830 and 2010 °F) isothermal phase diagrams of the Fe-Cr-C ternary alloy in Fig. 13.14. At both

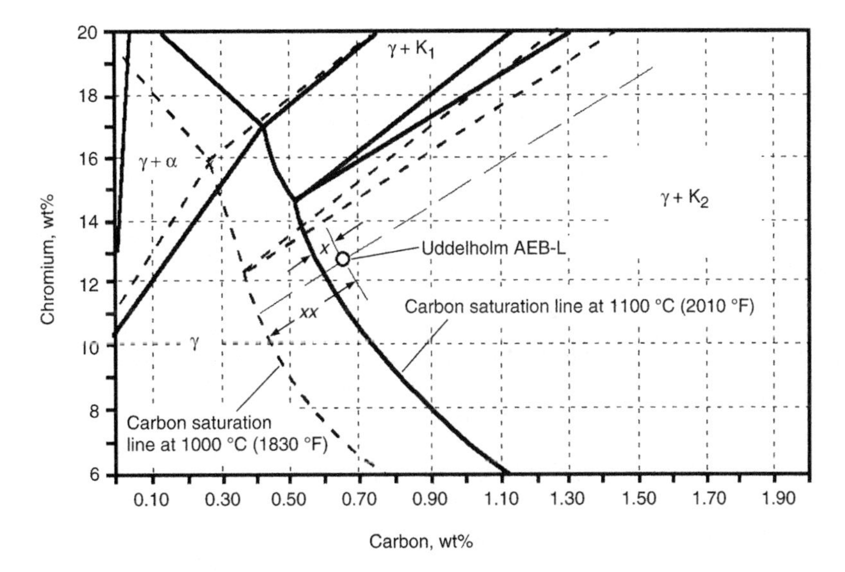

Fig. 13.14 The Fe-Cr-C isothermal sections at 1000 and 1100 °C (1830 and 2010 °F)

temperatures, the alloy composition lies in the austenite plus K_2 carbide region, and two-phase austenitization at both temperatures is expected to produce carbides in the austenite, which would give martensite + carbides on quenching. Note, however, that the distance, *x*, from the overall composition to the carbon saturation line is smaller at 1100 °C (2010 °F) than at 1000 °C (1830 °F), a distance marked *xx*. Therefore, a higher volume fraction of carbides is expected at the lower austenitization temperature.

Strips of the alloy were austenitized for 4 min in a resistance-heated furnace with an inert atmosphere (helium gas) and then forced-air cooled to 90 °C (200 °F) in 70 s, followed by water quenching. Metallographic samples were examined in the scanning electron microscope, and Fig. 13.15 is a typical micrograph showing the distribution of carbides in the martensitic matrix. The diameter of the carbides averaged approximately 0.6 μm (2 × 10^{-5}in.), which is somewhat finer than the secondary carbides found in 440 stainless steels. No primary carbides were observed. The absence of the coarse primary carbides should allow edge sharpening to a finer radius of curvature without carbide pullout, which is more likely to occur with the coarse primary carbides of 440C steels. The carbide density was measured as the number of carbides per square inch on micrographs of 2000× magnification, and the results are presented in Fig. 13.16. As expected from the previous discussion of Fig. 13.14, it is seen that the

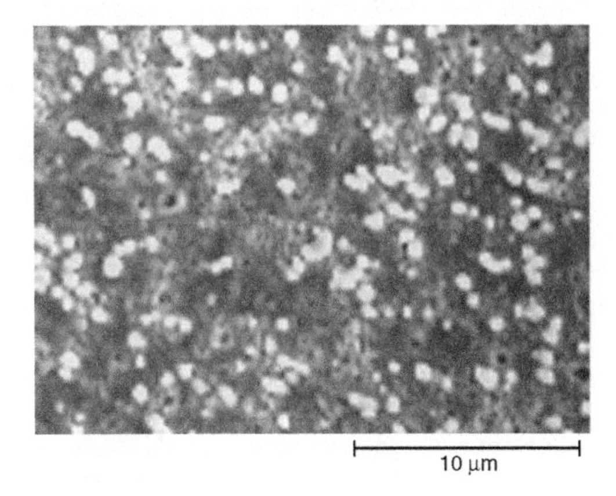

Fig. 13.15 SEM micrograph showing K2 carbides (white particles)

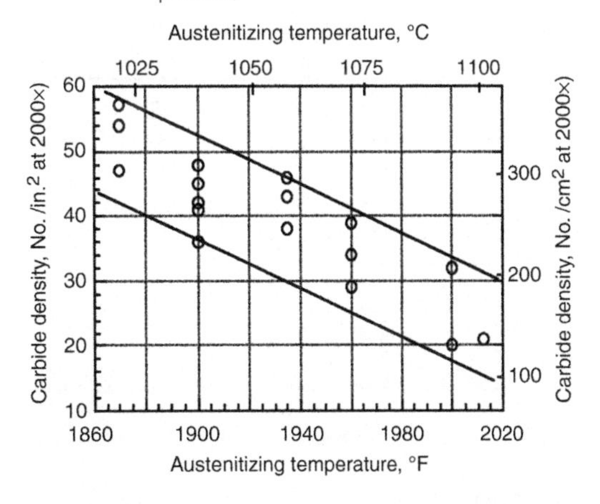

Fig. 13.16 Carbide density as a function of austenitizing temperature

carbide density increases as the austenitization temperature decreases.

The amount of retained austenite in the quenched samples was measured using standard x-ray diffraction techniques, and the results are shown in Fig. 13.17. The higher austenitizing temperatures resulted in a significant amount of retained austenite. As seen on the isotherms in Fig. 13.14, the higher austenitization temperature of 1100 °C (2010 °F) increases both the %C and the %Cr in the austenite (the points where the tie line meets the carbon saturation lines). This, in turn, lowers the M_s-M_f range, which results in more retained austenite for a room-temperature quench.

Mixtures of liquid nitrogen and isopentane (also called 2-methyl butane) were used to study the effects of cold quenching at temperatures of −60 and −96 °C (−75 and −140 °F). In samples quenched from both 1090 and 1040 °C (2000 and 1900 °F), the amount of retained austenite was reduced to zero at the –96 °C (−140 °F) quench temperature and to approximately 0.4% at the −60 °C (−75 °F) quench temperature. The latter temperature is the recommended cold quench temperature (Ref 13.13).

Because the steel strip was so thin, it was necessary to measure hardness with a microhardness tester and then convert the diamond pyramid hardness values to R_c values. The hardness was found to depend mainly on the amount of retained austenite. Samples quenched to −96 °C (−140 °F) from either 1040 or 1090 °C (1900 or 2000 °F), which had no retained austenite, both gave hardnesses of approximately R_c = 63.5. Samples quenched to room temperature, which had the amount of retained austenite shown in Fig. 13.17, had hardnesses of R_c = 62 and 60

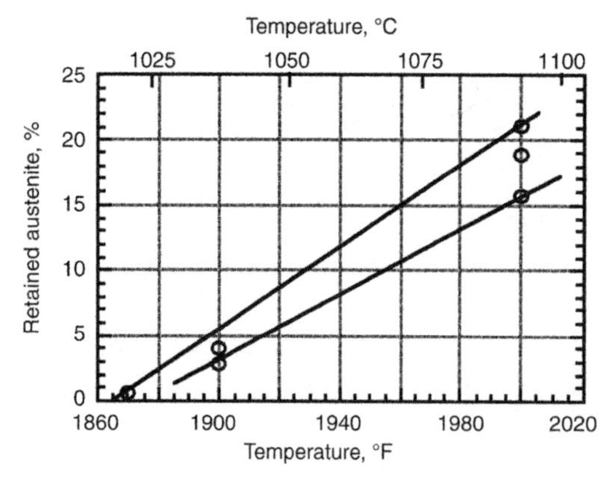

Fig. 13.17 Dependence of percent retained austenite on austenitizing temperature for room-temperature quench

when quenched from 1040 and 1090 °C (1900 and 2000 °F), respectively. Tempering these samples at 190 °C (380 °F), which is near the maximum recommended tempering temperature (Ref 13.12), reduced the hardness of both steels by approximately 1.7 R_c points.

Results of some stabilization experiments are presented in Table 13.9. A sample quenched to room temperature from 1090 °C (2000 °F) produced 21.1% retained austenite (RA). A similar sample quenched immediately to −96 °C (−140 °F) produced 0% RA, but the original sample quenched to −96 °C (−140 °F) after a three-day hold at room temperature produced 10.2% RA. Therefore, the three-day hold had stabilized 10.2% RA, just under half of the original 21.1% RA. For the 1040 °C (1900 °F) austenitization sample, all 4.3% of the RA was stabilized by the three-day hold. Also, in a 1090 °C (2000 °F) austenitized sample (bottom row of table) that had been tempered before the three-day hold, all of the original 19% RA was stabilized, illustrating that heat plus holding increases stabilization of the austenite. These results illustrate the importance of immediate cold quenching prior to tempering to maximize the elimination of retained austenite.

The measurement of retained austenite with x-ray diffraction analysis is a somewhat complicated and nonstandard technique. Reference 13.12 presents a simple technique that can be used as a guide to qualitatively evaluate the amount of retained austenite in the steels of Table 13.8 and thereby estimate if the austenitization temperature is near the desired level. Immediately after the room-temperature quench and prior to any tempering, blank samples are cold quenched to at least −35 °C (−31 °F) and their hardness compared to that of the room-temperature quenched samples. The cold quench will cause the hardness to increase due to loss of retained austenite. The guidelines (Ref 13.12) are as follows: (1) If hardness increases by more than 1.5 R_c points, the austenitization temperature was too high (too much retained austenite). (2) If

Table 13.9 Percent austenite in stabilization experiments

Austenitization temperature			
°C	°F	As-quenched room temperature, %	Quenched −96 °C (−140 °F) after 3 days, %
1090	2000	21.1	10.2
1040	1900	4.3	4.3
1090	2000	19	19(a)

(a) Sample tempered 30 min at 195 °C (380 °F)

hardness increases by less than 0.5 R_c points, the austenitization temperature was too low.

The K_1 and K_2 carbides present in the stainless steels of Table 13.8 are significantly harder than the cementite carbide (M_3C) of alloy steels. Table 13.10 presents some measurements of Knoop microhardness values that have been converted to estimated R_c values in the right column. The higher hardness of the K_1 and K_2 carbides has only a small effect on increasing the as measured R_c values of the stainless blades. As seen in Fig. 13.15, the carbides are well isolated from each other, and the depth of the diamond indentation of the hardness machine therefore measures the resistance to flow of the surrounding martensite matrix, with only a small effect from the carbides. However, the presence of the K_1 and K_2 carbides at the surface is effective in increasing the wear resistance over that of a pure martensitic matrix or of a martensite containing the M_3C carbide of alloy steels such as AISI 52100.

Wear tests on both of the steels in Table 13.8 (Ref 13.12, 13.13] show that these steels have significantly improved wear resistance over conventional steels, such as AISI 1095, at the same hardness levels of approximately $R_c = 60$. The improved wear produced by the fine arrays of the K_1 and K_2 carbides, combined with the lack of coarse primary carbides and the stainless corrosion resistance, makes the two steels of Table 13.8 among the best choices of steels for most knife applications. See Appendix B, "Stainless Steels for Knifemakers," for a further discussion of the popular stainless steels used to make knives.

Austenitic Stainless Steels

As the name implies, austenitic stainless steels are fully austenite at room temperature. The austenite is stabilized at room temperature by the addition of fairly large amounts of nickel to the steels. The steels are considered ternary alloys of Fe-Cr-Ni. Just as with the ferritic stainless steels, carbon is present in these steels at very low levels, and the corrosion resistance is enhanced by reducing the %C in them. The average compositions of the common austenitic stainless steels are presented in Table 13.11. As the %Cr increases, so does the %Ni. Both nickel and austenite are face-centered cubic structures, and, as may be expected, the addition of nickel tends to stabilize austenite to lower and lower temperatures.

The reason why the nickel content must increase as the %Cr increases can be seen by studying the 1100 °C (2010 °F) isothermal section of the Fe-Cr-Ni phase diagram shown in Fig. 13.18. To understand how compositions are plotted in this diagram, consider the 304 stainless composition that is shown in the diagram. The %Ni compositions run across the bottom of the diagram, and, because 304 contains 10% Ni, its composition must lie on the line going up from 10% Ni. However, two lines go up from 10% Ni on the bottom axis. The nickel lines are those going up toward the right, as shown by the arrows next to the legend on the bottom axis. (The other lines going up to the left are iron lines coming in from the left axis, as shown by the arrows on that axis.) Similarly, the chromium lines are the horizontal lines as shown by the arrows on the right axis next to its legend. Therefore, the composition of the 304 alloy is located at the intersection of the horizontal 19% Cr line and the 10% Ni line projecting up and to the right from the bottom axis. Alloy compositions that are 100% austenite (γ) at 1100 °C (2010 °F) must lie in the lower shaded region labeled γ in Fig. 13.18. Thus, the diagram requires that if %Cr is increased in the 304 alloy, it is necessary to also increase %Ni in order to remain in the 100% austenite region.

In order to understand how the phases may change on cooling, it is helpful to look at vertical sections through the ternary phase diagram. Figure 13.19 is the vertical section made along the line in Fig. 13.18 labeled "70% Fe vertical section." This vertical section is a cut made

Table 13.10 Hardness of some carbides

Phase	Knoop hardness	R_c
Martensite	846	65
Cementite (Fe_3C)	1150	70
K_1 carbide ($M_{23}C_6$)	. . .	73
K_2 carbide (M_7C_3)	1820	79

Source: Ref 13.14

Table 13.11 Austenitic stainless steels

AISI No.	%Cr	%Ni	Max %C
302	18	9	0.15
304	19	9.3	0.08
304L	19	10	0.03
308	20	11	0.08
309	23	13.5	0.20
310	25	20.5	0.25

Note: All contain approximately 2% Mn, 1% Si

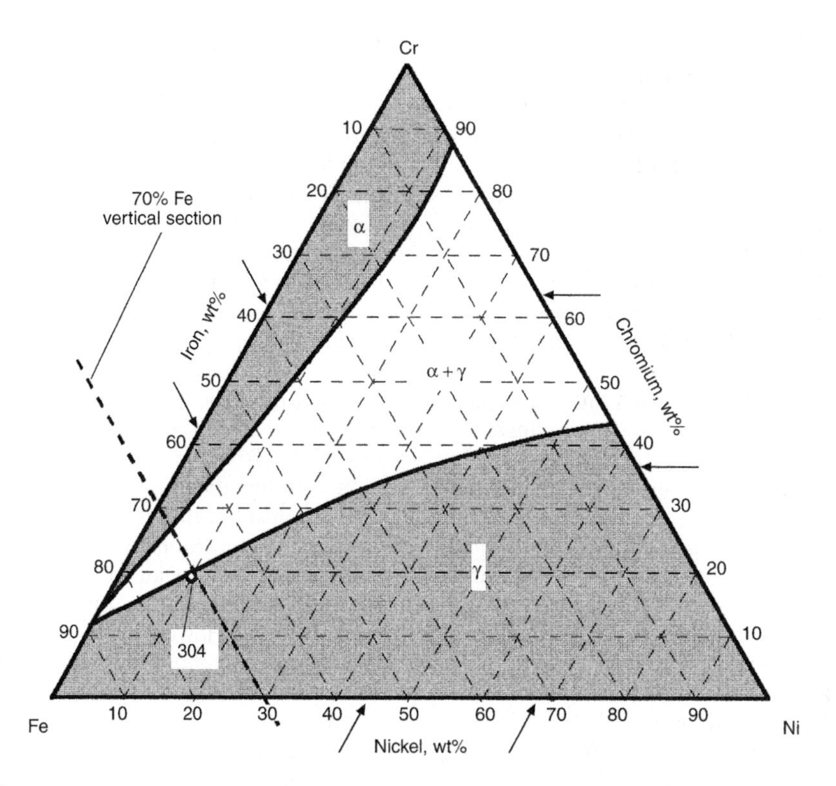

Fig. 13.18 Isothermal section of Fe-Cr-Ni phase diagram at 1100 °C (2010 °F). Source: Ref 13.4

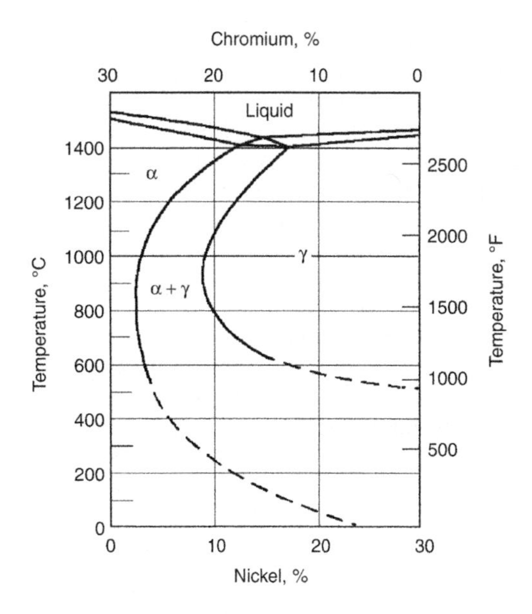

Fig. 13.19 Vertical section of Fe-Cr-Ni phase diagram at 70% Fe. Source: Ref 13.15

through the ternary diagram along alloy compositions that all contain 70% Fe, which means that the sum of the %Cr + %Ni compositions must equal 30%.

The 70% Fe vertical section (Fig. 13.19) shows that an alloy of 10% Ni + 20% Cr (close to 304) will be 100% austenite at 1100 °C (2010 °F), but on cooling below approximately 790 °C (1450 °F), it should become a mixture of austenite plus ferrite (γ + α). Notice that the lines below approximately 600 °C (1110 °F) are dashed. This is done to show that the austenite will remain stable below these temperatures without the formation of ferrite, even though this vertical section predicts that ferrite will form in the austenite. The dashed curves are used because it has been found that if the steel is cooled to a much lower temperature, or if it is mechanically deformed, particles of martensite will form in the austenite. Hence, the room-temperature austenite is said to be metastable.

Phase diagrams do not show where the metastable phases form. However, the metastable phase diagram for the austenitic stainless steels has been determined and is shown in Fig. 13.20. The regions on this diagram labeled A_M map the composition of alloys that form metastable austenite on rapid cooling to room temperature from approximately 1100 °C (2010 °F), meaning that particles of martensite may form in the austenite on cold quenching or on severe mechanical deformation at low temperatures. The common austenitic stainless steels of Table 13.11 are plotted on this diagram; all of the austenitic stainless steels lie in the metastable region at room temperature. The fact that the austenitic stainless steels are actually metastable alloys at room temperature is not important from a practical point of view, because, as long as severe cooling is avoided, they remain austenitic at room temperature and elevated temperatures with no problems of ferrite formation.

The austenitic stainless steels are generally considered to have the best corrosion resistance of the three types of stainless steels. It is customary to partition corrosion into three types, as shown in Table 13.12. As illustrated in the rightmost column of the table, corrosion resistance is enhanced by three main things: (1) decreasing the %C in the steel, (2) increasing the %Cr in the steel, and (3) adding a small amount of molybdenum to the steel. The Table 13.11 data show that the AISI 308, 309, and 310 series of austenitics employ increased %Cr to achieve improved corrosion resistance.

Table 13.13 presents some additional popular austenitic stainless steels with approximately

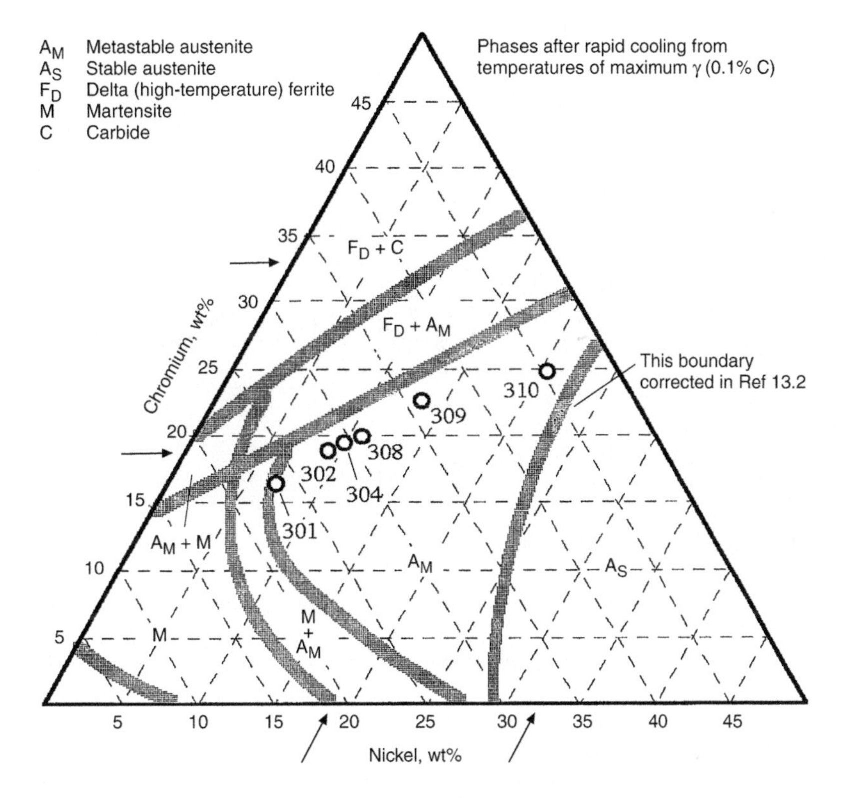

Fig. 13.20 Metastable phase diagram for austenitic stainless steels quenched from temperatures near 1100 °C (2010 °F) (the temperature of the isothermal section in Fig. 13.18). Source: Ref 13.3

Table 13.12 The three main types of corrosion in stainless steels

Type of corrosion	Description	To avoid(a)
Intergranular	As previously discussed, this type of corrosion results from the precipitation of the K_1 carbide, usually on grain boundaries, of either ferrite or austenite.	%C less than approximately 0.02
Pitting (crevice)	Small pits develop holes in the passivating film, which set up what is called a galvanic cell, producing corrosion.	%Cr greater than 23–24 %Mo greater than 2
Stress-corrosion cracking	Localized points of corrosion allow stresses initially unable to crack the steel to concentrate sufficiently to now do so. Details of the mechanism are complex and not well understood. The presence of the chlorine ion (a component of the salt in sea water) makes this type of corrosion a problem in salt waters.	%Cr greater than 20 %Mo greater than 1

(a) Source: Ref 13.5, Chap. 15

Table 13.13 Additional austenitic stainless steels

AISI No.	%Cr	%Ni	Max %C	Other
316	17	12	0.08	2.5 Mo
316L	17	12	0.03	2.5 Mo
321	18	10.5	0.08	Ti(5 × C_{min})
347	18	11	0.08	Nb(Cb)(8 × C_{min})

Note: All contain approximately 2% Mn, 1% Si

18% Cr that are modified to enhance corrosion resistance. The popular AISI 316 steel has molybdenum added, and 316L has the molybdenum addition plus a reduced %C to improve resistance to intergranular corrosion. The 321 and 347 alloys employ a small addition of a scavenger element to reduce intergranular corrosion. Both titanium and niobium (columbium) are very strong carbide-forming elements. When they are present in austenite, they will form carbides with carbon that is dissolved in the austenite. Essentially, they scavenge the carbon from the austenite, leaving less of it to form the K_1 carbide. Notice that the amount of the scavenger element added is related to the minimum amount of carbon in the steel. For example, in 321, the titanium level is five times the minimum carbon level.

The resistance to intergranular corrosion will be reduced significantly by heat treatments that cause the K_1 carbide to precipitate. Figure 13.21 is a type of IT diagram that maps the temperature-time conditions needed to cause the K_1 carbide to start to precipitate on the austenite grain boundaries. The diagram also contains the start curve for the intergranular corrosion that is caused by the carbide precipitation. Reference 13.5 has data for AISI 304 stainless that give a temperature-time at the nose of 850 °C (1560 °F) and 20 s for onset of grain-boundary precipitation and 690 °C (1275 °F) and 20 min for onset of corrosion. It also recommends annealing temperatures for the austenitic stainless steels, and, on cooling from these temperatures, the recommendation is to cool at a rate sufficient to drop the temperature from 870 to 425 °C (1600 to 795 °F) in no more than 3 min. Figure 13.21 shows that by following this recommendation, the possibility of promoting intergranular corrosion is avoided.

Figure 13.22 illustrates why it is advisable to maintain the carbon level of the austenitic stainless steels as low as possible. Notice how sensitive the minimum start time for K_1 precipitation is to the %C in the steels. As the %C decreases from 0.08 to 0.02% (a factor of 4), the minimum start time for K_1 precipitation increases from

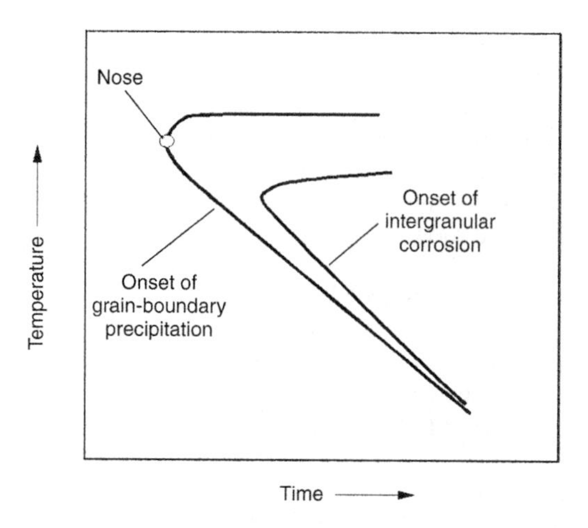

Fig. 13.21 Start times versus temperature for both K_1 precipitation and intergranular corrosion

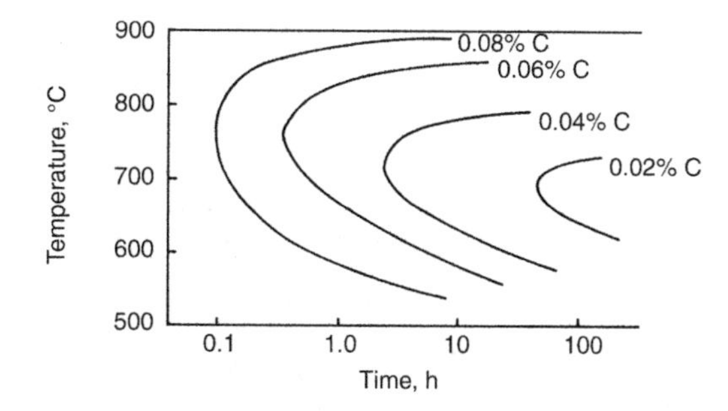

Fig. 13.22 Start time versus temperature for K_1 precipitation in 18Cr/10Ni stainless as %C changes. Source: Ref 13.16

0.1 to approximately 80 h (a factor of 800). These data illustrate why Ref 13.7 recommends care in forging to minimize carburization.

Although the austenitic stainless steels are generally considered to have the best corrosion resistance, the ferritic stainless steels are superior in their resistance to stress-corrosion cracking, particularly the superferritics that are low in nickel (Ref 13.1).

Austenitic stainless steels cannot be hardened except by cold working. During cold working, the strength (and hardness) rises while the ductility falls. Figure 13.20 shows that when the nickel content of 302 is reduced from its value of 9% to a value of 7% in AISI 301, the composition lies in the region where martensite may be expected to form. In 301, martensite does form during the cold-working operation, and this leads to increased strengths and reduced ductilities. Table 13.14 compares the tensile strength and ductility of 301 and 302 after a 40% cold reduction. The tensile strengths have been converted to estimated equivalent hardnesses (Table 5.3) to provide an idea of expected maximum hardnesses in these cold-worked austenitic stainless steels.

Hot ductility is an important concern in forging knives from stainless steels. Figure 13.23 shows the relative hot ductility of several types of stainless steels. Ferritic stainless steels have a nearly constant hot ductility throughout the forging temperature range. The hot ductilities of austenitic grades are much more sensitive to temperature. For example, the hot ductility of AISI 304 increases strongly with temperature and only becomes equal to the ferritic 430 at approximately 1250 °C (2290 °F). The falloff in these curves is dramatic and limits the maximum forging temperatures. The falloff is due to either formation of liquid or loss of grain-boundary cohesion. The significant reduction of hot ductility in AISI 316 and 347 is thought to be caused by inhibition of recrystallization produced by the additions of molybdenum and niobium (columbium), respectively, in these two steels (Ref 13.16).

Table 13.14 Mechanical properties after 40% cold reduction

Steel	Tensile strength		Elongation, %	Equivalent hardness, R_c
	MPa	ksi		
301	1324	192	5	42
302	1103	160	15	36

Source: Ref 13.2

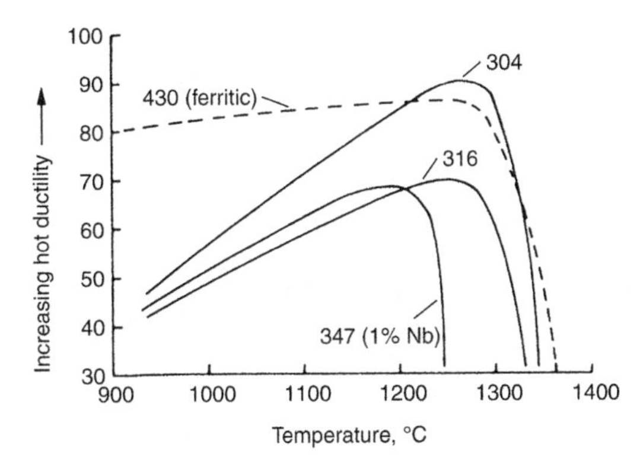

Fig. 13.23 Hot ductility of several stainless steels. Source: Ref 13.16

Precipitation-Hardening Stainless Steels

Recall that secondary hardening during the tempering of alloy steels occurs due to the precipitation of small carbide particles (Fig. 10.10). Precipitation-hardening stainless steels (PHSS) achieve the same type of hardening increase on heating to modest temperatures, but here it is due to precipitation of compounds that are not carbides. For example, when 17-7 PH stainless steel (Table 13.15) is heated to approximately 540 °C (1000 °F), the aluminum added to this alloy forms compounds with the nickel atoms of the stainless, such as Ni_3Al or NiAl.

There are three basic types of these steels, called martensitic, semiaustenitic, and austenitic. An example of each type is shown in Table 13.15. Some of these steels have been given an AISI number, which is seldom used because it has become customary to maintain the trade name used by the company that developed the steel. For example, Ref 13.7 describes the recommended heat treat practice for these steels without specifying the AISI numbers. Similar to a quench and temper heat treatment, these steels also are heat treated by a two-step process, with the final step being the precipitation anneal.

The steel type is related to the matrix that is present when the final precipitation-hardening treatment is carried out. For example, in the austenitic A-286 steel the precipitation occurs when the steel is fully austenite. The chromium-nickel compositions of the three steels in Table 13.15 are plotted in Fig. 13.24. The austenitic A-286 stainless steel lies in the stable austenite region, and the final step of the heat treatment causes precipitation to occur in an austenite matrix. The diagram shows that the martensitic 17-4 PH steel lies in the composition region where rapid cooling produces martensite, and in this steel, the final precipitation occurs when the steel has a martensite structure.

Table 13.15 Composition of three precipitation-hardening stainless steels

Type	AISI No.	Trade name	%Cr	%Ni	%C	Other(a)
Martensitic	630	17-4 PH Armco	16	4.2	0.04	3.4 Cu, 0.25 Nb
Semiaustenitic	631	17-7 PH Armco	17	7.1	0.07	1.2 Al
Austenitic	600	A-286 Allegheny-Ludlum	14.8	25.3	0.05	Low levels: Mo, Al, Ti, V, B

(a) Low levels of manganese and silicon are also present (Ref 13.5)

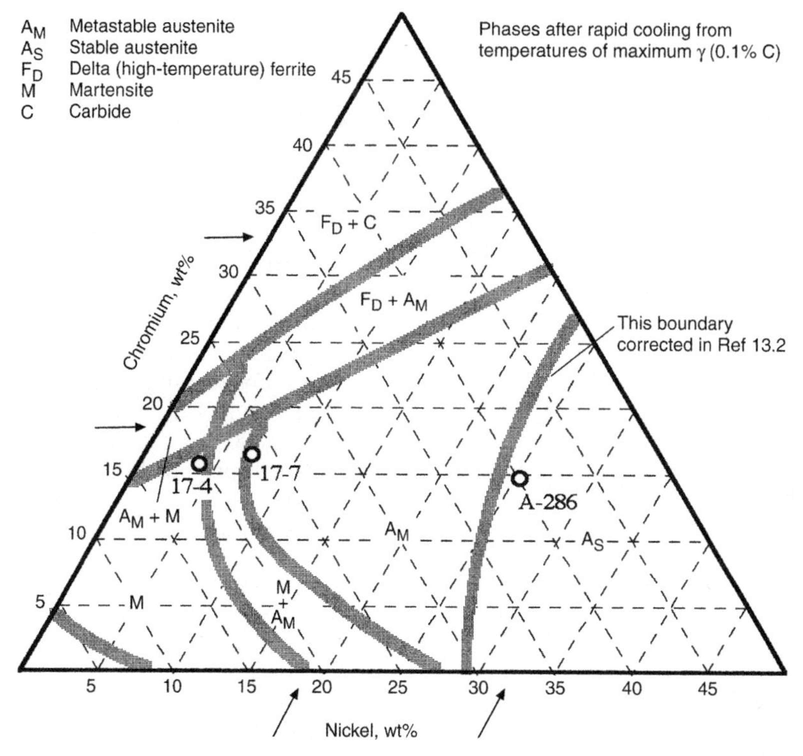

Fig. 13.24 Compositions of the three precipitation-hardening stainless steels in Table 13.15 plotted on the metastable phase diagram shown in Fig. 13.20

The semiaustenitic steels are the most complex of the three. It is helpful to realize that 17-7 is essentially the same steel as AISI 301, with a slightly lower carbon composition plus the addition of aluminum needed to form the precipitate. Figure 13.24 shows that its composition is right on the edge of the metastable austenite region. As described in Ref 13.7, a double heating is employed prior to the final precipitation anneal. The first heat treatment, usually done at the steel mill, produces an austenite matrix with minor amounts of delta ferrite. The second heating forms chromium carbides along the delta ferrite grain boundaries and changes the composition of the austenite matrix, so that on cooling it now becomes a martensite matrix. The final precipitation in the semiaustenitic stainless steels occurs in a predominately martensite matrix.

In the heat treating of these steels, it is common to specify the various steps in the process with the term *condition*. For example, a second step of heating to 790 °C (1455 °F) for 1.5 h and cooling to room temperature within 1 h and holding for 30 min is called condition T. If this is followed by the precipitation anneal of heating to 565 °C (1050 °F) for 1.5 h and air cooling, the final condition is called TH1050, whereas if the precipitation temperature is 510 °C (950 °F), the final condition is called TH950. Reference 13.7 gives information on the hardness achieved for the various conditions of 17-7 PH. Condition TH1050 produces a hardness of R_c = 38 to 44, while condition TH950 produces R_c = 42 to 48.

It is helpful to realize that the martensite structure of these steels is a low-carbon martensite with hardnesses well below that of the martensite of either alloy steels or the martensitic stainless steels. The final hardness ranges of the three steels after the heat treatments specified in Ref 13.7 are given in Table 13.16. These alloys are stainless steels with improved toughnesses at their level of hardness over the martensitic stainless steels.

A helpful hint in working with stainless steels is to remember that the austenitic form of

Table 13.16 Final hardness in three precipitation-hardening stainless steels

Steel	Final hardness, R_c
17-4 PH	42–44
17-7 PH	34–49
A-286	24–45

Source: Ref 13.7

iron is not magnetic and therefore will not be attracted to a magnet, so fully austenitic stainless steels will not be attracted to a magnet. Austenitic stainless steels that contain a small amount of ferrite will be weakly attracted to a magnet, with the attraction force dependent on the volume fraction ferrite in the steel. As seen in Fig. 13.19, forging some of the common austenitic stainless steels at higher temperatures will result in formation of the high-temperature delta ferrite. Use of a magnet on fully austenitic and ferritic steels as standards can help gage levels of residual ferrite.

There is a class of stainless steels not discussed here, called duplex stainless steels, that consist of a mixed ferrite-austenite structure. These steels contain a chromium-nickel composition of 25-5. According to Fig. 13.19, if steel with a duplex composition were forged at approximately 900 °C (1650 °F), it would consist of a mixture of ferrite and austenite. Refer to Ref 13.17 for a brief discussion of these stainless steels as well as all the other stainless steels.

Summary of the Major Ideas in Chapter 13

1. The outstanding feature of stainless steels is their excellent resistance to corrosion. The resistance is due to a thin chrome-containing oxide film on the surface of the steel that inhibits reaction with water solutions and moisture. When an effective corrosion-inhibiting film is present, the steel is said to be *passivated*. For a steel to be considered stainless, it must contain a minimum %Cr level of just under 12.
2. The corrosion data in Fig. 13.1 illustrates the significance of the 12% Cr composition. It also illustrates that stainless steels are only corrosion resistant in oxidizing aqueous solutions, such as nitric acid. In reducing acid solutions, such as sulfuric acid, the passive film is reduced to the point where it is not effective at inhibiting corrosion.
3. Table 13.12 presents the three types of corrosion that occur with stainless steels. It shows the three major compositional changes that lead to improved corrosion resistance: (1) minimizing the %C in the steel, (2) increasing %Cr up to the 20 to 25% range, and (3) adding 1 to 2% Mo to the steels.
4. Stainless steels are generally partitioned into four major types: ferritic, martensitic, austenitic, and precipitation-hardened stainless steels.
5. *Ferritic stainless steels* are Fe-Cr-C alloys. At room temperature, they have a ferrite (body-centered cubic) structure and are magnetic. The room-temperature ferrite contains a fine dispersion of chrome carbide, called the K_1 carbide, which has the chemical formula $(Fe+Cr)_{23}C_6$, often abbreviated to $M_{23}C_6$.

   As shown by the Fig. 13.4 phase diagrams, these steels will contain some austenite at forging temperatures (1040 to 1150 °C, or 1900 to 2100 °F). The high chromium level produces excellent hardenability (Fig. 13.5), so that steels air cooled after forging will contain martensite unless annealed adequately at approximately 700 °C (1300 °F). In addition, two forms of embrittlement can occur in these steels when subjected to prolonged heating at low temperatures, as illustrated in Fig. 13.6.

   The ferritic steels have AISI numbers in the 400 range. As shown in Table 13.1, the popular grades have three chromium contents: 12, 17, and 25% Cr, with increasing corrosion resistance as the %Cr increases. New grades of these steels, not given AISI numbers and called superferritics, have been developed. As shown in Table 13.2, these steels exhibit improved corrosion resistance with small additions of molybdenum and nickel and by reduction of the level of both carbon and nitrogen.
6. Similar to the ferritic stainless steels, the *martensitic stainless steels* are Fe-Cr-C alloys that have been assigned 400 AISI numbers (Table 13.3). Whereas it is usually desirable to have the %C as close to zero as is economically possible in the ferritics, it is necessary to increase %C in the martensitics in order to increase the hardness of the martensite formed in these alloys (Table 13.4). The steels have high hardenability and can be hardened with air cooling but are generally quenched to avoid problems with embrittlement and reduced corrosion resistance due to K_1 carbide precipitation. They need to be tempered to produce adequate toughness for most uses, and their tempering response is similar to alloy steels (Fig. 13.9).

   Similar to plain carbon and alloy steels, it is necessary to increase the %C to approximately 0.6 to obtain martensite with hardnesses in the low Rockwell 60 range.

However, it cannot be assumed that alloys containing 0.6% C or above will have both adequate hardness and corrosion resistance. The %Cr/%C level in the martensite is controlled by the level in the austenite formed at high temperature before quenching. However, the %Cr/%C level in the austenite may not be the same as the overall composition in alloys. For example, an overall composition of 12% Cr/0.6% C results in a 1100 °C (2010 °F) austenite of this same composition and leads to as-quenched R_c values in the low 60s, while maintaining the minimum 12% Cr for adequate passivation. However, if the %Cr is above 12 in order to improve corrosion resistance, the austenite formed at 1100 °C will contain %C below 0.6. Alternatively, if the %C is increased above 0.6, then the austenite formed at 1100 °C will contain more carbon, but the %Cr will fall below 12 and reduce corrosion resistance.

The best compositions for knife steels have %Cr/%C levels at or just above 12% Cr/0.6% C, and Table 13.8 presents two such steels. Optimal heat treatment of these steels is complicated, because it involves the two-phase austenitization discussed in Chapter 11. Austenitization temperature and time control martensite hardness, volume fraction of carbides, and amounts of retained austenite. An example illustrating these factors on heat treatment of Uddeholm AEB-1 stainless steel is described.

7. *Austenitic stainless steels* have significant amounts of nickel added to stabilize the austenite phase at room temperature. These alloys may be regarded as Fe-Ni-Cr alloys with carbon impurities. They have 300 AISI numbers (Table 13.11). As explained with the help of Fig. 13.18 and 13.20, the %Cr/%Ni compositions are balanced to ensure that the face-centered cubic austenite phase is stable at room temperature. Because the austenite is nonmagnetic, these alloys will not be attracted to a magnet unless they contain some residual ferrite. As shown in the Fe-Cr-Ni phase diagram vertical section (Fig. 13.19), an alloy of 10% Ni/20% Cr (close to AISI 304) will contain ferrite when forged above approximately 1100 °C (2010 °F), which illustrates that residual ferrite can be present in these steels, depending on prior heat treatment.

   The 300-series austenitic stainless steels have the best overall corrosion resistance. Corrosion behavior generally is classified into three types: intergranular, pitting (crevice), and stress corrosion. Table 13.12 summarizes the characteristics of each. It also shows how the compositions of %C, %Cr, and %Mo are varied to optimize corrosion resistance. The AISI 302 can be considered as the base steel, %Cr/%Ni = 18/9. Improved corrosion resistance is obtained in 304 and 304L by reducing the %C, in 308, 309, and 310 by increasing %Cr, and in 316 by adding molybdenum. The austenitic stainless steels can be work hardened to reasonably high hardness levels, as illustrated in Table 13.14. Their hot ductility is compared to that of ferritic steels in Fig. 13.23.

8. The *precipitation-hardening stainless steels* are heat treated in a two-step process similar to the quench and tempering of alloy steels, where, instead of tempering, the second heating produces an increased hardness through the precipitation of an alloy compound. Special elements, such as aluminum and copper, are added to these alloys to produce the precipitate. The alloys can be hardened to maximum values of 44 to 49 HRC and provide a stainless steel in this hardness range with improved toughness over the martensitic stainless steels.

   The %Cr/%Ni balance is altered in these steels to produce three types of steels, called martensitic, austenitic, and semiaustenitic. The names given to these steels are usually the trade names of the manufacturers, even though some of them have been assigned AISI numbers. Table 13.15 gives one example steel from each of the three types. The final precipitation reaction occurs in a martensitic matrix for both the martensitic and semiaustenitic types, and it occurs in an austenite matrix in the austenitic type. Figure 13.24 shows how the variation of the composition of these steels produces this effect for the three steels in Table 13.15.

9. A stainless steel is said to be *sensitized* if it has been held in the low-temperature range of approximately 870 to 425 °C (1600 to 795 °F). This treatment forms precipitates of the K_1 carbide on grain boundaries that remove chromium from the surrounding

matrix and leads to intergranular (grain-boundary) corrosion. It occurs in all types of stainless steels, particularly if the %C is high (Fig. 13.21, 13.22). The addition of the scavenging elements titanium or niobium (columbium) to the steel is sometimes employed to remove carbon from the austenite or ferrite as a carbide. Examples include the austenitic steels 321 and 347 (Table 13.13) and the superferritic steels 18-2 and 26-1 (Table 13.2).

10. Retained austenite occurs in steels when the M_s-M_f temperature range extends below room temperature, as is the case for the martensitic stainless steels. The retained austenite can be eliminated by cold quenching. Cold quenching is done in two steps: first to room temperature and then to a cold temperature. A time delay before the cold quench often results in an increase in the amount of retained austenite after the cold quench. The time delay has stabilized some of the retained austenite. *Stabilization* can produce quite subtle problems in martensitic stainless steels; see the discussion of Table 13.9.

REFERENCES

13.1. M.A. Streicher, Stainless Steels: Past, Present and Future, in *Stainless Steel '77*, R.Q. Barr, Ed., Climax Molybdenum Co., 1977, p 1–34.
13.2. *Source Book of Stainless Steels,* American Society for Metals, 1976, p 107
13.3. *Metals Handbook,* American Society for Metals, 1948, p 1261
13.4. *Metals Handbook,* 8th ed., American Society for Metals, 1973
13.5. D. Peckner and I.M. Bernstein, *Handbook of Stainless Steels,* McGraw-Hill Book Co., New York, 1977
13.6. A.E. Nehrenberg and P. Lillys, High Temperature Transformations in Ferritic Stainless Steels Containing 17 to 25% Chromium, *Trans. ASM*, Vol 46, 1954, p 1177–1213.
13.7. *Heat Treater's Guide: Practice and Procedures for Irons and Steels,* 2nd ed., ASM International, 1995
13.8. F.K. Bloom, Effect of Heat Treatment on, and Related Factors on Straight Stainless Steels, *Corrosion*, Vol 9, 1953, p 61
13.9. R.L. Rickett et al., Isothermal Transformation, Hardening and Tempering of 12% Chromium Steel, *Trans. ASM,* Vol 44, 1952, p 164
13.10. M. Hillert, Prediction of Iron-Base Phase Diagrams, *Hardenability Concepts with Applications to Steel,* D.V. Doane and J.S. Kirkaldy, Ed., Metallurgical Society of AIME, 1978, p 49
13.11. A. Omsen and L.G. Liljestrand, Reactions during Heating of a 13.5% Cr Razor Blade Steel, *Scand. J. Metall.*, Vol 1, 1972, p 241–246
13.12. P. Ericson, *Sandvik 12C27-Stainless Steel for Edge Tools,* Steel Research Centre, Sandvik, Sweden, 1981
13.13. "UHB Stainless AEB-L Technical Data," Uddeholm Strip Steel AB, Munkfors Sweden, Nov 1990
13.14. P. Leckie-Ewing, A Study of the Major Carbides in Some High Speed Steels, *Trans. ASM,* Vol 44, 1952, p 348
13.15. J.W. Pugh and J.D. Nisbet, *Trans. Metall. Soc. AIME,* Vol 188, 1950, p 273
13.16. F.B. Pickering, Physical Metallurgy Development of Stainless Steels, *Stainless Steels '84,* The Institute of Metals, London, 1985, p 1–28
13.17. W.F. Smith, *Structure and Properties of Engineering Alloys,* McGraw-Hill, New York, 1981

CHAPTER 14

Tool Steels

ONE OF THE CHARACTERISTICS of our species, *Homo sapiens*, has been the use of tools. Because our ancient ancestors used rock materials for their tools, implements from antiquity have been preserved, and archeologists are able to use the "tool kits" of ancient peoples to characterize differences between our *Homo sapiens* ancestors and other ancient species, such as Neanderthal man. These tool kits consisted of various shapes of rock material, including flints, that were used for such activities as cutting, hammering, punching, and so on.

Modern-day tool steels are essentially the evolved form of these 100,000 year old tool kits, because tool steels are the primary die and cutting materials that modern industry uses to shape other materials by such operations as forging, extrusion, shearing, and so on. Until the early 1800s, the main use of steel was for tools. Nowadays, however, tool steels represent only a small fraction of total steel production, but they continue to play a key role in man's industrial activities, as is evidenced by the fact that many industrial plants have small shops devoted to the tool steels of their machinery.

The first tool steel was developed in England by Robert Mushet in approximately 1868. He accidentally discovered that large additions of tungsten and manganese produced air-hardening steels. (These additions improved the hardenability of the steels to the point where air cooling rates were adequate to avoid formation of pearlite and bainite, thereby forming martensite directly on air cooling.) After systematic research, he developed a steel containing 2% C, 2.5% Mn, and 7% W that remained the tool steel of choice for approximately 25 years. It was called, appropriately, Robert Mushet's Specialty steel. Then, in the 1898 to 1900 time period, two Americans, Taylor and White at Bethlehem Steel, discovered what are now called high-speed tool steels. Their discovery was largely a processing discovery, which showed that by austenitizing high-tungsten steels at much higher temperatures than had previously been used, it was possible to produce very significant secondary hardening during tempering, similar to that shown in Fig. 10. 10.

The new steel was demonstrated at the 1900 World Fair in Paris, where it was shown that tools could operate for long periods at such high cutting speeds that a red heat (550 °C, or 1020 °F) was produced in the steels. Their patented steel contained 1.85% C, 3.8% Cr, and 8% W. Vigorous research in England and the United States quickly established that the %C level was too high and the %W too low for optimal performance. By 1910, the optimal composition had been adjusted to 0.68% C, 4.0% Cr, 18% W, 1.0% V, and 0.4% Mn (Ref 14.1). Taylor and White lost their patent claim in a controversial court case in 1908 in a story that is interesting to read (Ref 14.2). It is surprising how close the 1910 composition compares to one of the best modern-day high-speed tool steels, T1, whose composition is given in Table 14.1.

Tool Steel Classification

Tool steels have been classified into several types, as illustrated in the left column of Table 14.1. The various types are subdivided further according to a letter code followed by a number, as shown in the second column of this table. Chapter 6 of Ref 14.3 presents complete lists of all the tool steels along with an excellent discussion of selection criteria for these steels. There are many different steels for each letter code. For example, the W series listed in Ref 14.3 includes seven steels ranging from W1 to W7. To simplify this presentation, Table 14.1 lists just

Table 14.1 Classification of AISI tool steels showing weight percent composition of one alloy for each AISI type

Type	AISI	C	Mn	Si	Cr	V	W	Mo	Other	Total alloy
Carbon (water hardening)	W1	**1.0**	0.25	0.25	...	...	...	...	...	0.5
Low alloy	L6	0.7	0.5	0.25	0.9	...	...	0.3	1.6 Ni	3.6
Shock resisting	S2	0.47	0.40	**1.05**	...	0.22	...	0.45	...	2.1
Die steels for cold working	O2	0.9	**1.6**	0.25	0.22	**1.8**	...	0.3	...	4.2
	A2	1.0	0.62	0.25	**5.1**	0.32	...	**1.2**	...	7.5
	D2	**1.5**	0.3	0.25	**12**	0.6	...	0.95	...	14.1
Die steels for hot working	H13	**0.39**	0.35	1.0	**5.1**	1.0	...	**1.4**	...	8.9
	H21	**0.35**	0.30	0.3	3.5	0.5	**9**	...	...	13.6
High speed	M2	0.83	0.27	0.32	4.1	**2**	**6.1**	**5**	...	17.8
	T1	0.78	0.25	0.30	4.1	1.1	**18.2**	0.7	...	24.7
	T15	**1.55**	0.27	0.27	4.4	**4.9**	**12.4**	0.5	**5 Co**	27.7

Note: The P steels commonly used for molding plastics have not been included. Source: Ref 14.3

one steel for each letter type. Reference 14.4 presents a complete listing of the compositions of the AISI tool steels, which includes most of the steels in the more complete list of Ref 14.3. (Note: There are many tool steels in use that do not have an AISI classification, and Ref 14.3, p 230, presents a system for their classification. This system, which assigns a three-digit number to each steel, is not widely used.)

In the discussion of steel selection in Ref 14.3, the most important properties of the steels are considered to be wear resistance, toughness, and hot hardness. Table 14.2 ranks these three properties for each of the steels listed in Table 14.1, where a ranking of 1 to 10 is used, with 10 being the best possible value. In general, the main reasons for high values of the three properties listed in Table 14.2 are:

- *High wear resistance*: More carbides
- *High toughness*: Low %C in the steel
- *High hot hardness*: More alloy carbides

The various types of tool steels are briefly discussed by reference to the classification type shown in Tables 14.1 and 14.2.

W Steels. The "W" stands for water hardening. These steels are similar to the plain carbon 1095 steel and have very low hardenability. As shown in Table 14.2, the toughness is improved significantly if the steel is shallow hardened. This means the steel is quenched at a rate that produces martensite only near the surface, allowing the core to remain unhardened and therefore tougher. The poor hot hardness of these steels relative to all other steels is because the carbides in the other steels are alloy carbides, that is, M_3C rather than Fe_3C. The alloy carbides resist coarsening and dissolving at higher temperatures and give the improved hot hardness. The final column of Table 14.1 gives the total of the weight

Table 14.2 Relative values of the three most important properties of tool steels

Type	AISI	Wear resistance	Toughness	Hot hardness
Carbon (water hardening)	W1	4	7 (shallow hardened)	1
	(1095)	4	3 (through hardened)	1
Low alloy	L6	3	6	2
Shock resisting	S2	2	8	2
Die steels for cold working	O2	4	3	3
	A2	6	5	5
	D2	8	2	6
Die steels for hot working	H13	3	9	6
	H21	4	6	8
High speed	M2	7	3	8
	T1	7	3	8
	T15	9	1	9

Source: Ref 14.3

percent alloying elements in each steel. As one moves down the table, the amount of alloying addition generally increases, and therefore, the volume fraction of alloy carbides increases.

L Steels. The "L" stands for low alloy, and these steels have compositions similar to the AISI alloy steels discussed in Chapter 6, "The Low-Alloy AISI Steels." For example, comparing the composition of the L6 steel listed in Table 14.1 to the AISI 4340 steel of Table 6.3, it is seen that L6 is essentially a 4340 steel with %C increased from 0.4 to 0.7.

S Steels. The "S" stands for shock resistance. The high toughness needed for shock load resistance is obtained by reducing the %C in these steels to low levels, which also results in the low wear resistance and hot hardness of the S steels.

O Steels. The "O" stand for oil hardening. The key alloying elements for the various steels are shown in bold print in Table 14.1. For the O steels, manganese and vanadium are mainly responsible for its improved hardenability over the W steels, allowing them generally to be oil quenched to through hardness.

A Steels. The "A" stands for air hardening. Hardenability of these steels is improved to the point that they may be air hardened. The key alloying additions causing hardenability improvement are chromium and molybdenum.

D Steels. These steels are sometimes called high-carbon, high-chromium steels, and the "D" symbol has no obvious meaning, although it is usually associated with the term *direct hardening*. The steels can be air hardened. As illustrated in Table 14.2, the combination of high carbon and high alloy content results in excellent wear resistance, moderately good hot hardness, but poor toughness.

H Steels. The "H" stands for hot hardness. They are usually the steels of choice for the dies of hot extrusion operations. The combination of low carbon and moderate-to-high alloy content gives good toughness and hot hardness but only fair wear resistance.

M and T Steels. The "M" and "T" stand for the molybdenum and tungsten additions in these high-speed tool steels. The carbides of these two elements are stable to quite high temperatures. Therefore, the high level of these alloying elements produces large carbide volume fractions, which gives good wear resistance and hot hardness but poor toughness. The high-carbon T15 steel is included in Tables 14.1 and 14.2 to provide an example of a tool steel designed for maximum wear resistance and hot hardness at the sacrifice of toughness.

The Carbides in Tool Steels

It should be clear from the previous discussion that the carbides present in tool steels play a dominant role in the control of the mechanical properties desired in tool steels. The microstructure of most tool steels following heat treatment is similar to that of 52100 shown in Fig. 6.5. As may be seen from Fig. A2(c) of Ref 14.4 (p 548), the microstructure of A2 tool steel contains arrays of carbides that appear similar to Fig. 6.5 but have larger diameters. As with 52100, it is these arrays of carbides that give the tool steels their excellent wear resistance. The type of carbide present in the tool steels varies with the type of alloying elements in the steel. As may be seen from Table 14.1, there are four major alloying elements present in varying amounts in the tool steels: chromium, vanadium, tungsten, and molybdenum. As discussed earlier with the aid of Fig. 8.8, these elements all form very stable carbides.

Table 14.3 lists the six major types of carbides found in tool steels. As one goes down the table, it is seen that the hardness of the carbide type increases. (These hardness values, given as Rockwell C, are only estimates based on several sources.) Each of them is characterized by a different crystal structure that corresponds to a different chemical formula. The M_3C carbide can be thought of as having a chemical formula of $(Fe + X)_3C$, where X refers to different combinations of manganese and the four major alloying elements, chromium, vanadium, tungsten, and molybdenum. If X contains only iron and manganese, this is the cementite carbide found in plain carbon steels. Table 14.3 shows that in tool steels, the M in M_3C will be mostly iron, manganese, and chromium, with minor amounts of tungsten, molybdenum, and vanadium. The two chromium carbides that are dominant in stainless steels and discussed in Chapter 13, "Stainless Steels." $M_{23}C_6$ (called K_1) and M_7C_3 (called K_2), are also present in many tool steels and have M containing mainly chromium. The remaining three carbides, M_6C, M_2C, and MC, have M containing mainly molybdenum, tungsten, and vanadium.

On heating a carbide-containing steel to high temperature, austenite will form, and the carbides present in the original steel will begin to get smaller in size as the M atoms in the carbide dissolve into the austenite. The more difficult it is for the M atoms to dissolve, the hotter one can heat the steel and retain small carbides in the matrix. Figure 14.1 presents data showing the fraction of several alloying elements that have dissolved from the carbides into the austenite as the temperature of austenitization rises in an M4 high-speed tool steel. This steel contains three main carbides in the annealed condition: $M_{23}C_6$, M_6C, and MC. The figure shows that the predominantly chromium carbide, $M_{23}C_6$, dissolves at relatively low temperatures, while the M_6C and MC carbides, which contain mainly molybdenum, tungsten, and vanadium, will not

Table 14.3 The carbides in tool steels

		Element distribution in M	
Type	HRC	Most	Least
M_3C	70	Fe, Mn, Cr	W, Mo, V
$M_{23}C_6$ (K_1)	73	Cr	Mo, V, W
M_6C	75	Fe, Mo, W	Cr, V, Co
M_7C_3 (K_2)	79	Cr	. . .
M_2C	79	W, Mo	Cr
MC	84	V	W, Mo

dissolve until temperatures are quite high. The secondary hardening, which makes the high-speed tool steels have hot hardness, is due to precipitation of fine alloy carbides on tempering. However, this precipitation process requires that the alloying elements have been dissolved into the austenite during the austenitization step. It is apparent from Fig. 14.1 that quite high austenitization temperatures are required to dissolve the Mo-W-V alloy carbides of high-speed steels. This relationship between alloy content and recommended austenitization temperature is illustrated in Table 14.4, which has been made up using the recommended heat treat practice for the various steels given in Ref. 14.4. It is seen that the austenitization temperatures increase from values typical of plain carbon and low-alloy steels, approximately 845 °C (1550 °F), up to the extremely high temperatures of approximately 1230 °C (2250 °F) for the most highly alloyed steels, the high-speed tool steels.

As shown in Ref 14.3 (p 219), it is possible to characterize the different types of tool steels by the shapes of their hardness-versus-tempering temperature curves. Figure 14.2 presents the generalized curves. It is seen that as the alloying content of the tool steel increases, a well-developed secondary hardening peak is formed, which has

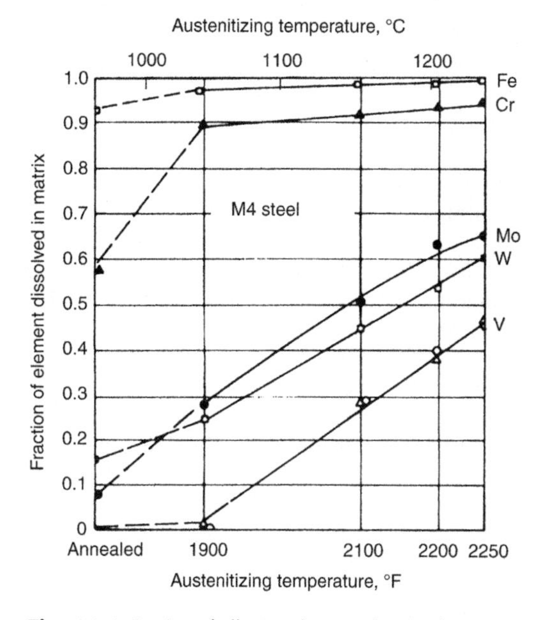

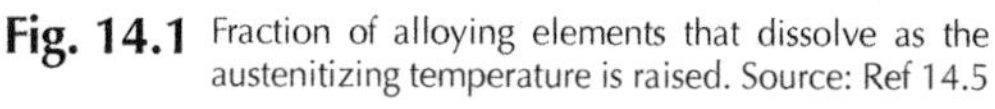

Fig. 14.1 Fraction of alloying elements that dissolve as the austenitizing temperature is raised. Source: Ref 14.5

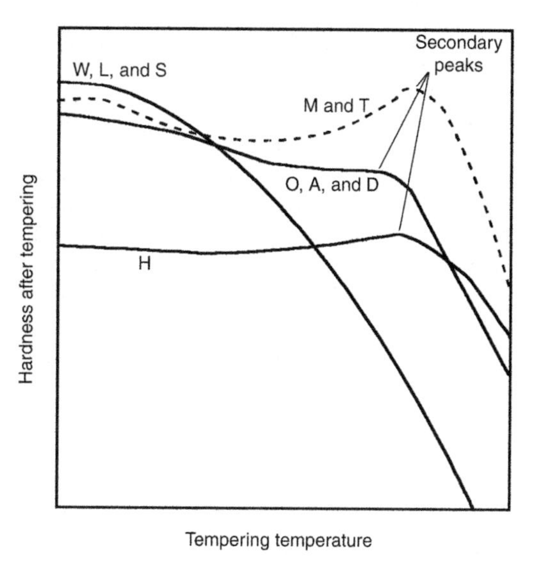

Fig. 14.2 Hardness versus tempering temperature of the various types of tool steels

Table 14.4 Recommended heat treat practice for a sampling of tool steels

		Austenitization temperature		Secondary peak			
					Temperature of max R_c		
Type	AISI	°C	°F	Max R_c	°C	°F	Number of tempers
Carbon (water hardening)	W1	760–845	1400–1553	None	None	None	Single
Low alloy	L6	790–845	1454–1553	None	None	None	Single
Shock resisting	S2	845–900	1553–1652	None	None	None	Single
Die steels for cold working	O2	760–800	1400–1472	57	260	500	Single
	A2	925–980	1697–1796	57	454	850	Double
	D2	980–1025	1796–1877	58–59	482	900	Double
Die steels for hot working	H13	995–1040	1823–1904	52–54	524	975	Triple
	H21	1095–1205	2003–2201	52–56	532	990	Triple
High speed	M2	1190–1230	2174–2246	64–66	543	1010	Double
	T1	1260–1300	2300–2372	65–67	527	980	Double
	T15	1205–1260	2201–2300	67	543	1010	Triple

Source: Ref 14.4

its maximum hardness for the high-speed tool steels. Reference 14.4 presents the hardness versus tempering temperature for most tool steels. Table 14.4 summarizes the temperatures of the secondary hardness peaks and the R_c hardness at the peak for the steels discussed here. When studying the data of Ref 14.4, it will be noticed that small changes occur in the tempering curves depending on both the quench type (air versus oil versus salt) and the austenitizing temperature. The higher values of the hardness ranges listed in Table 14.4 correspond to austenitization at the higher temperatures of the austenitizing ranges that are listed. Notice the much lower overall hardness of the H steels, shown in Fig. 14.2. This results from the low %C in these steels, which is required by the need for high impact strength in their use as hot work die steels for such applications as extrusion.

Special Heat Treatment Effects with Tool Steels

The heat treatment of tool steels involves many subtle effects that are not present in heat treating low-alloy and plain carbon steels. These fall into five general categories: two-phase effects, stabilization of austenite, conditioning of retained austenite, high austenitizing temperatures, and complicated heat treatment requirements.

Two-Phase Effects. Most of these steels are two-phase at their austenitization temperatures and therefore involve the complications previously discussed for this case in Chapters 11 and 13. The carbon composition and the alloy composition of the austenite are controlled by the austenitization temperature. In general, higher austenitization temperatures will increase both the %C and percent alloying elements in the austenite. Increasing %C leads to higher-hardness martensite but also an increase in percent retained austenite, unless subzero cooling is employed. Specific two-phase effects include:

- The austenitization time and temperature control the amount of carbon in the austenite, which in turn controls the as-quenched hardness, mainly by determining the amount of retained austenite in these high-carbon steels. (As an example, see p 623 and 625 in Ref 14.4 for data on T1 tool steel.)
- The austenitization temperature affects the secondary hardening peak. Higher temperatures put more alloying elements into the austenite and generally raise the secondary peak hardness. (Again, see p 625 in Ref 14.4 for T1 data.)

Stabilization of Austenite involves the following:

- Many of the tool steels need to be cold quenched to remove retained austenite. As explained in Chapter 13, "Stainless Steels," it is important to avoid holding the sample at room temperature between the quench to room temperature and immersion into the cold quench medium. Data on the effect of the room-temperature holding time versus percent retained austenite may be found for some of the tool steels in Ref 14.4; for example, p 625 presents data for T1.
- The higher-alloyed tool steels are air hardening. However, austenite stabilization effects produce more retained austenite for air-quenched steels than oil-quenched steels, just as was described for stainless steels in Chapter 13. Hence, the air-hardened steels end up with slightly lower hardness than oil-hardened steels, as is illustrated for D2 by the data on p 562 in Ref 14.4.

Conditioning of Retained Austenite. As discussed in Chapter 10, "Tempering," during stage 2 of the tempering process, the retained austenite decomposes into a mixture of carbides and ferrite. (This mixture is often called bainite, although it forms on up-quenching austenite, rather than the usual case of bainite formation on down-quenching austenite.) In the highly alloyed tool steels, it is found that this stage 2 decomposition of retained austenite into bainite is not complete, and some or all of the remaining retained austenite is said to be conditioned during the first temper. Furthermore, during cooling from the tempering temperature, the conditioned retained austenite will transform to fresh martensite, which, not being tempered, is brittle. (For reasons too complex to go into here, the conditioning has raised the M_s of the conditioned austenite, which then allows martensite formation on cooling.) Thus, in many tool steels, it is often recommended that the steels be given a double or even a triple temper to be sure all fresh martensite has been tempered. The number of temper treatments recommended for the various tool steels listed in Ref 14.4 are given in the rightmost column of Table 14.4.

High Austenitizing Temperatures, particularly for the high-speed tool steels, require careful atmosphere control to avoid surface degradation effects. Salt bath heating is an effective method of minimizing scaling and/or surface decarburization. An advantage of the T steels over the M steels is that they are less sensitive to decarburization. In controlled-atmosphere furnaces, the M steels will decarburize even in a reducing CO atmosphere. For example, in an 11% CO atmosphere at 1200 °C (2200 °F), the M steels decarburize significantly at the surface in just 5 min, whereas the T steels actually carburize at short times and then decarburize at longer times (Ref 14.3, p 713). Interestingly, as-quenched surface hardness can be decreased by both surface decarburization and carburization. In the latter case, the reduced hardness results from increased amounts of retained austenite, while in the former, it results from a reduction of the hardness of the martensite.

Complicated Heat Treatment Requirements. The recommended heat treatment practice for tool steels is much more complicated than that for the plain carbon and low-alloy steels. As evident from the previous discussion, austenitization atmospheres, temperature, and times must be carefully controlled to achieve proper amounts of %C and percent alloying element in the austenite. Control of retained austenite often requires the use of cold-temperature quenching, control of room-temperature hold time in this process, and the use of double or even triple tempering. In addition, for tool steel dies, it is often important to minimize any shape or size changes during the heat treat process. Also, because these steels are often fairly brittle, it is important to avoid any crack formation induced by thermal stresses on heat treating. Therefore, it is usually recommended that the steels be given a stress-relief anneal after rough machining to shape and also a preheat treatment after finishing machining. In the preheat treatment, the tool is held just below the A_1 temperature to minimize temperature gradients and thereby reduce any distortion or possibility of crack formation when the austenite begins to form above the A_1 temperature. Table 14.5 illustrates the complexity of recommended heat treat practice for most of the steels discussed in this chapter and is taken mainly from Ref 14.6. Reference 14.4 presents similar recommended practice, with more details given for each of the various steps.

Summary of the Major Ideas in Chapter 14

1. Tool steels are the modern version of the "tool kits" employed by our ancient ancestors to shape the relatively soft materials, such as wood and animal skins, used in their daily lives. Modern tool steels are used to shape, cut, punch, deform, and so on the much stronger and varied materials used by modern society. They are specialty steels, produced in relatively low volumes, that have various desirable combinations of a triad of desirable properties in tools: wear resistance, toughness, and hot hardness. Tables 14.1 and 14.2 list a selection of common AISI tool steels along with the relative values of the triad of properties, and the accompanying text presents a brief discussion of the various tool steel types.
2. The wear resistance generally desired in tool steels results from the various alloy

Table 14.5 Recommended full heat treat practice

Steel	Stress relief after rough machining, °C (°F)	Preheat after finish machining, °C (°F)	Austenitization temperature, °C (°F)	Quench media(a)	Low-temperature stabilization, °C (°F)	Temper(b)
W1	620 (1150)	565–650 (1050–1200)	760–840 (1400–1550)	W or B	Not required / Not required	To target R_c
S2	650 (1200)	650 (1200)	840–900 (1550–1650)	W or B	−75 to −196 (−105 to −320)	To target R_c
O2	650 (1200)	650 (1200)	790–815 (1460–1500)	O	−75 to −196(c) (−105 to −320)(c)	To target R_c
A2	675 (1250)	600–700 (1110–1290)	925–980 (1700–1800)	A	−75 to −196 (−105 to −320)(c)	To target R_c
D2	675 (1250)	600–700 (1110–1290)	980–1065 (1800–1950)	A	−75 to −196 (−105 to −320)(c)	To target R_c
H13	650 (1200)	815 (1500)	995–1040 (1820–1900)	A	−75 to −196 (−105 to −320)(c)	D or T at 500 °C (930 °F)
M2	750 (1380)	780–840 (1440–1550)	1100–1230 (2010–2250)	A or O	−75 to −196 (−105 to −320)(c)	D or T at 540 °C (1000 °F)
T1	700–750 (1290–1380)	815–870 (1500–1600)	1260–1300 (2300–2370)	A or O	−75 to −196 (−105 to −320)(c)	D or T at 570 °C (1060 °F)

(a) W, water; B, brine, A; air, O, oil. (b) D, double temper; T, triple temper. (c) Recommend prior 150 °C (300 °F) stress-relief treatment. Source: Ref 14.4, 14.6

carbides formed in these steels. The key alloying elements used in tool steels, chromium, molybdenum, tungsten, and vanadium, form the various carbides shown in Table 14.3. This table presents the relative hardness of the carbides. As shown in Fig. 14.2, most of the tool steels produce a strong secondary hardening peak during tempering that plays a key role in enhancing hot hardness. To successfully produce this peak, it is necessary to austenitize the tool steels at temperatures high enough to dissolve the carbide present in the steel as supplied from the mill. Figure 14.1 shows that steels high in molybdenum, tungsten, and vanadium must be austenitized hotter than steels high in chromium, and Table 14.4 presents recommended austenitizing temperatures that reflect this fact.

3. Heat treatment of tool steels is considerably more complicated than heat treatment of plain carbon and low-alloy steels. As illustrated in Table 14.5, possible additional steps involve (1) an initial stress relief after rough machining, (2) a preheat just below the A_1 temperature, (3) cold quenching to avoid retained austenite, and (4) double or triple tempering to avoid retained austenite. The reasons for the need for these various additional steps include two-phase effects, stabilization of austenite, conditioning of retained austenite, and high austenitizing temperatures.

REFERENCES

14.1. O.M. Becker, *High Speed Steels*, McGraw-Hill, New York, 1910
14.2. A.S. Townsend, Alloy Tool Steels and the Development of High Speed Steels, *Trans. Am. Soc. Steel Treat.*, Vol 21, 1933, p 769 (Note: This journal became *Trans. ASM.*)
14.3. G.A. Roberts and R.A. Cary, *Tool Steels*, 4th ed., American Society for Metals, 1980
14.4. *Heat Treater's Guide: Practices and Procedures for Irons and Steels*, ASM International, 1995
14.5. F. Kayser and M. Cohen, Carbides in High Speed Steels—Their Nature and Quantity, *Met. Prog.*, June 1952, p 79. See also Ref 14.3, p 744
14.6. R. Wilson, *Metallurgy and Heat Treatment of Tool Steels*, McGraw-Hill, New York, 1975

CHAPTER 15

Solidification

VIRTUALLY ALL STEEL PRODUCTS have been solidified from the liquid in the first stages of their production. (Even objects made from powders fit this category, because almost all steel powder is produced from molten metal.) Consequently, the solidification process has an influence on the properties of all steel products. This process exerts a strong influence on three important factors of cast metals:

1. Microsegregation of alloying elements, carbides, and inclusions
2. Microstructure (grain size, grain shape, and phases present)
3. Level of porosity in the cast metal

It is common for casting processes to produce severe microsegregation, large grains, and significant porosity, all of which lead to a reduction in mechanical properties. All three of these factors are improved by severe mechanical deformation, such as hot forging or rolling, so that wrought steels generally have improved mechanical properties over cast steels. However, even with wrought steels, mechanical properties can be influenced by the original casting operation, so it is helpful to have an understanding of the solidification process, even for those who work with wrought steels and not steel castings.

As a material solidifies from the liquid, there is a solid front that advances into the liquid. The shape of this solid front, sometimes called the solid-liquid interface, plays a strong role in controlling the three factors mentioned previously. Consider the solidification process occurring on the left in Fig. 15.1. It shows a crucible that has been filled with liquid steel and then allowed to cool until the solid-liquid interface has advanced inward approximately a third of the way to the center. Because heat is being removed from the bottom as well as the vertical walls, the solid grows upward approximately an equal distance from the bottom as it solidifies inward from the

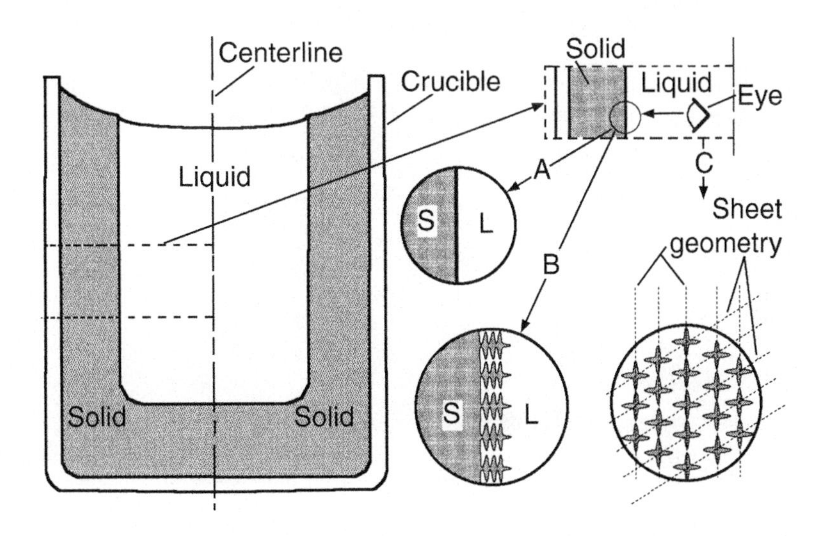

Fig. 15.1 Nature of the solid front during solidification of steels

mold walls. Because the solid is denser than the liquid, a volume contraction occurs upon solidification, causing the height of the solid to decrease as it freezes toward the crucible center. The inset on the right shows a section from the left side of the crucible, where the solid-liquid interface is advancing toward the right during solidification. Intuitively, one would expect the shape of the solid-liquid interface to be planar, as shown by the inset labeled "A." For very pure metals, this generally is correct, but not for steels and virtually all metal alloys. The interface actually consists of arrays of small branched structures that look like pine trees growing into the liquid, with branches coming off at right angles, as shown in the inset labeled "B." Each little treelike structure is called a dendrite (which is the Greek word for tree). If the liquid metal were transparent, the view looking perpendicular to the solid surface (the eye on the diagram) would appear as an array of little trees growing upward, as shown in "C." The view would be similar to the view when flying over a pine plantation and looking down on the treetops. However, the individual dendrites are very small and would require a magnifying glass to be seen.

The spacing of the dendrites depends on how fast the solid-liquid interface is moving, which, in turn, depends on how fast heat is being withdrawn from the liquid metal. Figure 15.2 shows the spacing of the main stalks of the dendrites as *d.* In a large steel casting, heat removal is slow, and dendrite spacing, *d,* can approach 1 mm (1000 μm). For continuous-cast steel, where heat is removed at high rates, the spacing is on the order of 300 μm, and in weld metal, where the small liquid pool allows even faster cooling rates, the spacings can be as small as 100 μm. The diameters of the dendrite main stalks (called primary dendrite arms) are smaller than their spacing, *d,* by approximately 10 times. The diameter of a human hair is approximately 50 μm (0.002 in.), so most dendrites have main stalk diameters of sizes comparable to or smaller than a human hair. These small dendrites play a strong role in controlling the first and third factors listed previously, but only a weak role in the second factor.

Microsegregation

Liquid iron can dissolve more of virtually all types of impurity atoms than solid iron can. Sulfur is an outstanding example. Whereas liquid iron can dissolve 31 wt% S at 1000 °C (1830 °F), solid iron can dissolve only 0.01% S at this temperature. This means that all of the sulfur impurities in a steel will be dissolved in the liquid

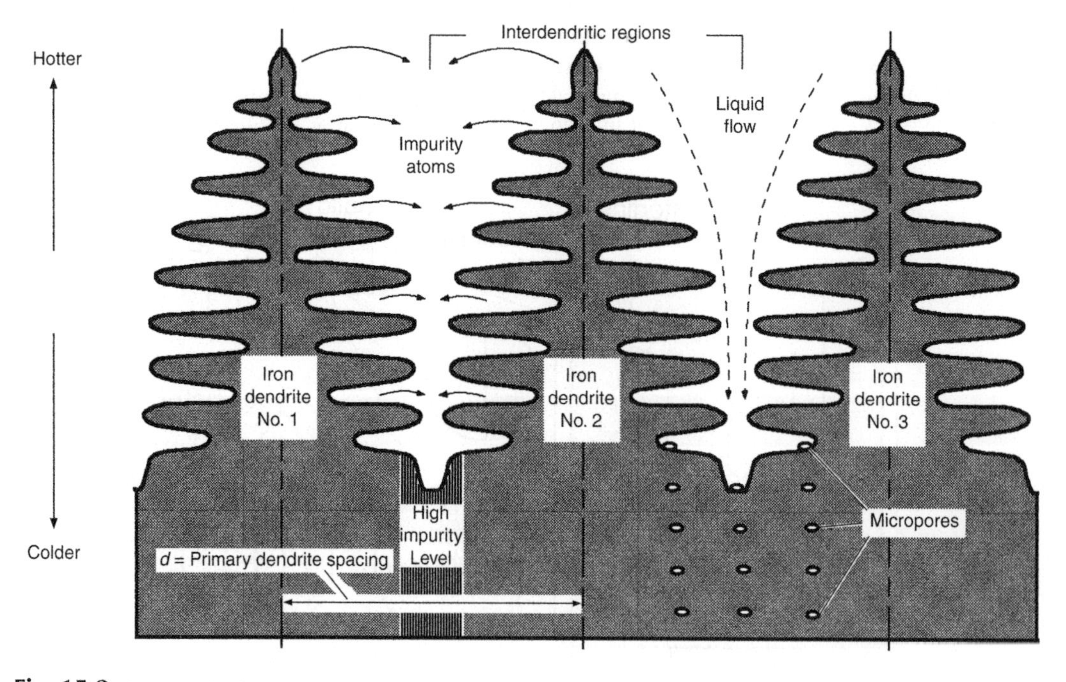

Fig. 15.2 Three iron dendrites growing vertically into the liquid during solidification

prior to solidification, but basically none of the sulfur will be dissolved in the iron after it freezes. The sulfur ends up in the solid in the form of small particles, called inclusions. The sulfide inclusions are a chemical compound of iron sulfide having the formula (MnFe)S. (The manganese present in all steels is contained in these sulfides in amounts that vary with the overall level of manganese in the steel and how fast the steel was cooled.) After solidification, the sulfide particles are distributed nonuniformly throughout the solid in what is called a microsegregated distribution. The dendrites cause the microsegregation, as demonstrated by examining dendrites No. 1 and No. 2 in Fig. 15.2. Because the solid dendrites can hold virtually no sulfur atoms, these atoms must be ejected from the growing dendrite arms into the liquid region between the two dendrites, as shown by the arrows on the diagram. The composition of sulfur in the interdendritic liquid is greater near the base of the dendrites at the bottom of the picture, and eventually, it becomes high enough that particles of the sulfide compound form. Depending on the chemical composition of the steel, the particles form from the liquid as either solid particles or immiscible liquid droplets that freeze to solid at temperatures lower than the temperature at the base of the dendrite arrays. Hence, all of the sulfide particles end up lying in the interdendritic regions, which are shown as the area labeled "high impurity level" in Fig. 15.2. The name *microsegregation* is used because the segregation spacing is on a microscopic scale, as dictated by the micron-level spacing, *d,* of the dendrites.

The noniron atoms deliberately added to the iron to make steels, such as manganese, carbon, and alloying elements, become microsegregated between the dendrites in the as-solidified steel. This phenomenon can lead to formation of particles of compounds in the austenite on cooldown that would not be expected. To understand how this occurs, consider the Fe-1.8%C alloy whose composition is shown in the full iron-carbon phase diagram in Fig. 15.3. The diagram shows that when this alloy is held at 1100 °C (2010 °F), it should consist of single-phase austenite (no particles present) having the composition of 1.8% C. Consider, however, the effects of the microsegregation of carbon between the dendrites. The phase diagram in Fig. 15.3 predicts that this alloy will begin to form solid when the liquid temperature falls to 1400 °C (2550 °F). This means that the tips of the primary dendrites in Fig. 15.2, which are the first-formed solid, will be positioned in the liquid, where the temperature is close to 1400 °C (2550 °F). The phase diagram also predicts that the composition of the first-formed solid will be only approximately 0.7% C. Thus, as the solid tips form, the %C

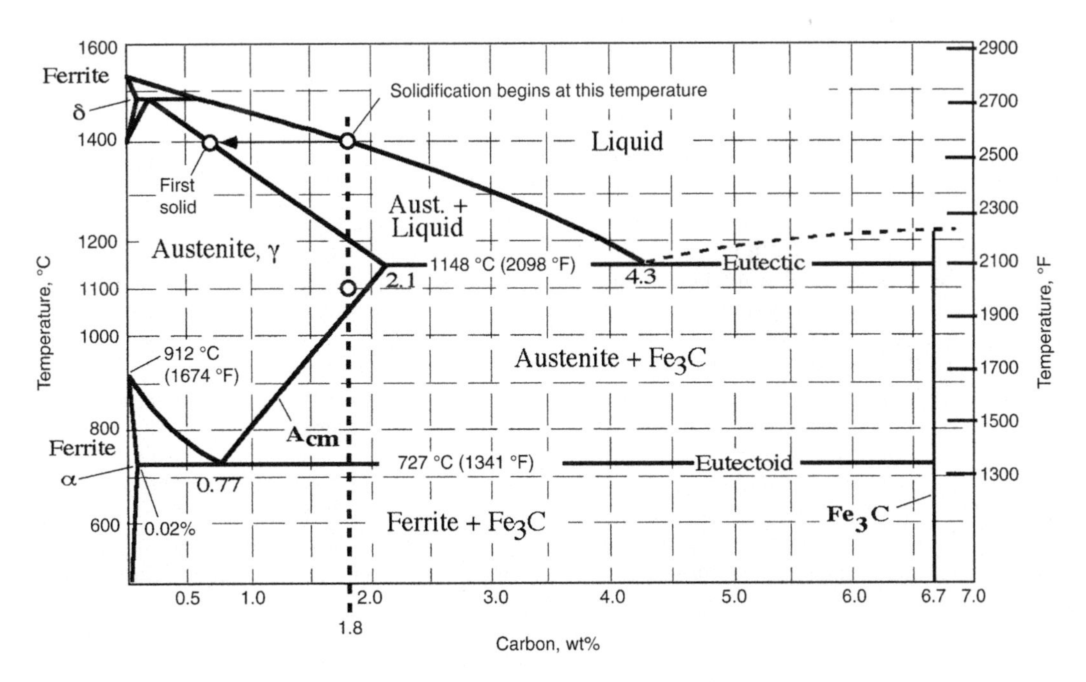

Fig. 15.3 Composition of a Fe-1.8% C alloy on the iron-carbon phase diagram

must drop from 1.8 in the liquid to 0.7 in the solid, and the difference of 1.1% C must be ejected into the liquid in the direction of the arrows shown between the tips of dendrites No. 1 and No. 2 in Fig. 15.2. As one moves toward the base of the dendrites in Fig. 15.2, the %C in the interdendritic regions must increase. Also, the temperature must decrease. When the liquid temperature decreases to 1148 °C (2100 °F), Fig. 15.3 shows that %C in the interdendritic liquid will reach the eutectic composition of 4.3, and all of the remaining liquid will solidify at 1148 °C (2100 °F) as a eutectic mixture of austenite containing cementite particles.

Now, consider two different cases of what may happen as the temperature drops further from 1148 °C (2100 °F) to the 1100 °C (2010 °F) temperature just discussed. The core regions of the dendrites initially form with %C values at approximately 0.7, well below 1.8%. Therefore, it is possible that the carbon in the freshly formed cementite particles, Fe_3C, will diffuse out into these cores, and the cementite particles will dissolve. In this first case, the result would be single-phase austenite at 1100 °C (2010 °F). Also, if the carbon diffusion coefficient were high enough and the steel were cooled slow enough, carbon would migrate by diffusion into the core regions of the austenite dendrites sufficiently to produce a homogeneous austenite having %C = 1.8 throughout at 1100 °C (2010 °F), as predicted by the equilibrium phase diagram in Fig. 15.3. This case is called *equilibrium freezing*.

In the second case, the diffusion rate is not adequate to dissolve all of the cementite particles formed in the solid eutectic at 1148 °C (2100 °F) as the temperature falls from 1148 to 1100 °C (2100 to 2010 °F). Now, the result is a mixture of austenite and Fe_3C particles at 1100 °C (2010 °F). Because the equilibrium phase diagram predicts that a 1.8% C alloy should contain no cementite particles at 1100 °C (2010 °F), this case is called *nonequilibrium freezing*. This second case shows how it is possible to end up with particles between the dendrites after freezing, even though the phase diagram predicts a single-phase (particle-free) austenite should form. In pure iron-carbon alloys, the carbon diffusion coefficient of the tiny carbon atoms is so large that equilibrium freezing usually occurs in cast alloys. Steels, however, contain manganese and alloying elements. The manganese and some of the alloying elements, such as molybdenum and chromium, are carbide formers, and the cementite carbide particles now become $(FeX)_3C$, where X is a combination of manganese, molybdenum, and chromium. To dissolve these carbide particles, the X element atoms as well as the carbon atoms must diffuse. As shown in Chapter 7, "Diffusion—A Mechanism for Atom Migration within a Metal" (Table 7.1), these elements diffuse thousands of times slower than carbon, and it is found that nonequilibrium freezing generally occurs in low-alloy steels as well as in tool and stainless steels.

When working with alloy steels, tool steels, and stainless steels, people often talk about *primary carbides* and *secondary carbides*. Consider the nonequilibrium freezing of the 1.8% C alloy just discussed. The cementite carbides formed out of the liquid between the dendrites are the primary carbides. After the alloy temperature falls below the A_{cm} temperature shown in Fig. 15.3, additional cementite (M_3C in an alloy steel) is expected to form, and this carbide is called secondary carbide. There is often a difference in the distribution of these carbides, because the primary carbides are distributed along the interdendritic regions, and the secondary carbides tend to form along the austenite grain boundaries. Notice that the primary carbides form out of the liquid, and the secondary carbides form out of the solid. There is usually a difference in the size of the primary and secondary carbides. In tool steels and stainless steels, the primary carbides that form from the liquid are generally much larger than the secondary carbides that form in the solid. The large primary carbides formed in tool steels and stainless steels can lead to extreme brittleness of as-solidified steels, because the carbides are brittle and form an interconnected network threading through the dendrite arrays in which they formed. For this reason, it is often necessary to hot forge tool and stainless steels to break up the network carbides and achieve adequate toughness for practical uses. By employing rapid solidification, one can reduce the size of the primary carbides and refine the networks. This is the reason tool steels are sometimes made from powders. Whereas there is no way to rapidly solidify a big ingot of steel, powders formed from small liquid droplets freeze very rapidly, thus reducing the dendrite spacing. After compaction and high-temperature sintering, one ends up with much finer primary carbides.

The phenomenon of banding in steels discussed in Chapters 4 and 5 (Fig. 4.9, 5.5, and 5.6) results from microsegregation of the alloying elements in steel. The mechanism that produces this dramatic effect on microstructure is a little complicated to understand, because one must first have a grasp of the concepts of hardenability discussed in Chapter 8, "Control of Grain Size by Heat Treatment and Forging." Consider a manganese steel such as AISI 1340 (Table 6.1), which is often used for train track rails. After solidification, the manganese is microsegregated into the interdendritic regions, which, as shown at position "C" in Fig. 15.1, tends to be aligned into a sheet geometry. The alignment shown at "C" would occur only within a single grain, with neighboring grains having interdendritic sheets aligned at various angles to those shown in "C." Figure 15.4 illustrates the variation of the dendrite orientation between neighboring grains in an as-cast steel. (As explained subsequently, a special etchant causes original austenite regions to appear black.)

Forging or rolling causes the sheet geometry of all the various grains to align in the plane of the metal flow produced during the deformation. After adequate forging or rolling flow to form a plate geometry, the result is microsegregated sheets of high- and low-manganese content aligned in the plane of the plate. The high-manganese sheets are located at the positions of the prior-interdendritic regions, and the low-manganese sheets are at the locations of the austenite dendrite regions. The banded microstructure forms as a result of the effect of manganese on hardenability. In the high-manganese bands, the ferrite start curve of the isothermal transformation diagram is pushed to longer times than in the low-manganese band regions. Hence, on adequately slow cooling from the

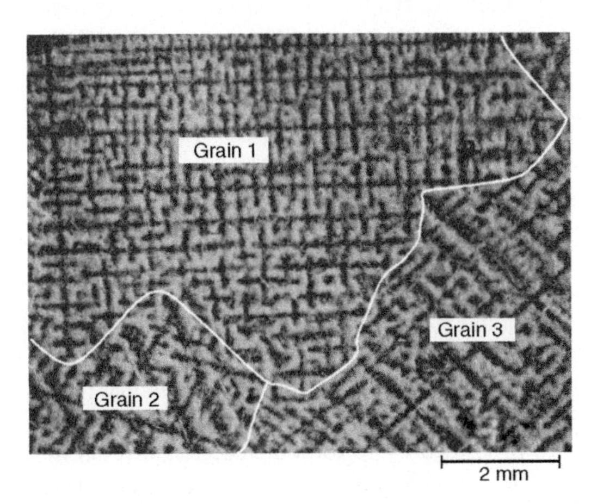

Fig. 15.4 Austenite dendrites in a 1.6% C as-cast steel revealed with Stead's etch

austenite temperature region, the ferrite nucleates first in the low-manganese band regions. Remember, ferrite is essentially pure iron. Therefore, as the ferrite grows out from the low-manganese band regions, it ejects carbon into the high-manganese band regions. Eventually, the %C in those regions becomes high enough that pearlite forms there, and bands of ferrite intersperse with bands of pearlite. This structure is called ferrite-pearlite banding. It is also possible for the microsegregation to form several other types of less common banding, such as ferrite-bainite, ferrite-martensite, pearlite-bainite, and bainite-martensite banding.

Figure 15.4 presents an illustration of the microsegregation caused by dendritic solidification. The steel had a phosphorus impurity level of 0.07%. Similar to the elements discussed previously, the phosphorus atoms segregate into the interdendritic liquid between the dendrites during solidification. Early in the 20th century, John Edward Stead developed a special chemical etch that is sensitive to variation in phosphorus levels even when the steels contain only tiny amounts of phosphorus (Ref 15.1). When a polished steel surface is etched with Stead's etch, regions containing higher concentrations of phosphorus appear white relative to the low-phosphorus regions. Because most of the phosphorus has microsegregated to the interdendritic regions, Stead's etch makes the interdendritic regions white and the dendrite stalks dark, as shown in Fig. 15.4.

The common Damascus steel knives of modern bladesmiths are made using the technique of pattern welding. In this method, alternate sheets of two different types of steel are forge welded together in bundles that are folded back upon themselves several times to produce blades having approximately 200 to 400 layers of the alternating sheets. The surface patterns on these blades result from the fact that one of the layers will etch darker than the other, thereby revealing the pattern in which the layers intersect the blade surface. Pattern-welded steel originated long ago, in approximately 200 A.D. The name Damascus steel originates from a type of steel quite different from these pattern-welded steels that was developed in approximately the same time period. These blades are called genuine Damascus blades, simply because the name originated from them. Genuine Damascus blades are named after the city of Damascus in modern-day Syria, because these Damascus swords were first encountered by Western Europeans in the

Crusades, and it was thought that they were produced in Damascus.

Almost all of the Damascus steel swords found in the arms and armor sections of large modern museums are of the genuine Damascus type. Figure 15.5(a) presents the surface pattern of a museum-quality Damascus sword. Genuine Damascus steels are made from small ingots of fairly high-purity iron containing approximately 1.5 to 1.7% C. The original small steel ingots were manufactured on a large scale in India, and this steel has come to be called wootz steel in the English-speaking world. The method of making these ingots and the mechanism causing the surface pattern on forging were not known until recent times (Ref 15.2). The surface pattern on genuine Damascus blades results from a type of carbide banding, as illustrated by the longitudinal section of the sword of Fig. 15.5(a) shown in Fig. 15.5(b). The micrograph is shown at an original magnification of 90×, and the dark bands consist of arrays of Fe_3C particles that have clustered together into bands during the forging process. The surface pattern results from a type of carbide banding, and recent experiments (Ref 15.3) show that the banding is caused by the microsegregation of low levels of carbide-forming elements, such as vanadium and manganese, that are present in the original wootz ingots. During the repeated heating and cooling cycles of the forging process, the carbides present in the original casting are continually being partially dissolved and reformed. The microsegregated vanadium and/or manganese atoms cause the reforming carbides to slowly become positioned primarily at their locations on the interdendritic planes. Just as with pearlite-ferrite banding discussed previously, the interdendritic planes are rotated into the forging plane during the forging deformation. As seen in Fig. 15.5(b), the alignment of the carbide bands in the final blade is so good that it looks like the carbides may have been aligned by mechanical placement, as is the case for the bands of pattern-welded Damascus blades.

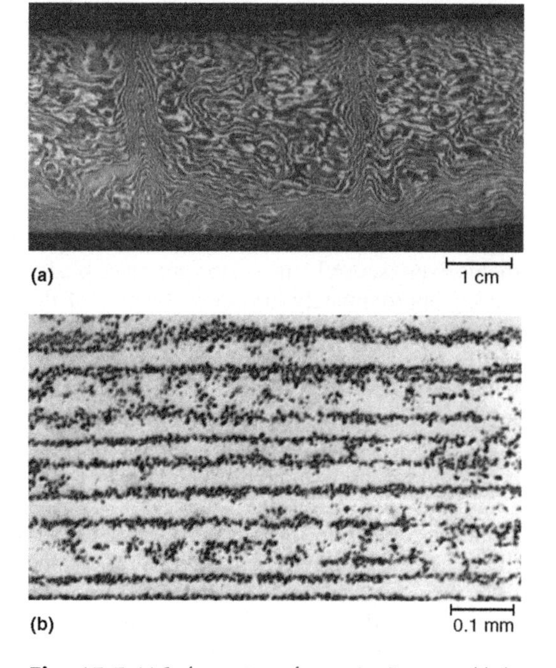

Fig. 15.5 (a) Surface pattern of a genuine Damascus blade, 17th century (Moser blade No. 9). (b) Longitudinal section through the blade. Etched to make cementite particles dark and the pearlite matrix white

Grain Size and Shape

Here, the main interest is to understand the grain shapes that are produced by the solidification process. After the liquid metal is poured into the crucible shown in Fig. 15.1, it cools most rapidly at the walls of the crucible, and it is on the walls that the first solid forms out of the liquid. It is common terminology to say that the solid *nucleates* on the walls. Little particles of solid may nucleate at thousands of sites per square inch of wall, and each of these little particles can grow out into the liquid and become individual grains. The growth front of each little particle contains arrays of aligned dendrites, such as those shown at the bottom of Fig. 15.2, and those particles with dendrite stalks pointing to the right will grow toward the crucible centerline the fastest and crowd out their neighbors. Consequently, near the wall in the region labeled "N_w" in Fig. 15.6, there is a growth competition, and the winning grains grow into long columnar shapes extending inward from the wall. Meanwhile, the liquid ahead of the advancing columnar grains is cooling, and it is possible that small particles of solid can nucleate within this cooling liquid, especially if small foreign particles are present in this liquid. The foreign particles can come from either mechanical addition or by precipitation from the liquid, for example, sulfide, oxide, carbide, or boride compounds. (Also, in some cases, solid particles of the alloy itself are produced by liquid convection currents that break off side arms of the dendrites on the columnar growth front and carry them out into the liquid. When this occurs, the dendrites play an important role in controlling grain size and shape, but otherwise they do not.)

Figure 15.6 also shows small particles of the solid alloy nucleating in the liquid in the region labeled "N_L." These particles can grow out in all directions, and each particle gives rise to a grain called an equiaxed grain, because it has approximately equal axial lengths in all directions. When the long columnar grains hit the equiaxed grains formed in the liquid, the columnar grains stop growing, and the remaining liquid solidifies as equiaxed grains. Hence, it is common to find columnar grains near the crucible wall and equiaxed grains in the center. The morphology is often called the *columnar-to-equiaxed transition* (CET), as shown in Fig. 15.6.

In metal alloys in which the as-solidified grain structure is directly inherited by the room-temperature casting, the presence of the long columnar grains will generally lead to low toughness. As previously discussed, fine grain size is desired to enhance toughness. Depending on the melting point of an alloy, its thermal conductivity, and how fast the crucible wall conducts heat, the radial position of the CET in the crucible will change. If the CET does not occur until near the center of the crucible, the metal is said to be a skin-forming alloy, for the obvious reason that it freezes by forming a solid skin that progressively thickens. If the CET occurs at or very close to the mold wall, the metal is said to be a mushy-forming alloy, because the liquid becomes a mush containing a mixture of small solid particles and liquid that completes solidification by growth of the arrays of small particles. This distinction is important for designing molds in the foundry industry to produce castings with minimized porosity, and the details are too complex to pursue here.

When using industrial foundry molds (sand molds of low thermal conductivity), steels are skin-forming alloys, while cast irons, copper alloys, and aluminum alloys are mushy-forming alloys. However, when making the ingots used for production of wrought alloys, most alloys become more skin-forming due to higher solidification rates resulting from the higher-thermal-conductivity metal ingot molds and continuous casting molds. Because the columnar grains are broken up on hot forging or rolling of these ingots and slabs, their inherent brittle nature is not a problem. However, columnar grains are a common feature of weld metal, brazing metal, and solder metal joints. So, in many brazing alloys and aluminum alloys, tricks are employed to produce equiaxed grains. These tricks are aimed at avoiding columnar grains by using additives that nucleate particles in the molten bead to enhance equiaxed grain formation.

For steels, the formation of columnar grains is not as much of a problem, because the solid phase that forms from the dendrites is not

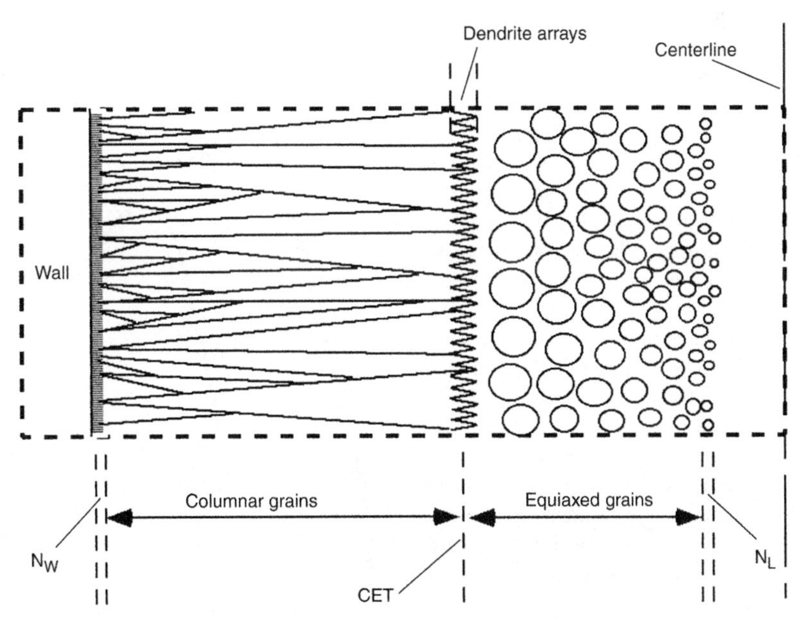

Fig. 15.6 Development of columnar grain structure and equiaxed grain structure during solidification. CET, columnar-to-equiaxed transition

directly inherited in the room-temperature product. Such is not the case for most copper and aluminum alloys. Figure 15.3 illustrates that for steels of composition greater than 0.5% C, the dendritic solid will be the face-centered cubic austenite, which is replaced by a mixture of body-centered cubic ferrite and Fe_3C, usually in the form of ferrite + pearlite, on cooling to room temperatures. Thus, the long columnar austenite grains formed from the liquid will be subdivided into finer ferrite + pearlite grains at room temperature. For steels with %C less than 0.5, the columnar grains will be the high-temperature body-centered cubic form of ferrite, called delta (δ) ferrite, which is twice replaced on cooling. As Fig. 15.3 shows, on cooling to room temperature, delta ferrite will change first to austenite and then to ferrite or a mixture of ferrite+Fe_3C at room temperature. Nevertheless, control of the CET in the continuous casting of steel is an important consideration for plant managers. By preventing the CET from getting too close to the center of the continuous cast billet, both porosity and segregation at the center of the billet are reduced significantly.

Porosity

When a given volume of liquid metal solidifies, the solid formed from it will occupy a smaller volume (solids are more dense than liquids), which leads to the development of porosity during the solidification process. The volume reduction is commonly called the *shrinkage volume*. If this shrinkage volume is filled with more of the liquid metal, there is no porosity, but if it is filled by air or some other gas, porosity is often the result. The reduction of height in the metal shell formed at the crucible wall in Fig. 15.1 is due to shrinkage volume. When the final liquid freezes along the centerline of the ingot, one would expect to observe a shrinkage pipe going down the centerline, as is familiar to anyone who has cast little lead soldiers. Suppose the walls of the crucible were slanted as shown in Fig. 15.7. As the solid metal advances in from the walls, it will form a bridge at the top center before all of the liquid has frozen, and it will trap a large pocket of liquid metal in the ingot center. After this trapped liquid freezes, a large internal pore remains. This type of pore is referred to as a *cavity porosity*. It is a problem in the foundry industry, where special feeders (or risers) are designed into the molds to ensure that the final liquid to freeze is never below the surface of the part being cast.

There is a second type of porosity produced by the solidification process that is called *microporosity*. This microporosity results because of the geometry of the dendritic solidification front, and the mechanism of formation is a little complicated. Consider dendrites No. 2 and No. 3 in Fig. 15.2. As shown by the two downward-slanted dashed arrows between these dendrites, there must be a flow of the liquid metal to fill in the shrinkage volume generated as the dendrites solidify. In order to produce this liquid flow, the pressure in the liquid at the dendrite tips must be higher than at the dendrite base. Because the pressure at the tips will be close to room pressure, this means that the liquid at the dendrite base must be below room pressure. Therefore, the interdendritic liquid at the base of the dendrites is under a partial vacuum.

To better understand what this implies, consider the following experiment. A low-melting-point metal such as lead or tin is heated in a small glass beaker. When the metal is melted, the top of the beaker is attached to a vacuum pump that pulls the air out of the beaker. This reduces the pressure above the liquid metal, creating a vacuum there. What one observes next is that small gas bubbles begin to form on the walls of the glass container. The gas in these bubbles consists of gaseous molecules such as oxygen (O_2), nitrogen (N_2), or maybe

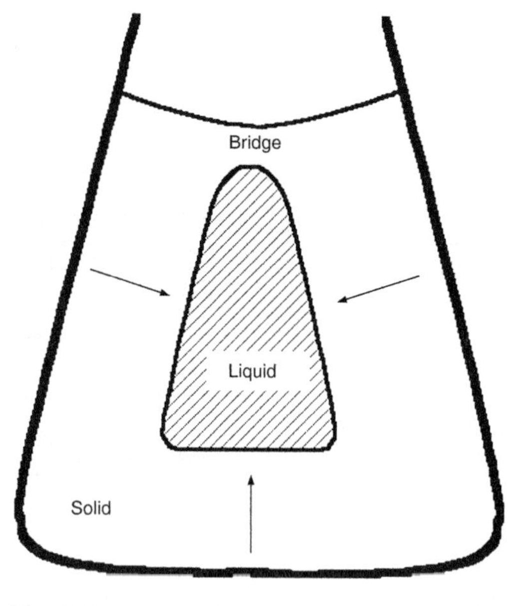

Fig. 15.7 Trapped liquid forms a cavity porosity

carbon monoxide (CO) that had been dissolved in the liquid metal. Therefore, because the interdendritic liquid at the base of the dendrites is under a vacuum, it is possible that small gas bubbles will form there, particularly if the dendrites are long and if the metal contains significant amounts of dissolved gaseous elements. Because the gas bubbles produce quite small pores, this form of porosity is called microporosity. It is too small to see with the eye on the polished surface of a metal. Typical diameters are approximately 5 to 10 μm for columnar grains and approximately 25 μm for equiaxed grains. They are easily seen in an optical microscope.

The porosity in as-solidified metals can be removed by mechanical deformation, as long as the pore walls have not been oxidized. In general, micropores will not be oxidized, and cavity pores will only be oxidized if they have an open path to the atmosphere. For example, the cavity formed in Fig. 15.7 will not be exposed to air unless the ingot cracks, whereas the pipe formed at the center of an unbridged ingot will oxidize. Most porosity formed by the solidification process can be removed by deformation sufficient to cause the pore walls to close on themselves. They simply weld together under the deformation forces.

Steel is a rather special case when it comes to porosity formation during solidification. Because the molten metal always contains some level of both oxygen and carbon, it is common to form carbon monoxide (CO) gas as the molten metal cools. This is an age-old problem that steel producers must deal with. The evolution of CO gas during steel solidification is generally managed by controlling the level of oxygen dissolved in the liquid steel. Steel producers control the level by adding elements to the liquid steel that selectively form oxides and reduce the dissolved oxygen level of the liquid steel. The resulting oxide particles end up as small inclusions that are relatively harmless. The common deoxidizers added are aluminum and silicon, which form the alumina (Al_2O_3) and silica (SiO_2) inclusions that collect in the interdendritic liquid of steels. For example, if suppliers refer to a steel as an AK steel, they mean that aluminum was added to "kill" the steel. The added aluminum reduced the oxygen level such that no CO bubbling occurred.

It is common to have a certain level of microporosity in as-solidified steels, which is not a problem because most steels are used in the wrought form, and the porosity has been closed up. One interesting result of the presence of this microporosity is that steel ingots can form with no apparent shrink cavity along their centerline. When the volume of the microporosity equals that of the shrinkage volume, the solid solidifies with the same shape as the liquid. This occurs for the small wootz ingots used to make Damascus steel.

Summary of the Major Ideas in Chapter 15

1. In general, as-cast metals have inferior strength and toughness properties compared to wrought metals, that is, metals prepared by forging, rolling, or other processes that deform the as-cast structure.
2. There are three primary factors of the solidification process that contribute to this difference in mechanical properties:
   a. Microsegregation of alloying elements, carbides, and inclusions
   b. Microstructure (grain size, grain shape, and phases present)
   c. Level of porosity in the cast metal
3. As illustrated in Fig. 15.1, when steel solidifies inward from a container wall, the advancing solid-liquid interface does not have a planar shape, as shown at location "A," but consists of arrays of dendrites, as shown at locations "B" and "C" and in the blowup in Fig. 15.2. The treelike dendrite structures have very small main trunk diameters, on the order of the diameter of a human hair. These dendrite arrays cause the first and third factors listed in summary point 2 and can influence the second factor.
4. *Microsegregation* results because solid steel cannot dissolve nearly as much of the impurity elements or alloying elements as liquid steel can. Hence, these impurities tend to segregate into the liquid between the dendrites. Because the dendrites are spaced on the order of microns (fractions of a millimeter), the segregation is spaced on this small length, which requires a microscope to see. It is called microsegregation.
5. The microsegregation can cause several effects in steels. First, *primary carbides* are not expected to form in steels with less than 2.1% C. Primary carbides are those carbides that form from the liquid, as opposed to *secondary carbides* that form from austenite on cooling below the A_{cm} or A_1 temperatures. Microsegregation can cause primary

carbides to form at compositions well below 2.1% C, particularly in alloy steels.

Second, microsegregation produces the banded structures common in slow-cooled steels and illustrated in Fig. 4.9, 5.5, and 5.6. These structures form by a rather complicated mechanism involving the effect of the microsegregated alloy or impurity elements on hardenability.

Third, the attractive structure formed on the surface of genuine Damascus blades, as opposed to pattern-welded Damascus blades, also results from a banding due to microsegregation.

6. As shown in Fig. 15.6, grains that grow in from the wall have an elongated shape and are called *columnar grains*. It is also possible for grains of solid to nucleate in the liquid ahead of these columnar grains. These grains grow with a spherical shape and are called *equiaxed grains*. Columnar grains are often present in weld and brazing metal and can be a problem there, because columnar grains reduce toughness. They are not as much of a problem in wrought metals, because hot mechanical deformation causes them to be replaced with new sets of smaller recrystallized grains, as explained in Chapter 8, "Control of Grain Size by Heat Treatment and Forging." For as-cast steels, the problem is also reduced, because, on cooling, the original austenite or ferrite dendrites are replaced by new ferrite grains due to the A_1 transformation in steels. Such is not the case in many aluminum and copper alloys, where similar transformations do not occur on cooling.
7. As explained with Fig. 15.6, metal castings often display columnar grains growing in from the mold walls and approximately spherical-shaped (equiaxed) grains in the casting center. The transition point from columnar to equiaxed is called the CET. If this CET occurs near the casting centerline, the casting mode is called skin-forming, and the grains are columnar. If it occurs near the mold wall, the casting mode is called mushy-forming, and the grains are equiaxed. For castings that must retain their as-cast grain structure, the mushy mode is preferred if toughness is important.
8. The solidification process gives rise to two kinds of porosity that result from the fact that a solid formed from a given volume of liquid will occupy a smaller volume.

Cavity porosity results from the final solidifying liquid being fully enclosed in a solid shell, often when the top of the liquid is bridged over by solid, as shown in Fig. 15.7. Cavity pores generally are larger than a millimeter and can be avoided by proper casting design.

Microporosity is caused by problems in feeding the liquid metal into the tiny spaces between freezing dendrites, as explained with Fig. 15.2. This type of porosity is characterized by sizes in the very small 10 to 25 μm diameter range. The density of cavity porosity can be reduced by removing dissolved gas from the liquid metal or by casting under pressure. It is most prominent in mushy-forming alloys. It can be so severe in some aluminum and bronze alloys that as-cast parts cannot easily be made vacuumtight.

Both types of porosity can be removed by hot deformation adequate to cause the walls of the pores to close in on themselves, provided the walls have not oxidized by exposure to air. The unoxidized walls simply weld together in the same fashion that hot deformation welds together the alternating sheets in pattern-welded Damascus blades.

REFERENCES

15.1. G.F. Vandervoort, *Metallography, Principles and Practice*, McGraw-Hill, New York, 1984

15.2. J.D. Verhoeven, The Mystery of Damascus Blades, *Sci. Am.,* Vol 284, 2001, p 62–67

15.3. J.D. Verhoeven, A.H. Pendray, and W.E. Dausch, The Key Role of Impurities in Ancient Damascus Steel Blades, *JOM,* Vol 50, Sept 1998, p 58–64. http://www.tms.org/pubs/journals/JOM/9809/Verhoeven-9809.html (accessed May 23, 2007)

CHAPTER 16

Cast Irons

CAST IRONS DIFFER from high-carbon steels in several aspects, including the following two:

- They cannot be converted to 100% austenite during heat treatment.
- The other phase present in the austenite is generally graphite as opposed to the cementite (Fe_3C) present in steels.

To understand the first point, consider again the solidification of an Fe-18% C alloy, as discussed with the aid of Fig. 15.3 in Chapter 15, "Solidification." It was explained that the interdendritic liquid will increase its %C to the eutectic value of 4.3 at the base of the dendrites, where the temperature drops to 1148 °C (2100 °F), and the cementite carbide (Fe_3C) will form from the liquid as a eutectic mixture of austenite + Fe_3C.* However, on cooling to 1100 °C (2010 °F), this Fe_3C can dissolve into the austenite dendrites so that a single-phase austenite may form at 1100 °C (2010 °F), as shown at the circle in Fig. 15.3. If, however, the %C in the original alloy is raised to values greater than 2.1, it is no longer possible to produce 100% austenite on cooling. Notice that an alloy of 2.5% C, for example, on cooling to 1100 °C (2010 °F) (open circle in Fig. 16.1) must lie in the two-phase region and consist of austenite+Fe_3C. As with cast irons, this alloy cannot be converted to 100% austenite on heating to its melting point, 1148 °C (2100 °F).

The second point says that graphite will form in most cast irons. Graphite is a crystalline form of pure carbon that is quite soft and very easily fractured. Its presence generally leads to a dramatic reduction in toughness in gray cast iron but not in nodular cast iron, as is discussed later. As mentioned in Chapter 10, "Tempering" the iron-carbon phase diagram presented in Chapter 3, "Steel and the Iron-Carbon Phase Diagram," and as Fig. 15.3 is not really an equilibrium diagram. Strictly speaking, the diagram in Fig. 15.3 should be called the Fe-Fe_3C diagram, because it predicts that at high %C values, the Fe_3C compound will form in austenite. There is a second phase diagram, called the iron-graphite diagram, that is, in fact, the equilibrium phase diagram.

Figure 16.1 shows the superposition of both of these diagrams, where the dashed lines represent the iron-graphite diagram. Notice that, whereas the Fe-Fe_3C eutectic occurs at 1148 °C (2100 °F), the iron-graphite eutectic occurs at 1154 °C (2110 °F), 6 °C higher. This means that it is possible to form graphite from iron-carbon liquid at temperatures from 1054 to 1149 °C (2110 to 2100 °F) but not Fe_3C, and therefore, graphite is a more stable phase. Thus, in iron-carbon alloys, Fe_3C is actually a metastable phase. On cooling pure, high-carbon austenite, the graphite phase can form at the higher temperatures (given by the A_{gr} line) than can the Fe_3C cementite phase (compare to the A_{cm} line). Thus, the more stable graphite would be expected to form as a high-carbon steel, such as 1095 or 52100, and is cooled below the A_{gr} and A_{cm} lines. In steels, however, it is rare to form graphite; the metastable Fe_3C virtually always forms, particularly if carbide-forming alloying elements such as chromium are present. The motion of atoms needed at the atomic level to form graphite, either from the liquid or from austenite, is more difficult than is needed to form Fe_3C, and consequently, the metastable

* When liquids of eutectic compositions freeze, the two solids at the ends of the horizontal eutectic line (austenite at 2.11%+Fe_3C here) usually grow side-by-side as a coupled pair from the liquid. If, however, one solid forms first and then the other, it is called a divorced eutectic.

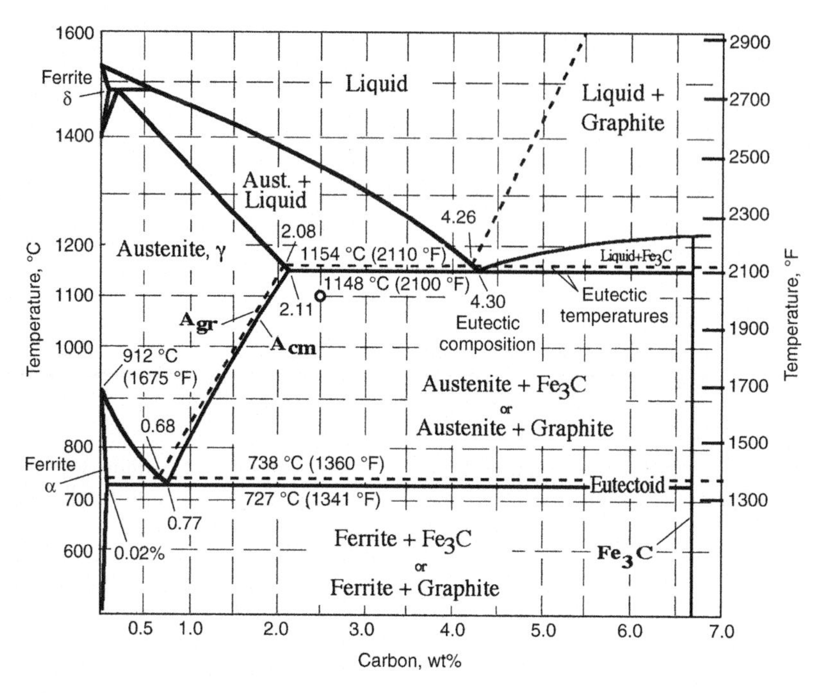

Fig. 16.1 Combined iron-graphite and Fe-Fe_3C phase diagrams. Source: Ref 16.1

Fe_3C forms first and is stable in most industrial applications. It is possible for graphite to form in steels, in which case the steel is said to have graphitized. Graphitization sometimes occurs when steel is held at high temperatures for months or years, as in steam lines. So, graphite formation in steels is not of general practical importance, but it is in the production and heat treatment of cast irons.

Gray and White Cast Irons

Cast irons are ternary alloys of Fe+C+Si. There are four main types of cast irons, and their composition ranges are specified in Table 16.1. Gray and white cast irons are discussed first, followed by malleable and ductile cast irons. In cast irons, a eutectic reaction occurs at the base of the austenite dendrites.

In gray iron, the graphite eutectic occurs, and the result is carbon in mostly the graphite form, while in the white irons, the Fe_3C eutectic occurs, and the result is carbon mostly in the form of Fe_3C, as shown in the far right column of Table 16.1. The addition of silicon allows graphite to form more easily, particularly its formation from the liquid. Perhaps it may be more correct to say that the addition of silicon makes it more difficult to form Fe_3C. To explain, consider the data in Fig. 16.2, which show that silicon additions lower the temperature of the austenite-Fe_3C eutectic while raising that of the austenite-graphite eutectic. To understand how this affects graphite formation in cast irons, it is helpful to review the discussion of the meaning of the A_3 and A_{r_3} temperatures from Chapter 3 (Fig. 3.10). The A_3 line on the phase diagram gives the temperature at which ferrite is expected to precipitate (form) from austenite on cooling. In reality, however, ferrite always forms at a lower temperature, which is termed the A_{r_3} temperature. The data in Fig. 3.10 for a 1018 steel show that ferrite did not form (precipitate) until the austenite had cooled

Table 16.1 The four main cast iron types

Iron	%C	%Si	%Mn	%S	%P	Carbon form
Gray	2.5–4.0	1–3	0.4–1	0.05–0.25	0.05–1.0	Graphite
White	1.8–3.6	0.5–2	0.2–0.8	0.06–0.2	0.06–0.18	Fe_3C
Malleable	2–2.6	1.1–1.6	0.2–0.8	0.06–0.2	0.06–0.18	Graphite
Ductile	3–4	1.8–2.8	0.2–0.9	0.03 max	0.1 max	Graphite

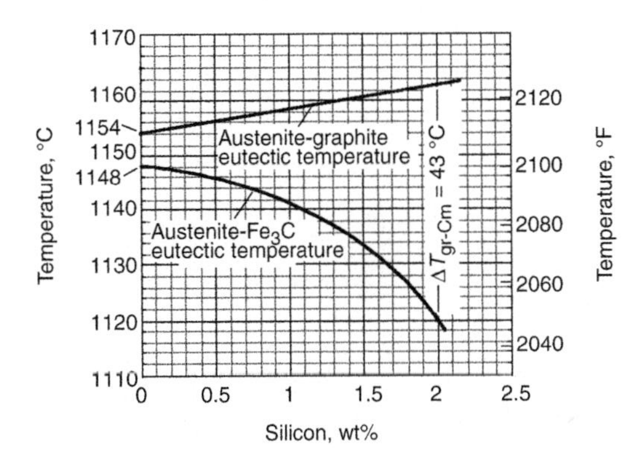

Fig. 16.2 Variation of the graphite and Fe_3C eutectic temperatures as %Si increases from 0 to 2. Source: Ref16.2

62 °C (110 °F) below the A_3 temperature. This temperature difference of 62 °C = (A_3–A_{r_3}) is often called supercooling, because it determines how far the austenite must be cooled below the equilibrium transformation temperature given on the phase diagram by the A_3 line in order to start ferrite formation. By cooling faster, the A_{r_3} temperature decreases, meaning that more supercooling is needed to cause ferrite to form, and if cooling slower, the opposite occurs.

Now, consider the solidification of an Fe-3.5 %C alloy containing 2% Si. Figure 16.2 shows that there are two eutectic temperatures for this alloy: 1163 °C (2125 °F) if graphite forms and 1120 °C (2050 °F) if Fe_3C forms. As solidification proceeds, the liquid in between the dendrites continues to become enriched in carbon as its temperature drops until one of these two eutectics forms. There is a competition between which one will form first. The competition involves three important factors:

1. Both eutectics require some supercooling below their eutectic temperatures shown in Fig. 16.2 before they will form from the interdendritic liquid.
2. Just as with the supercooling required to form ferrite from austenite, the amount of supercooling required increases in both cases as the cooling rate increases.
3. The amount of supercooling required to form the austenite-Fe_3C eutectic is smaller than that required to form the austenite-graphite eutectic.

Suppose the interdendritic liquid is cooled to 1130 °C (2070 °F), and neither eutectic has formed (nucleated). Then, the liquid will be supercooled by 33 °C (from 1163 to 1130 °C for formation of the graphite eutectic, and there will be no supercooling for the Fe_3C eutectic, because the liquid temperature has not fallen below 1120 °C (2050 °F). Hence, the Fe_3C eutectic cannot form. If the 33 °C supercooling is adequate to make the graphite eutectic form, then the result is a gray cast iron.

Now, imagine cooling faster, such that the amount of supercooling required to form the graphite eutectic increases to 53 °C. This means that the graphite eutectic will not form in the interdendritic liquid until it reaches 1110 °C (1163 − 53 °C). However, at 1110 °C the liquid is supercooled 10 °C below the Fe_3C eutectic (from 1120 to 1110 °C). Because little supercooling (less than 10 °C) is needed to form the Fe_3C eutectic, the cast iron will freeze as a white cast iron. This example illustrates why a given cast iron composition will form as a white iron on fast cooling and as a gray iron on slow cooling. It also illustrates why the addition of silicon favors formation of gray irons. The value of $\Delta T_{gr\text{-}Cm}$ in Fig. 16.2 drops from 43 °C with 2% Si to only 6 °C (from 1154 to1148 °C) with no silicon addition. Without silicon, it is much more likely that the interdendritic liquid will supercool below the white eutectic temperature before the graphite eutectic has formed.

Foundrymen do a simple experiment that dramatically demonstrates the effect of cooling rate on white iron formation. The molten metal is poured into a small sand mold that has a cast iron plate on its bottom. The setup is called a chill-test casting because the plate on the bottom causes solidification to occur rapidly at the bottom where the melt contacts the cast iron. As freezing proceeds upward, the solidification rate decreases, and a transition from a white iron to a gray iron can be observed by simply fracturing the sample along a vertical plane. For reasons to be discussed later, the fracture surface of a white iron is white and that of a gray iron is gray. As illustrated in Fig. 16.3, the transition from white to gray is easily seen on the fracture surface, and the height above the chill surface is called the chill depth. Note that in the transition zone, there is a mixture of white and gray; such regions are referred to as mottled.

As the silicon content of the cast iron increases, the chill depth decreases. By adding certain chemical elements, called inoculants, to the liquid iron ladle just before pouring, the chill depth can also be decreased. The inoculants form small solid particles in the liquid that make it easier for the graphite eutectic to form,

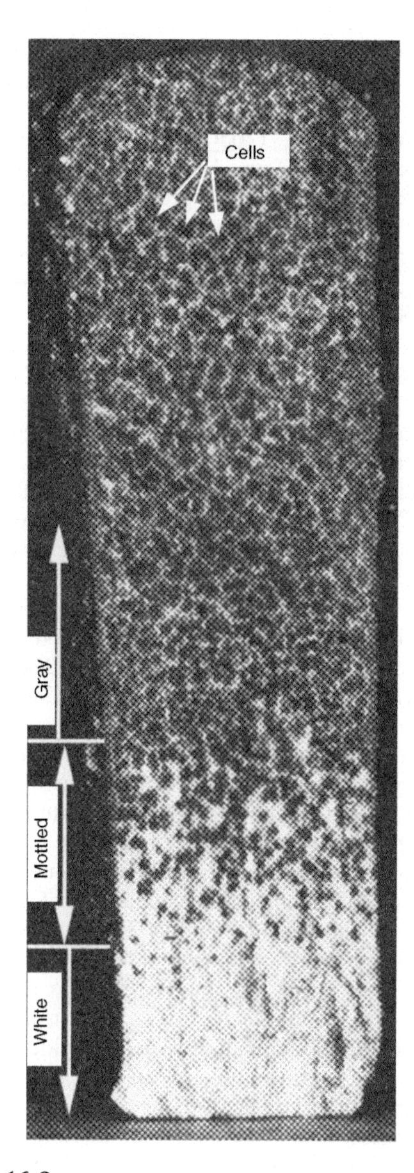

Fig. 16.3 Chill casting showing white, gray, and mottled areas

thereby reducing the supercooling needed for its formation and allowing it to form at faster cooling rates. The most effective inoculating element is calcium, and it is likely, but not certain, that its effectiveness results from formation of small sulfide particles by reaction with the sulfur impurities present in the liquid cast iron.

Figure 16.4 presents a graphical plot of the %C+%Si listed in Table 16.1 for the various types of cast irons. People who make cast irons often refer to the carbon equivalent (CE) of a specific iron. The carbon equivalent is defined as:

$$CE = \%C + \tfrac{1}{3}\%\ Si$$

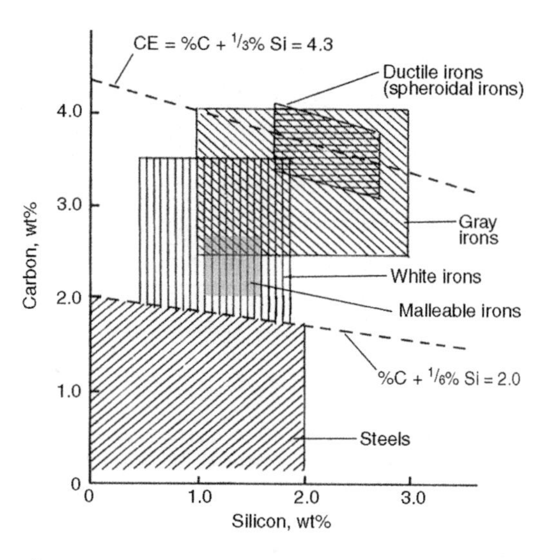

Fig. 16.4 Carbon and silicon composition ranges for various cast irons and silicon-containing steels

Notice that the dashed line at the top of Fig. 16.4 shows the position of any cast iron whose CE = 4.3%. Suppose such an iron had %Si = 0. It would consist of only Fe + C, and its location could be plotted on Fig. 16.1 at %C = 4.3. Figure 16.1 shows that this composition is right at the eutectic composition of Fe-Fe_3C alloys and very close to the eutectic composition of iron-graphite alloys. The dashed line in Fig. 16.4 is a good approximation of how these eutectic compositions change as silicon is added to iron-carbon alloys. Generally, if a cast iron has a composition close to the eutectic, the fraction of austenite dendrites will be very small. This means that as the CE of a cast iron drops more below the 4.3 value, there will be a larger volume fraction of solid forming from the liquid as austenite dendrites. Similarly, as the CE approaches 4.3, there will be a larger fraction solid forming from the liquid as the eutectic mixture of either austenite + graphite in gray irons or austenite + Fe_3C in white irons.

The geometric shapes of the eutectic solids that form in gray and white cast irons are now discussed. Assume that the cast iron has CE values below 4.3, so that the first solid to form consists of austenite dendrites. The top row in Table 16.2 presents a simple schematic summary in the two left columns showing how the microstructure develops during solidification of gray cast irons. The austenite-graphite eutectic solid nucleates (forms) at various locations within the interdendritic liquid. It then grows outward with an overall spherical shape. As

Table 16.2 Schematic summary of microstructure development during solidification of cast irons having C.E. below 4.3

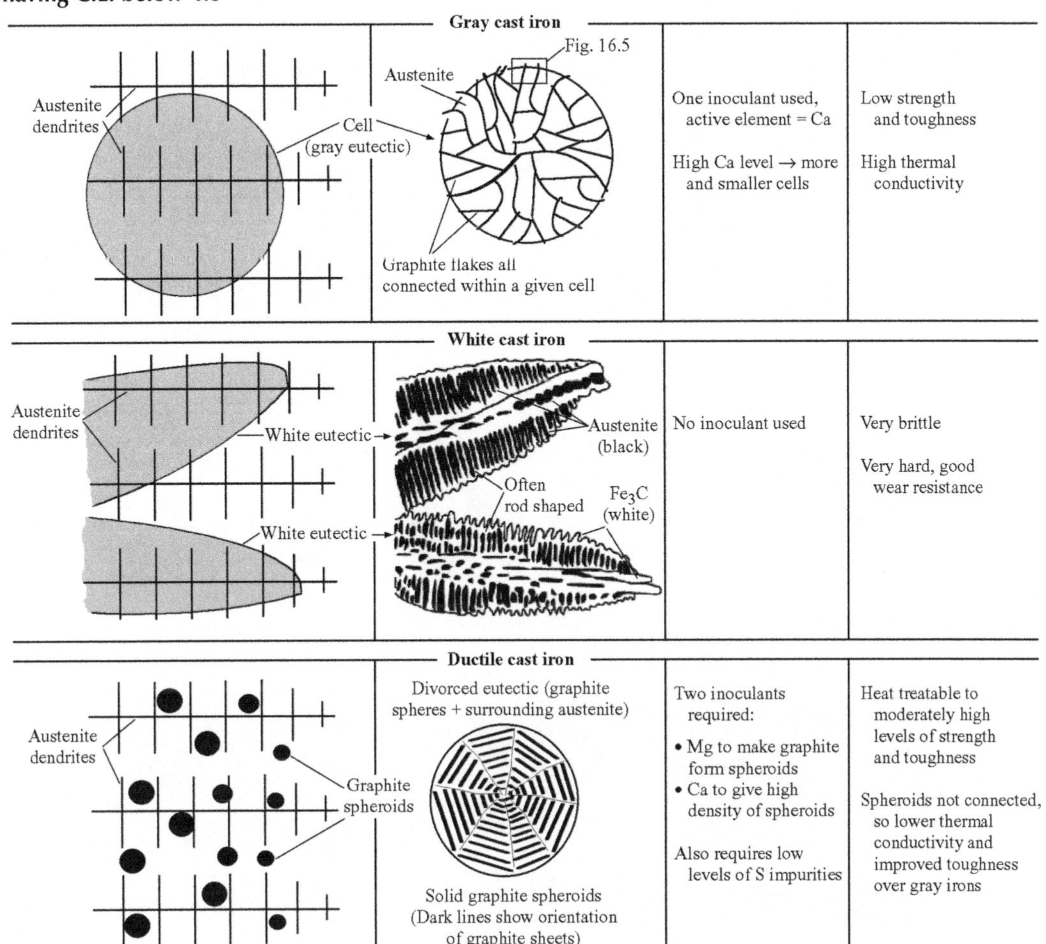

Cast iron	Inoculation	Properties
Gray cast iron	One inoculant used, active element = Ca High Ca level → more and smaller cells	Low strength and toughness High thermal conductivity
White cast iron	No inoculant used	Very brittle Very hard, good wear resistance
Ductile cast iron	Two inoculants required: • Mg to make graphite form spheroids • Ca to give high density of spheroids Also requires low levels of S impurities	Heat treatable to moderately high levels of strength and toughness Spheroids not connected, so lower thermal conductivity and improved toughness over gray irons

shown in the top left column, the spheres grow around and encase the austenite dendrites. They generally grow to sizes larger than the dendrite spacing. Notice that in the mottled region in Fig. 16.3, many of these spherical gray regions, which have become trapped in the white iron that surrounds them, are visible. In fully gray cast irons, the spherical regions have grown outward until they contact one another, and each of these regions is called a cell. Several cells are located by the arrows at the top of Fig. 16.3. At the growth front between the spherical cells and the liquid, both flake-shaped graphite and the austenite grow together into the liquid as a type of coupled eutectic growth. Figure 16.5 shows an actual growth front preserved by quenching, and the flakes are covered by austenite nearly up to their protruding tips. Because this sample was made by quenching, the liquid surrounding the spherical growing cells solidified rapidly as white iron and preserved the solid austenite-graphite eutectic growth front.

The micrograph in Fig. 16.5 illustrates the typical shape of what is called type A graphite in gray cast irons. This is probably the most common shape of graphite in sand cast gray irons. Gray irons that are solidified more rapidly can have finer distributions of graphite, and these have been classified into types B through E by ASTM International. A comparison of structures of all of the types, A to E, can be found in Ref 16.3 (p 820, 874). Figure 16.6 presents SEM micrographs of a type A and a type D gray iron. These are the most common gray irons, with type D forming in thin sections that cool fast, and type A forming in fatter sections that cool slower. The samples were first polished to a mirror finish and then immersed in

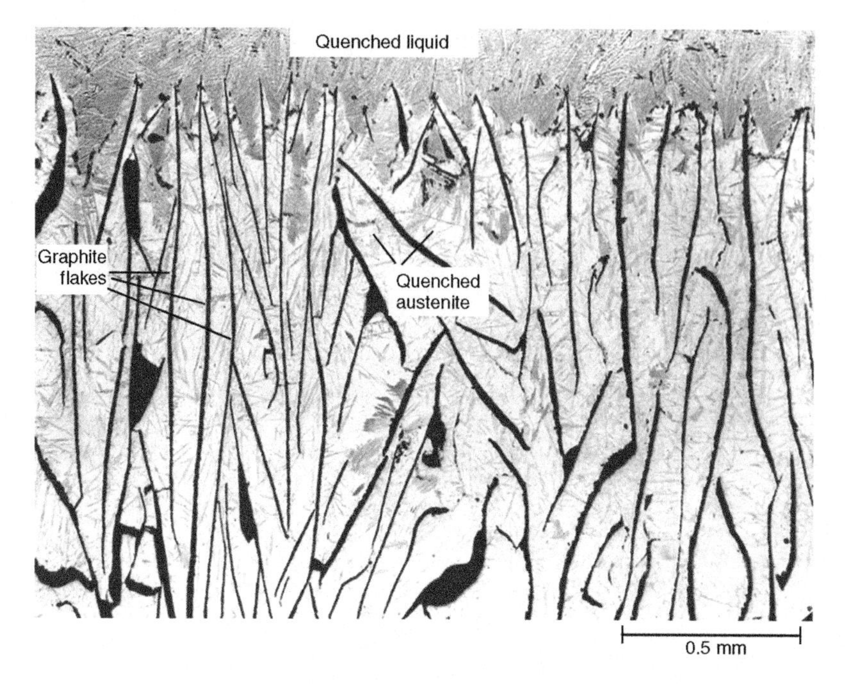

Fig. 16.5 Optical micrograph showing growth front of austenite-graphite eutectic into the liquid at the cell-liquid interface. Original magnification: 40×

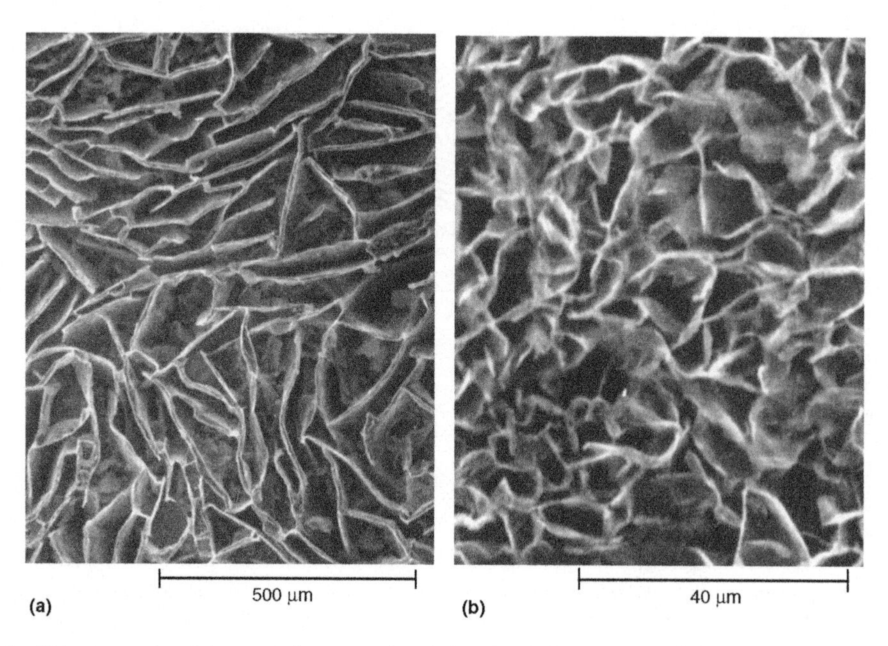

Fig. 16.6 SEM micrographs of deep-etched gray cast iron (3.5% C, 2% Si). (a) Type A (flake). Original magnification: 75×. (b) Type D. Original magnification: 1000×. Source: Ref 16.4

hydrochloric acid, which dissolved away the iron lying between the graphite flakes. So, the micrographs illustrate how the graphite extends below the original polished surface. Notice that the original magnification of the type D graphite in Fig. 16.6(b) is 13 times larger than the type A graphite in Fig. 16.6(a). This illustrates the smaller spacing and the thinner thicknesses of the flakes in type D graphite.

Graphite is a relatively soft material, and its presence in gray iron produces some useful properties in these irons. These properties include good resistance to sliding wear, excellent machinability, and outstanding damping capacity, from 25 to 100 times better than a 1080 steel (Ref 16.5, p 435). An important characteristic of gray irons is that all of the graphite flakes in a given cell are connected, the way the branches on a tree interconnect. This feature is illustrated fairly well in the micrographs of deeply etched irons shown in Fig. 16.6. Because of the special properties of graphite, the interconnectedness of the graphite has a significant effect on the thermal and mechanical properties of gray irons. The thermal conductivity of graphite in the directions along the planes of the flakes is very high, on the order of that in high-purity copper. Consequently, the thermal conductivity of gray iron is higher than that of steels. The low mechanical strength of the graphite flakes, particularly in directions at right angles to the plane of the flakes, has a large effect on the mechanical properties of gray cast irons. Gray irons have limited ductility and tend to fracture under impact loading. The fracture surface has a distinctly gray color because it is basically all graphite, except where it penetrates old austenite dendrites.

These features are illustrated in the SEM pictures of the fracture surfaces of type A graphite shown in Fig. 16.7(a) and (b). Micrograph (a) shows how the relatively large flakes have delaminated into bundles of parallel sheets under the fracture stress. Graphite is similar to mica in that it tends to delaminate into sheets along certain crystal planes, called basal planes, and these planes lie along the plane of the flakes in type A irons. In micrograph (b), the fractured region is located at a remnant iron dendrite. (Note: The two graphite flakes at the center illustrate nicely how these graphite flakes are composed of stacks of sheets, analogous to the structure of mica. It is this sheet structure that causes the strength of the flakes to be so low in directions at right angles to the flakes.) At the dendrite fracture surface in Fig. 16.7(b), the white surface of the iron dendrite is characteristic of a microvoid coalescence fracture surface previously discussed in reference to Fig. 5.12 of Chapter 5, "Mechanical Properties." Figure 16.7(b) also illustrates how the flake graphite has managed to grow around the dendrite branches. The overall fracture surface appears gray to the eye because the fractured flakes protrude above the fractured iron dendrite surfaces, and because the dendrites occupy only a minor volume fraction of the gray iron. Figure 16.7(c) is the fracture surface of a type D iron. It appears essentially similar to the type A fracture surface, except that all of the fractured graphite features reflect the finer graphite spacings and flake thicknesses of these irons, which result from faster solidification.

Gray cast irons are often classified according to an ASTM code that runs from 20 to 60, as shown in the left column in Table 16.3. The

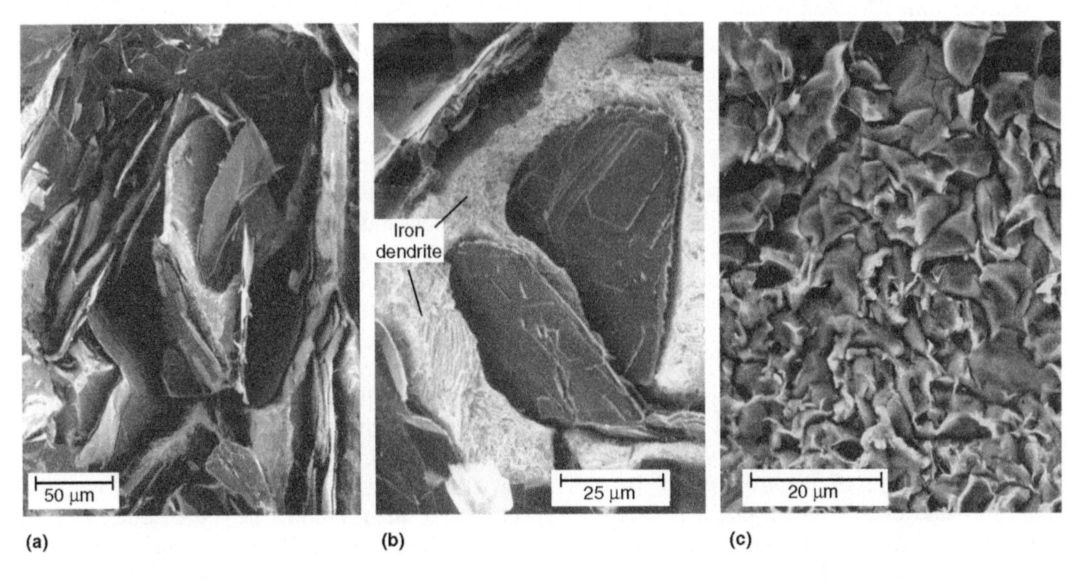

Fig. 16.7 Fracture surfaces of gray cast iron (3.5% C, 2% Si) revealed in SEM micrographs. (a) Type A (flake). Original magnification: 220×. (b) Type A (flake) at a dendrite. Original magnification: 600×. (c) Type D. Original magnification: 1000×. Source: Ref 16.4

Table 16.3 Typical properties of gray cast iron test bars

ASTM (A 48 class)	Tensile strength		Compressive strength		Hardness, Brinell	Hardness (equivalent R_c)
	MPa	ksi(a)	MPa	ksi(a)		
20	152	22	572	83	156	(2)
30	214	31	752	109	210	(16)
40	293	42.5	965	140	235	22
50	362	52.5	1131	164	262	27
60	431	62.5	1296	188	302	30

(a) The "k" stands for 1000s, so 22 ksi = 22,000 lb/in.^{2}. Source: Ref 16.5

code number approximates the tensile strength. The hardness of gray cast irons is virtually always given with the Brinell scale. The large ball indenter used with the Brinell test provides a much better average hardness over the graphite and iron structure than do the much smaller Rockwell B or C indenters. Using Table 5.3, the measured Brinell hardness numbers have been converted to equivalent R_c values in the final column. This is helpful for people who routinely use the C scale as a measure of hardness and find it difficult to evaluate relative hardness using other scales. The data in Table 16.3 illustrate that gray cast irons are much stronger in compression than in tension, which is a direct result of the low strength of the graphite flakes and the interconnectedness of the flakes in gray irons. This is why gray irons are most useful in applications that are subject mainly to compressive stresses and not to tensile stresses. Often, they are used in massive structures such as lath beds, rolling and forging bodies, engine blocks, and so on.

The strength variation of the irons in Table 16.3 results mainly from variation of the microstructure of the iron phases and the volume fraction of iron dendrites. As the carbon equivalent of an iron falls further below the 4.3 value of the eutectic, the volume fraction of iron dendrites increases and raises the strength. Iron microstructures that are mainly pearlitic will be stronger than mainly ferrite microstructures. Gray irons are generally not heat treated, and when they are (Ref 16.3), the heat treatment usually is an anneal, which decreases strength. Therefore, the variation of iron microstructures in the irons in Table 16.3 arises during the cooldown in the casting process.

To understand how the foundryman controls the microstructure of the iron-matrix part of both gray and ductile cast irons, it is useful to understand Fig. 16.8. This figure shows how the phase diagram in Fig. 16.1 for pure iron-carbon alloys is changed when 2% Si is added. It is a vertical section through the ternary Fe-C-Si system at a constant value of 2% Si and is the same type of diagram as those in Fig. 13.4, which were used in discussing the Fe-C-Cr stainless steels. As in Fig. 16.1, both the stable graphite lines and the metastable Fe_3C lines are shown, but in this case, the stable graphite lines are solid and the metastable Fe_3C lines are dashed. Notice that the diagram illustrates the lower temperature of the Fe_3C eutectic relative to the graphite eutectic, previously shown to be $\Delta T_{gr\text{-}Cm}$ = 43 °C in Fig. 16.2.

Consider now an iron of 3.5% C+2% Si. This iron would have a CE of 4.17, and its composition appears in Fig. 16.8 at the vertical 3.5% C line. The development of the microstructure of this alloy as it forms in a mold after pouring can be divided into two stages as cooling occurs along this vertical line. Stage I involves the solidification of the liquid metal as the temperature falls from point 1 (~1250 °C) to point 2 (~1060 °C). During stage I for a gray iron, the structure forms as shown in Table 16.2. At completion, when the temperature has fallen to point 2, the iron will consist of a small volume fraction of austenite dendrites surrounded by large cells of graphite flakes in an austenite matrix. Stage II involves cooling in the mold from temperature 2 to temperatures of 3 (~720 °C) and below. Consider a gray iron. As the temperature falls to the eutectoid temperatures, the austenite component must continuously reduce its carbon composition, as required by the A_{gr} line on the diagram. During this time, the carbon atoms ejected from the austenite are deposited on the graphite flakes, causing them to thicken. When the temperature falls to the eutectoid region of the diagram, just as with the eutectic reaction in the liquid, there is a competition between the formation of graphite versus Fe_3C. Again, it is necessary to supercool more below the graphite eutectoid than below the Fe_3C eutectoid to cause either to form. If the cooling rate is sufficiently fast in the mold, or if low levels of pearlite-stabilizing elements are present, particularly tin or antimony, the graphite eutectoid

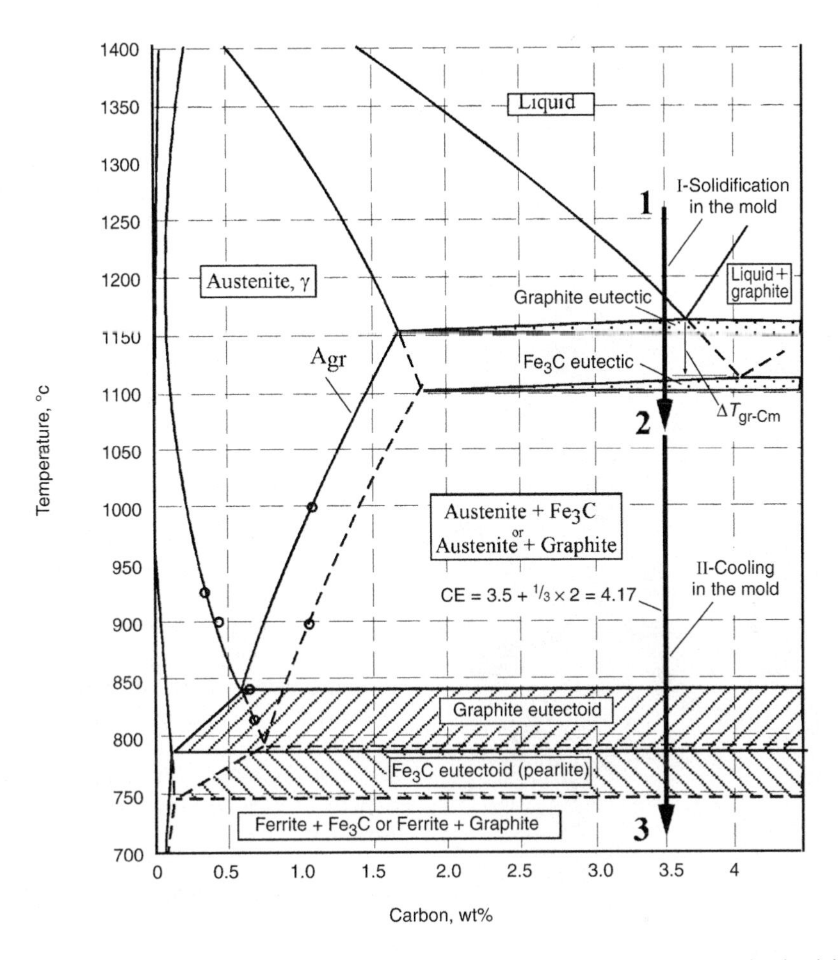

Fig. 16.8 Vertical sections of Fe-C-Si phase diagram at 2% Si. The stable graphite system is represented with solid lines and the metastable Fe_3C system with dashed lines

will not nucleate. In this case, all of the austenite surrounding the graphite flakes will transform to pearlite, similar to what happens in steels, as explained in Chapter 3, "Steel and the Iron-Carbon Phase Diagram." The resulting iron is called a pearlitic gray cast iron, which consists of the flake graphite surrounded by grains that are fully pearlitic.

If, however, the cooling rate is extremely slow, it can happen that no pearlite will form. In this case, the graphite eutectoid reaction occurs in preference to the Fe_3C eutectoid, but it occurs in a divorced manner, and the remaining austenite transforms into rims of ferrite around the flakes. As the ferrite rims grow outward, essentially all of the carbon that was in the austenite diffuses back toward the flakes and is deposited on them. The result is graphite flakes surrounded by ferrite in what is called a ferritic cast iron. Figure 16.9 shows the intermediate case where some of the graphite flakes are surrounded by rims of ferrite, but the remainder of the iron matrix is pearlite. In this case, the divorced eutectoid graphite reaction (called divorced because the graphite and ferrite do not form side-by-side as a coupled pair) occurred first, and then, as the temperature dropped below the pearlite eutectoid, the remaining austenite converted to pearlite (which is the more common coupled eutectoid, with Fe_3C and ferrite plates forming side-by-side as a coupled pair). The greater the fraction of pearlite in a cast iron, the higher its tensile strength. The foundryman controls the fraction of pearlite by controlling the cooling rate in the mold and with alloy additions that stabilize the pearlite. The most common heat treatment of gray irons is annealing, which converts some or all of the pearlite to ferrite and reduces tensile strength and improves machinability. See Ref 16.3 for discussion of the heat treatment of gray irons.

As shown in Fig. 16.4, white cast irons have lower carbon and silicon compositions than gray irons. Graphite does not form in these irons, so a structure consisting of Fe_3C (cementite) and pearlite results. Table 16.2 illustrates that both white cast irons and gray irons having a CE below 4.3 begin solidification from the liquid with the formation of austenite dendrites. In white irons, the interdendritic liquid solidifies as the eutectic mixture of austenite+Fe_3C, which forms between the dendrites and grows out around them, as shown schematically in Table 16.2. The white eutectic has an overall planar shape, with a central spine that consists of Fe_3C (white in Table 16.2) with rods of austenite (black) growing out perpendicular to the planes. Fig. 16.10 shows an optical micrograph of the eutectic component of a white iron. The Fe_3C also appears white in this micrograph, and the austenite, which has transformed to pearlite, is black. Because the Fe_3C is very hard and very brittle, these irons are also hard and brittle. The high hardness produces very good wear resistance. Because cast irons can be made to freeze with the white structure at fast cooling rates, and gray iron at slower cooling rates, as illustrated in Fig. 16.3, it is possible to make items such as large rollers with a very hard white iron surface and a softer gray iron core, with the core having adequate toughness for many industrial applications. Improved toughness may be achieved in white irons by alloying with moderate amounts of nickel and chromium, and a nice discussion of these irons may be found in Ref 16.6. The fracture surface of a white iron runs along the brittle Fe_3C component of the iron. This component fractures by a cleavage mechanism and produces a bright white surface, and thus the name for these irons.

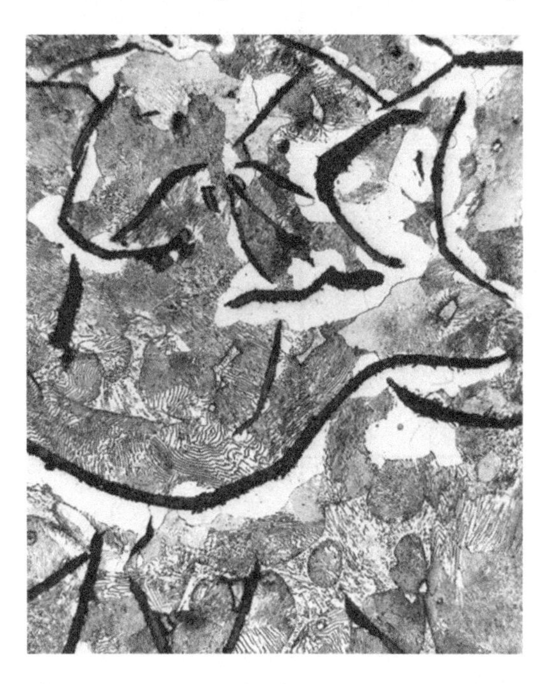

Fig. 16.9 Gray iron with a ferrite-pearlite matrix. Original magnification: 250×

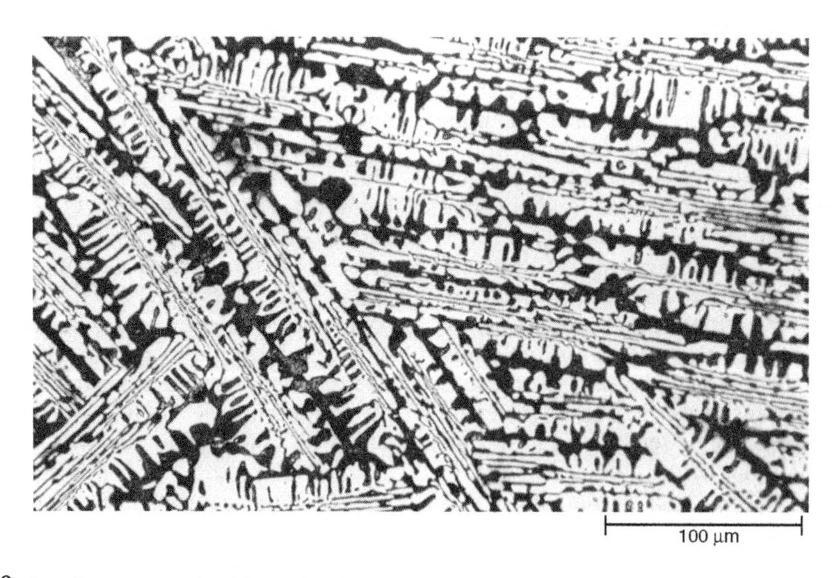

Fig. 16.10 Eutectic component of a white cast iron

Ductile and Malleable Cast Iron

Ductile cast irons represent a triumph of 20th century metallurgical research. These irons were developed independently in approximately 1948 at the International Nickel Company (INCO) in the United States and at the British Cast Iron Research Association (BCIRA) in England. Both groups discovered that by keeping the sulfur and phosphorus levels low and adding very small amounts of a key chemical element, the shape of the graphite could be changed from the interconnected flakes of gray irons into isolated spheres (usually called spheroids) of graphite. (Ductile cast iron is also called nodular cast iron and spheroidal cast iron.) The INCO team showed that the effect was produced by the addition of only 0.02 to 0.1% Mg, and the BCIRA team by the addition of only 0.02 to 0.04% Ce (the rare earth metal of atomic number 58; see the periodic table on the inside back cover). Current foundry practice in the United States uses the magnesium addition method to produce ductile irons.

As illustrated in the bottom row of Table 16.2, the graphite spheroids grow directly out of the interdendritic liquid and are not connected to each other. Their formation is an unusual type of divorced eutectic growth, with the other component of the eutectic, the austenite, forming on the surrounding dendrites of austenite. The spheroids are 100% C in the form of graphite, with the characteristic graphite sheets rotated around so as to be nearly parallel to the surface of the spheroids, as shown in Table 16.2. The spheroid diameter of a ductile iron is much smaller than the cell diameter of a gray iron; the typical volume of a single gray iron cell would be occupied by approximately 200 spheroids in a ductile iron. The mechanical properties of a ductile iron are comparable to a high-carbon steel filled with very small and isolated spherical voids. The fact that the graphite spheroids are not connected and that they have smooth surfaces means that they do not act as crack initiation sites. Hence, they reduce mechanical properties only in approximate proportion to their very small volume fraction. Consequently, as their name implies, ductile cast irons have fairly good toughness, and, similar to steels, they are often used in the heat treated condition. See Ref 16.3 (p 824) for a discussion of heat treatment.

Prior to 1948, a type of cast iron similar to ductile cast iron and called malleable cast iron was available. The malleable cast iron produced in the United States was an American invention, having its origin in the 1826 work of Seth Boyden, an iron founder from New Jersey. It was called black heart malleable to distinguish it from iron produced by an inferior European process called white heart malleable. The industry was shrouded in secrecy and the process in mystery until the early 20th century, and Ref 16.7 presents an interesting discussion of those times. Malleable irons are produced in a two-step process, where the first step is the production of a white cast iron. In the second step, the white iron is heated to approximately 950 °C (1740 °F), held for a day or two, and then slow cooled over a day or two. As seen in Fig. 16.8, at 950 °C (1740 °F), the prolonged hold is in the temperature range of the austenite + Fe_3C or austenite + graphite region of the phase diagram. Therefore, at this temperature, the white structure of austenite+Fe_3C decomposes into the more stable austenite+graphite structure. Furthermore, if the proper mix of impurity elements, heating rate, and holding temperature and times are employed (Ref 16.8, p 656) the graphite forms as discrete, isolated clumps, similar to the spheroids of ductile iron.

The similarity of the microstructures of ductile and malleable irons is shown in the micrographs in Fig. 16.11. The micrographs on the right were made in an SEM after the polished sample surface had been deep etched to allow the graphite to emerge. These micrographs show nicely the difference in the graphite shape in the two irons. The surface of the graphite in ductile iron is fairly smooth and spherical, while it is more jagged and clumpy in the malleable iron. The malleable cast iron industry has largely been replaced by the ductile industry, although malleable irons continue to be produced and have some limited advantages, especially for thin samples (Ref 16.10, p 57).

Similar to gray irons, the matrix of ductile and malleable cast irons can be controlled by the cooling process, the addition of alloying elements prior to casting, and heat treatment subsequent to casting. In the as-cast condition, a ductile iron will be ferritic, pearlitic, or, more often, some mix of ferritic and pearlitic. Fig. 16.12 presents a micrograph of a ductile cast iron that has a matrix with a pearlite-ferrite mix that is predominantly white ferrite. Ductile irons are graded or classified as shown in Table 16.4. The three numbers of the specification refer to

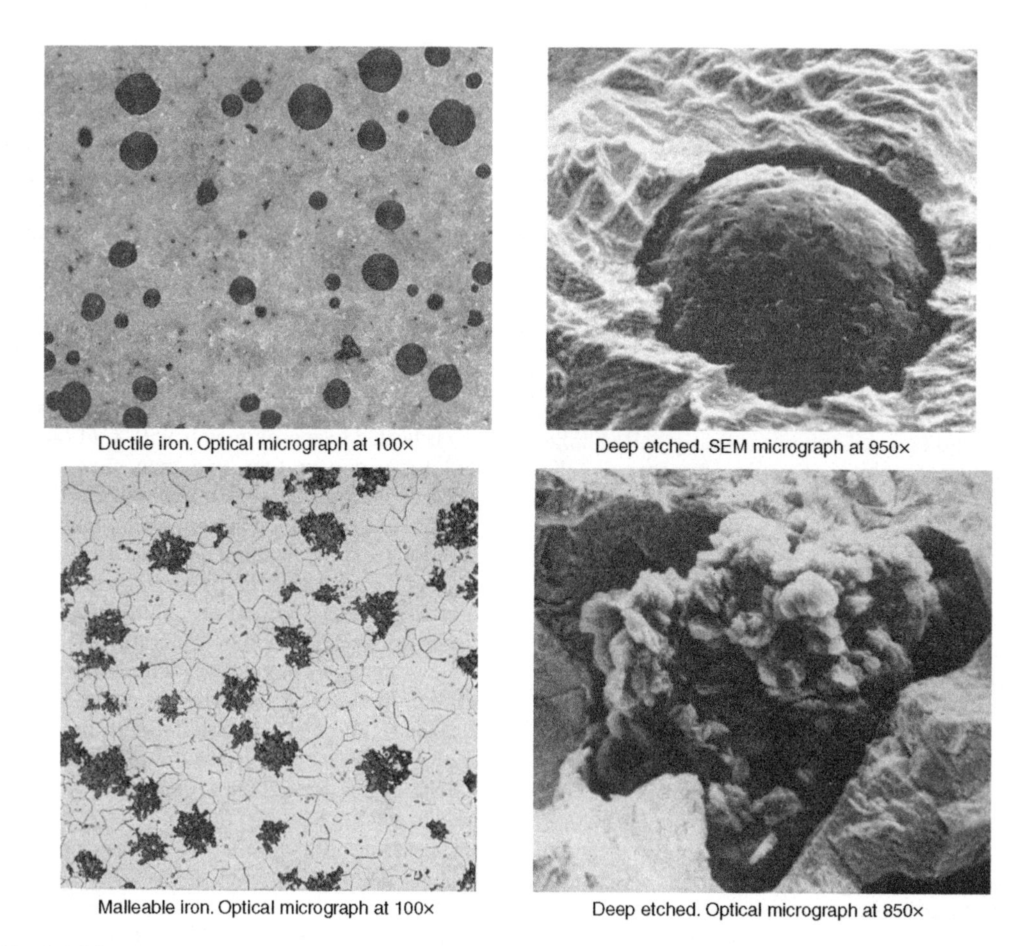

Fig. 16.11 Microstructures of ductile cast iron (top) and malleable cast iron (bottom). Source: Ref 16.9

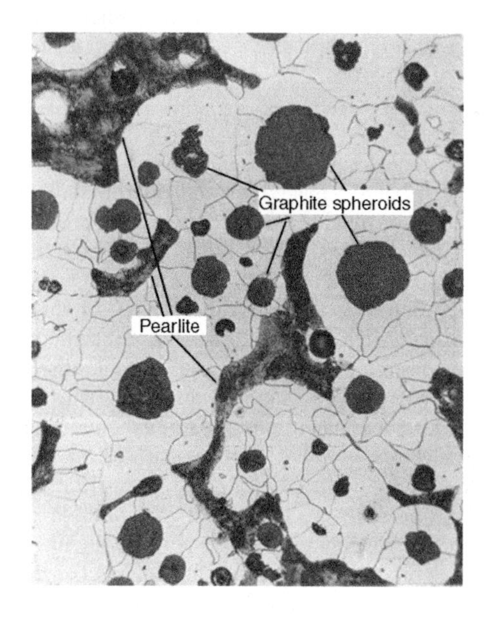

Fig. 16.12 Ductile cast iron with a matrix that is predominantly ferrite. Original magnification: 250×

approximate estimates of the three mechanical properties: tensile strength, yield strength, and percent elongation, with strength units of ksi (thousands of pounds per square inch). For example, the 60-40-18 ferrite ductile will have approximate tensile and yield strengths of 60 and 40 ksi (414 and 276 MPa), respectively, and an approximate elongation at failure of 18%. The fully ferritic structure is produced by an annealing heat treatment. The next two irons in the table (grades 65-45-12 and 80-55-06) are produced in the as-cast condition by composition control in the casting process, as was described earlier for gray irons. The fully pearlitic structure is often produced by a normalization heat treatment (normalization means air cooling). The final row in Table 16.4 (grade 135-100-8) refers to a special form of ductile iron that has become widely used because of its attractive mechanical properties for a relatively cheap cast iron. It is widely referred to as austempered ductile iron (ADI). As discussed in Chapter 12,

Table 16.4 Types of ductile cast iron

Grade or class	Matrix microstructure	Heat treatment after casting	Structure controlled by
60-40-18	All ferrite	Anneal	Heat treatment
65-45-12	Ferrite+small % pearlite	None (as-cast)	Composition control
80-55-06	Ferrite+large % pearlite	None (as-cast)	Composition control
100-70-03	All pearlite	Normalized	Heat treatment
120-90-02	Tempered martensite	Quench and temper	Heat treatment
135-100-8(a)	Bainite	Austempered	Heat treatment

(a) Austempered ductile iron (ADI)

Table 16.5 Grades of malleable cast iron

Grade	Matrix	Tensile strength/yield strength/elongation	ASTM No.
Ferritic malleable	Ferrite	53-35-18	35018
Pearlitic malleable	Pearlite+ferrite	65-45-10	45010

"Quenching," the austempering produces a bainite structure that, particularly at the high carbon levels of a cast iron, has excellent toughness compared to quenched and tempered martensite. These irons have found wide use in crankshafts and as a cheaper replacement for carburized gears.

Table 16.5 is presented to show the specification system for malleable irons. The two most common irons are presented in the table, along with their ASTM numbers. A column is included that gives the tensile strength, yield strength, and elongation in units comparable to that of Table 16.4. It is apparent that the mechanical properties of the two irons are comparable with the same type of ferrite-pearlite microstructure.

A relatively new form of cast iron, called *compact graphite* cast iron, has been developed in the past 20 to 30 years. These irons develop their graphite structure in cells, similar to that shown for gray irons in the top row of Table 16.2, except the graphite has a more cylindrical or wormlike shape. They are sometimes called vermicular graphite irons. Through research, composition controls have been developed that allow these irons to be produced reliably (Ref 16.6, p 22). They have mechanical properties somewhat intermediate between gray and ductile irons, and the interconnectedness of the graphite produces machinability and thermal conductivity superior to ductile irons.

Summary of the Major Ideas in Chapter 16

1. There are four main types of cast irons. As shown in Table 16.1, carbon is present mainly as graphite in gray, malleable, and ductile cast iron, while it is present as Fe_3C in white cast iron. The final microstructures of gray, white, and ductile irons form directly in the mold, while in malleable iron, it is produced by a special heat treatment of a white iron.
2. Cast irons are ternary alloys of Fe-C-Si. The addition of silicon makes it more difficult for Fe_3C to form in these alloys, and instead, graphite forms in the gray, ductile, and malleable irons.
3. There are dramatic differences in the properties of the cast irons. The white irons are hard but brittle due to the relatively large volume fraction of Fe_3C in them. The graphite in gray iron is largely interconnected, which produces excellent machinability, damping capacity, and thermal conductivity but low strength and toughness. As shown in Table 16.3, strength in tension is much less than in compression. The graphite in ductile and malleable cast irons is present as isolated spheres or clumps, respectively. These irons can achieve mechanical properties approaching steels and are often heat treated, as shown in Table 16.4 for ductile iron.
4. The top two rows of Table 16.2 show that both gray and white irons begin solidification by formation of austenite dendrites. In gray irons, the interdendritic liquid then solidifies as the gray eutectic in the form of large spherical cells, with graphite flakes and austenite growing outward at the spherical growth front, as shown in Fig. 16.5. In white irons, the interdendritic liquid solidifies as the white eutectic in the form of plate-shaped structures composed of Fe_3C + austenite having the shapes shown in Fig. 16.10. As explained with the aid of Fig. 16.1 and 16.2, a competition occurs between the formation of the gray eutectic and the white eutectic during solidification after the dendrites have formed. As illustrated by Fig. 16.3, fast cooling allows the white eutectic

to win the competition. The bottom of this casting froze fast in contact with a metal chill plate, and the top froze slowly in contact with a sand mold.

5. The composition ranges of the four cast irons are shown in Fig. 16.4. Foundryman talk of the carbon equivalent (CE) of cast irons, which equals %C + ⅓% Si. This number may be calculated for any iron. The farther the CE value is below 4.3, the larger will be the volume fraction of austenite dendrites in the iron.
6. The fracture surface of a gray iron appears gray because it is covered with graphite. Figure 16.7 shows how the graphite flakes pull up and cover the fracture surface nearly completely in these irons. The fracture surface of a white iron appears white because the brittle Fe_3C fractures by cleavage, giving a smooth surface that reflects light well. Notice that in Fig. 16.3, one can actually see the spherical gray cells of the gray iron encased within the white iron in the mottled region, where both forms have solidified.
7. Like gray and white irons, ductile iron begins solidification by formation of austenite dendrites (bottom row of Table 16.2). The addition of a minute amount of magnesium to an iron low in sulfur and phosphorus impurities changes the form of the graphite formation in the interdendritic liquid to small discrete spheroids of 100% graphite. The isolated spheroids detract little from the mechanical properties, and thus, these irons are called ductile cast iron. They are often heat treated, as shown in Table 16.4. Austempering to lower bainite produces excellent mechanical properties, and these irons, called austempered ductile iron (ADI) have wide industrial use.
8. In both gray and ductile irons, the matrix between the graphite that is present after solidification can be ferrite, pearlite, or a mixture of the two. As explained with Fig. 16.8, a competition occurs between the formation of the pearlite eutectoid and the graphite eutectoid on cooling. High cooling rates and/or the addition of special alloying elements will enhance the formation of the pearlite eutectoid. Such an iron is called a pearlitic gray or pearlitic ductile iron, as opposed to a ferritic gray or a ferritic ductile iron.
9. Malleable cast iron is produced by a prolonged heat treatment of white cast iron. The graphite distribution is similar to that of ductile irons, except that, as shown in Fig. 16.11, the isolated clumps of graphite (often called temper carbon) are more irregular. Malleable irons, developed in approximately 1830, have now been largely replaced by ductile irons, which were developed in the late 1940s and are cheaper to manufacture.

REFERENCES

16.1. *Metallography, Structures and Phase Diagrams,* Vol 8, *Metals Handbook,* 8th ed., American Society for Metals, 1972

16.2. I.C.H. Hughes, A Review of Solidification of Cast Irons with Flake Graphite Structures, *The Solidification of Metals,* Publication 110, ISI London, 1967, p 184–192

16.3. *Heat Treater's Guide: Practices and Procedures for Irons and Steels,* 2nd ed., ASM International, 1995

16.4. J.S. Park, "Microstructural Evolution and Control in the Directional Solidification of Fe-C-Si Alloys," Ph.D. thesis, Parks Library, Iowa State University, 1990

16.5. *Cast Irons,* ASM Specialty Handbook, ASM International, 1996

16.6. R. Elliott, *Cast Iron Technology,* Butterworth, 1988, p 33

16.7. H.A. Schwartz, *American Malleable Cast Iron,* The Penton Publishing Co., 1922

16.8. R.W. Heine, C.R. Loper, Jr., and P.C. Rosenthal, *Principles of Metal Casting,* McGraw-Hill, 1967

16.9. Broadsheet No. 138, British Cast Iron Research Association, U.K., 1976

16.10. *Properties and Selection: Iron and Steels,* Vol 1, *Metals Handbook,* 9th ed., American Society for Metals, 1978

CHAPTER 17

Surface Hardening Treatment of Steels

THERE ARE MANY APPLICATIONS where a very hard surface for wear resistance is desirable, for example, at the surface of a gear tooth, at the surface of a crankshaft along the bearing journals, or at the edges of cutting tools and knives. Because high hardness leads to low toughness, maintaining high hardness from the surface to the center of a steel part will make it too brittle for many applications. This problem can be overcome by producing a hard surface layer on a softer and tougher core material. There are several different techniques that have been developed over the years for this purpose. Descriptions of these techniques can be found in ASM International publications (Ref 17.1–17.4), and a brief introduction to some of these is presented here.

The techniques can be partitioned into two broad categories:

1. Techniques that add a surface coating to the steel. Common examples include chromium plating to increase surface hardness or corrosion resistance and the hard gold-colored layers of titanium nitride commonly applied to drill bits.
2. Techniques that modify surface layers of the steel either by selective surface heat treating or by diffusing atomic elements into the surface, usually carbon or nitrogen. This chapter focuses on surface modification techniques.

Surface Heat Treatments

Surface heat treatment (mentioned as part of the second technique) involves heating a steel part very rapidly above the A_{c_3} temperature (Fig. 3.11), resulting in a thin layer of austenite on the surface. The part is then quenched, which produces a thin layer of martensite on the surface. If the %C in the bar is in the 0.4 to 0.6 range, the martensite will be of the more ductile lath type and be very hard. At higher %C values, one will obtain the more brittle plate martensite with increasing amounts of soft retained austenite. Consequently, the most common steel parts treated with this technique have carbon contents of approximately 0.4%. In addition, when the martensite forms, it will expand in volume. The core material will restrain this expansion and pull the iron atoms closer together than they want to be, so that the surface layer ends up having a compressive residual stress. As explained with Fig. 5.13 and 5.14, this condition is highly favorable for avoiding fatigue failure in rotating parts.

The common, traditional techniques for surface heat treating are either flame hardening or induction hardening. More sophisticated and expensive techniques involve heating the surface by laser beams or by electron beams (Ref 17.2). Figure. 17.1 presents a polished and etched cross section of a 2.5 cm (1 in.) diameter bar of 1045 steel that was induction hardened, and the hardened martensitic case (layer) is clearly visible. The depth of the layer affected by the surface hardening treatment is commonly referred to as the *case depth*. This depth is typically defined as the distance below the surface where the hardness has fallen to HRC = 50. As illustrated in Fig. 17.2 by the hardness profile taken on the bar with a microhardness tester, the case depth here is 2.5 mm (0.1 in., or 100 mils). As is the case here, the core material is often a pearlite-ferrite mix.

Figure 17.3 illustrates how the hardened case develops during surface heat treatment. The top of this figure shows a hypothetical temperature rise near the surface of the bar at a time just prior to quenching. If one locates the A_{c_3} temperature on this plot as shown, the region I to

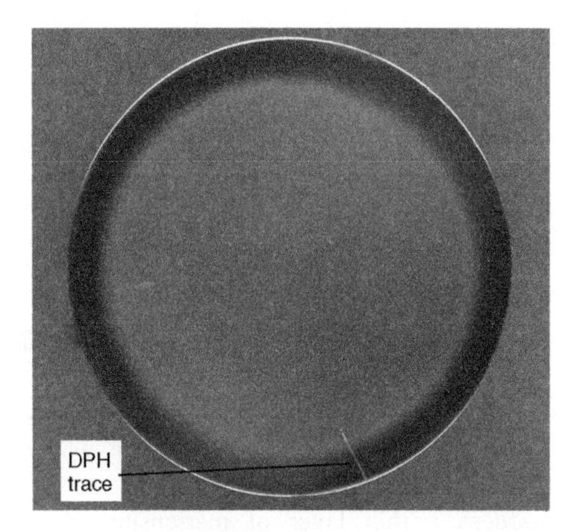

Fig. 17.1 Section view of a 25 mm (1 in.) diameter induction-hardened 1045 steel. The trace created by diamond pyramid hardness (DPH) indentations is visible

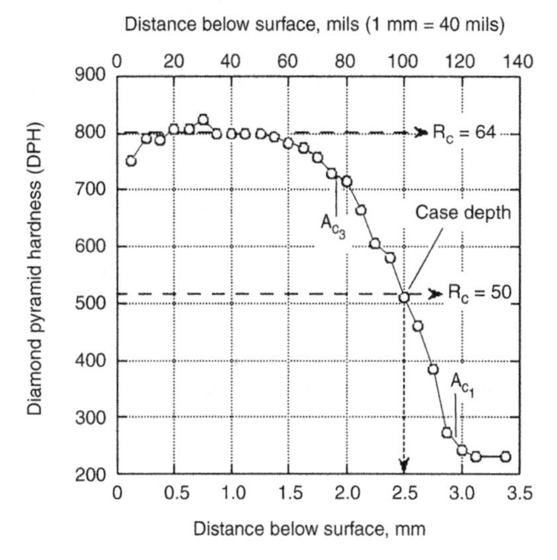

Fig. 17.2 Microhardness trace through case of Fig. 17.1. (See Table 5.3 to convert DPH to R_c.)

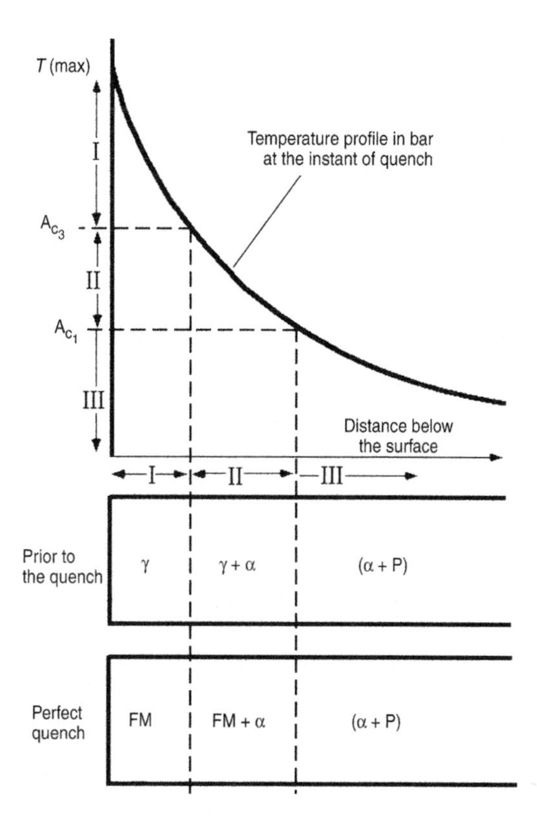

Fig. 17.3 Microconstituents present before and after quenching, illustrating the correlation between the temperature profile and the microconstituents. FM, fresh martensite

the left of this location must consist of austenite (γ). With a perfect quench, all of this austenite will transform to fresh martensite, shown as FM in the bottom diagram. At the A_{c_1} temperature, all of the pearlite will have dissolved, so that in region II there will be a mixture of austenite (γ) and ferrite (α) just prior to quench and a mixture of fresh martensite and ferrite with a perfect quench. Region III, the core material, will be a mixture of ferrite and pearlite. Upon examination of the bar under higher magnification in an optical microscope (Fig. 17.1), the locations were found where the original pearlite disappeared (corresponding to A_{c_1}) and where the original ferrite disappeared (corresponding to A_{c_3}). These locations are shown in Fig. 17.2. Notice that the R_c50 case depth occurs between these two locations. The 1045 bar in Fig. 17.1 was quenched in a water spray, and even with this fast quench, due to the poor hardenability of 1045 steel, very small clumps of the reaction products ferrite and bainite formed during the quench in the region from the A_{c_1} location to the point where the hardness topped out at diamond pyramid hardness (DPH) = 800.

It is quite fascinating to observe the induction hardening operation, because the surface temperature of the bar is raised to a bright orange color in just a matter of seconds. The reason that induction hardening is able to raise the surface temperature and not the interior temperature is because the induced alternating current in the steel bar is forced to flow very near the bar surface by a trick of electrical engineering. The current is induced in the bar from a close-fitting coil wrapped around the bar, which has a very high-frequency current flowing through it, and the trick is the high frequency. This induced current will only penetrate the bar surface to a distance called the skin depth, *S*, which is inversely proportional to the frequency; the higher the frequency, the lower the skin depth. For the experiment that resulted in the bar shown in Fig. 17.1, the frequency used was 450 kHz (450,000 cycles/s). For the coil used and with this frequency, the skin depth is calculated to vary as shown in Fig. 17.4.

The curves show that when the surface is still ferromagnetic ferrite at 700 °C (1290 °F), the skin depth is extremely small, under 100 μm (0.004 in., or 4 mils), and even after heating sufficiently to transform to nonferromagnetic ferrite plus austenite at 800 °C (1470 °F), the skin depth remains less than 1 mm (0.04 in.). Analysis of the power input equations shows that 98.2% of the power per unit area being induced into the steel by the coil is produced by the current flowing at a depth of only two times the skin depth. By using high-frequency current, it is possible to cause the temperature to rise very steeply only near the surface. As this analysis using Fig. 17.3 has shown, the case depth will occur at approximately the midpoint of region II. Therefore, by controlling the power input, one can control the temperature profile shown in Fig. 17.3 and thereby adjust the case depth to various desired levels. There are different types of power converters used in surface induction hardening, with frequencies ranging from as low as 3 to the 450 kHz used here. The lower-frequency units are used for deeper case depths. An excellent review of induction heat treating of steels is found in Ref 17.5.

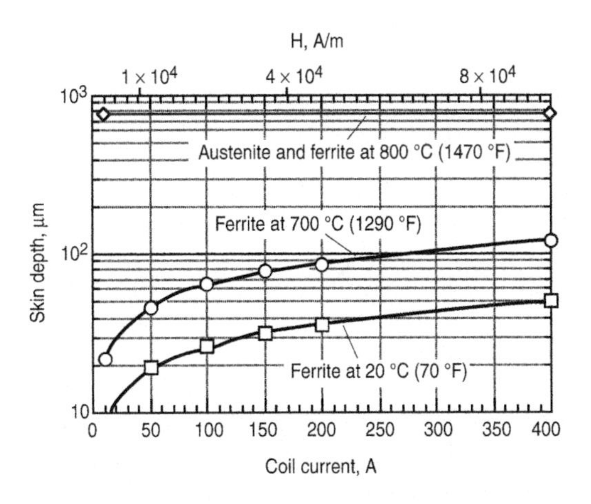

Fig. 17.4 Induction heating skin depths for ferrite at 20 and 700 °C (70 and 1290 °F) and for austenite and ferrite at 800 °C (1470 °F) as a function of the current through the work coil

An interesting feature of induction hardening is that the surface hardness can achieve values in excess of that obtained by conventional furnace heating and quenching. As shown in Fig. 4.13, the hardness of fresh martensite in a conventionally heat treated 1045 steel is expected to be HRC = 60 (DPH = 697, see Table 5.3). The actual hardness found at the surface of the induction-hardened bar shown in Fig. 17.3 runs significantly higher, approximately HRC = 64. This phenomenon is well known, and the increment in hardness obtained with induction hardening is generally called superhardness. As illustrated in Fig. 17.5, the superhardness increment produces an increase in the HRC values that range up to 5 HRC points. It apparently is related to the compressive residual stress produced by induction hardening.

Surface Diffusion Layers

There are a large variety of techniques that achieve selective hardening at the surface of steel parts by diffusing an atomic element into the outer surface layers, which produces an increased surface hardness either simply by its presence or by its effect on hardening during a subsequent heat treatment. The names of these techniques relate directly to the type of atom diffused into the surface. Adding carbon is called carburizing, adding nitrogen is called nitriding, adding combinations of carbon and nitrogen is either nitrocarburizing or carbonitriding, and adding boron is called boronizing. All of these techniques are discussed briefly in Ref 17.1, and an excellent introductory summary is presented in Ref 17.2 (p 259). A short review of some of these techniques is presented here.

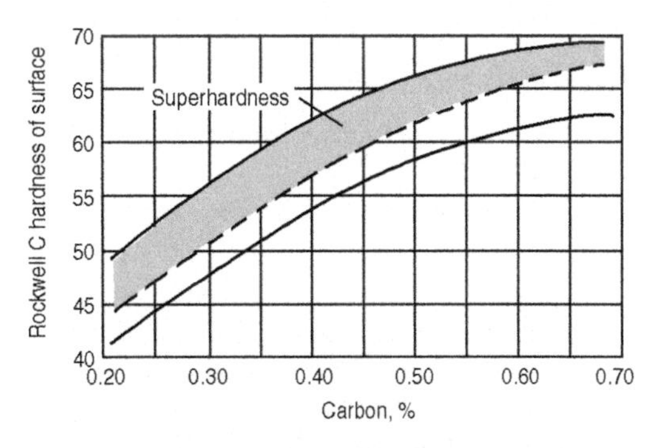

Fig. 17.5 Hardness for water-quenched steels as a function of carbon content. The lower curve is for furnace hardening. The upper two curves give a range of increased hardnesses for induction hardening. Source: Ref 17.4, p 179

Carburizing

This technique was discussed briefly in Chapter 7, "Diffusion—A Mechanism for Atom Migration within a Metal." It involves supplying carbon to the steel surface at high temperatures, where the carbon atoms diffuse into the steel and produce an increased carbon content locally under the surface, as shown schematically in Fig. 7.2. Subsequent heat treatment then produces a hardened surface layer, because the increased carbon content at the surface increases the hardness of the martensite formed there on quenching, as predicted by Fig. 4.13. As discussed with reference to Eq. 7.2, the carbon atoms are carried to the steel surface in a gas containing a mixture of carbon dioxide (CO_2) and carbon monoxide (CO). Figure 7.7 and its discussion explain how engineers have used the science of thermodynamics to calculate the $\%CO_2$ in the *endo gas* used commercially in U.S. heat treat facilities to control the so-called *carbon potential* of endo gas. The %C in a 1095 steel is 0.95. Therefore, the line labeled 1095 in Fig. 7.7 is the 0.95 carbon potential line. For example, according to Fig. 7.7, at 1000 °C (1830 °F), an endo gas having $\%CO_2 = 0.067$ will have a carbon potential of 0.95% C. Therefore, a 1095 steel placed into this gas under these conditions will neither carburize nor decarburize. In the discussion of Fig. 7.7, this gas was called a neutral gas for the 1095 steel.

Now, suppose a low-carbon steel such as a 1018 is placed into a furnace at 1000 °C (1830 °F) with the endo composition set to the 0.95 carbon potential value. The steel would immediately start to carburize, the surface composition would increase to the 0.95% value, and carbon would start to diffuse into the 1018 bar. Figure 17.6 is a schematic diagram showing how the %C would have increased in the bar after 5 min, 30 min, and 1h. The carbon transport occurs in two steps:

1. Carbon atoms move from the gas to the steel surface.
2. Carbon atoms migrate into the steel by diffusion.

At short times, step 1 is slower than step 2, and the rate of carburization is limited by step 1. During this time, the composition of the steel at the surface rises from the initial 0.18% C value to the carbon potential value of 0.95% C, after which time it remains fixed at this value, as set by the gas composition. At longer times, the rate of carburization is controlled by the diffusion process, which was discussed in Chapter 7. If the bar is quenched after the 1 h hold time and if its hardenability is adequate, fresh martensite forms near the surface, and Fig. 4.13 can be used to determine the value of the hardness at the surface. It can be seen that the surface hardness should be approximately HRC = 66. Figure 4.13 also predicts that when the %C decreases to approximately 0.28, the hardness will have fallen to HRC = 50. The construction in Fig. 17.6 shows how the R_c50 case depth is set by the composition profile produced by the carburization process.

In the previous example, where the source of carbon is supplied to the steel from the surrounding gas, the process is called gas carburizing. Obviously, control of the carbon potential of the gas is very important in controlling this carburization process. Figure 17.7 is a more generalized way of presenting the carbon potential information at a constant temperature than was presented in Fig. 7.7. Usually, conventional carburization is done at lower temperatures than the 1000 °C (1830 °F) value of the previous example in order to avoid grain growth of the austenite. A more common temperature is 927 °C (1700 °F), which is the temperature chosen for Fig. 17.7. This diagram can be used with any mixture of CO_2/CO gas. In endo gas, the %CO remains near 20, and the vertical line shown on the diagram applies. The diagram shows nicely one of the neat features of endo gas. By adjusting

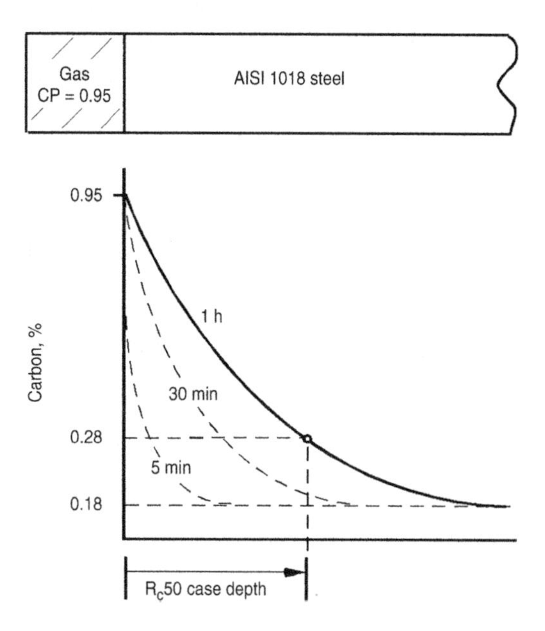

Fig. 17.6 Carbon profile produced by carburizing. CP, carbon potential

the air/natural gas input to the endo gas generator, the $\%CO_2$ in the gas can be controlled, thereby setting the carbon potential to any desired level up to the maximum possible for austenite, which is given by the A_{cm} composition, for example, 1.3% C at 927 °C (1700 °F). For reasons to be explained subsequently, it is desirable to fix the surface composition of carburized parts at approximately 0.8% C. Therefore, the $\%CO_2$ level should be set at approximately 0.17.

Steel was first made in the 2nd millennium B.C. by carburization of iron. The carbon was supplied to the iron surface from hot charcoal surrounding the iron, with the air flow to the charcoal forge adjusted by a clever blacksmith to achieve a carbon potential suitable for carburization rather than decarburization. Adding carbon to a steel surface from surrounding charcoal is called *pack carburizing*. It is rarely used in current times because it suffers from, among other things, lack of control of the carbon potential. However, pack carburizing can be done very cheaply, because all one needs to do is to pack a piece of steel in charcoal inside a closed iron (or better, Inconel) pipe and heat it to the carburizing temperature for a few hours. It is advisable to add an *energizer* to the charcoal, which helps generate the CO gas needed for carburization from the solid charcoal. It is common to add barium carbonate as the energizer, but such additives as molasses will also work well. Figure 17.8 shows the results of the pack carburization of a rectangular bar of 1018 steel heated to 927 °C (1700 °F) for 3 h and then water quenched. One-half of the bar is shown after sectioning, polishing, and etching in nital, with the location of a microhardness trace identified. A plot of the microhardness trace is shown in Fig. 17.9. It is seen that the R_c50 case depth is 0.92 mm (0.037 in., or 37 mils).

The carbon potential in the gas generated by pack carburizing is generally so high that the austenite becomes saturated with carbon. This means that the carbon potential reaches the value of the A_{cm} (or the A_{gr}) value for the carburizing temperature. As shown in Fig. 16.1, at 927 °C (1700 °F) this value will be approximately 1.3% C. By referring to Fig. 4.13, it can be seen that as the %C increases above approximately 0.95, the hardness of fresh martensite begins to drop below values of approximately 66. The cause of this decrease is the formation of retained austenite in the quenched steel. Figure 17.10 is a micrograph of the sample in Fig. 17.8 at the location between the surface and the first microhardness indentation. Dark plate martensite

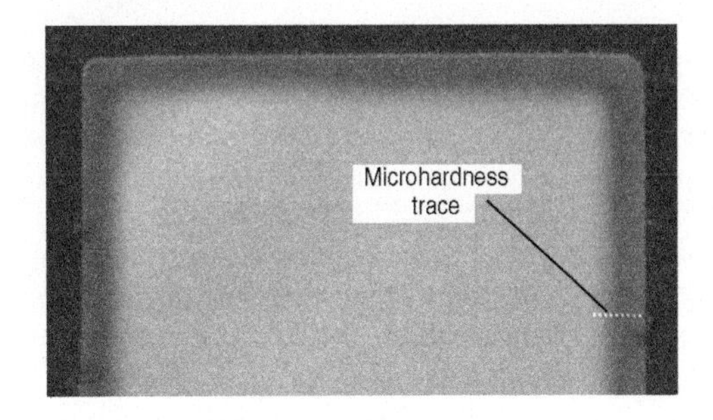

Fig. 17.8 Section of a pack-carburized square bar of 1018 steel

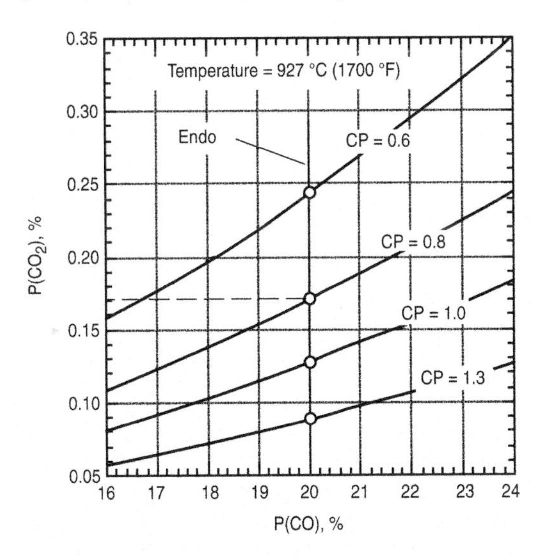

Fig. 17.7 Plot of carbon potentials in a CO_2/CO gas mixture at 927 °C (1700 °F) for carbon potential (CP) values varying from 0.6 wt% up to the saturation value of the A_{cm}, 1.3 wt%

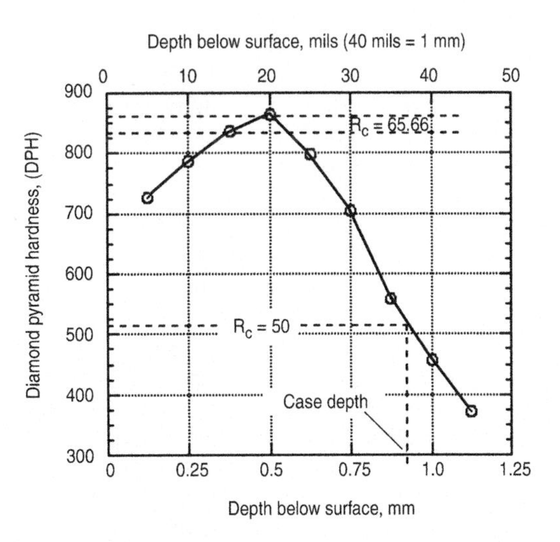

Fig. 17.9 Microhardness profile of trace shown in Fig. 17.8

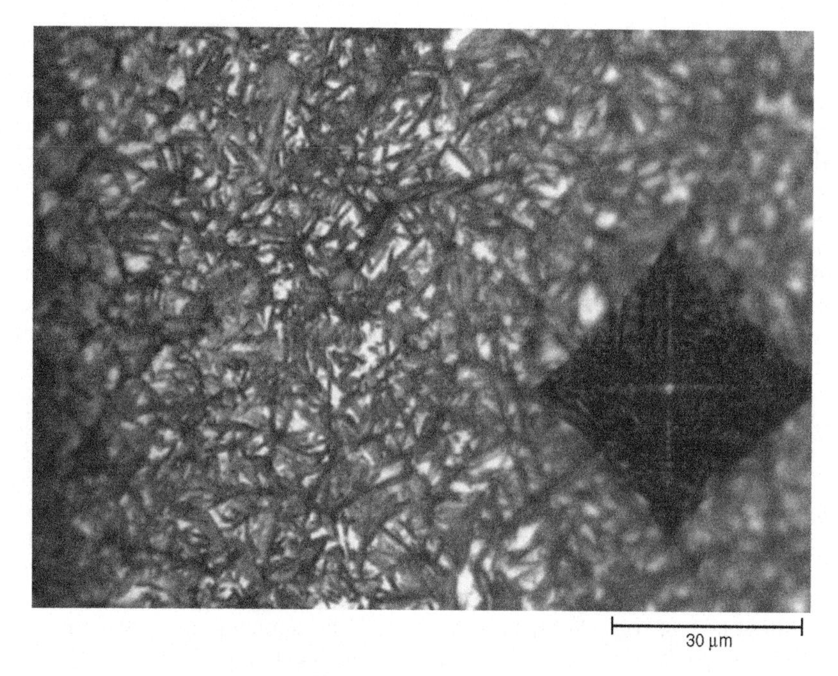

Fig. 17.10 Region of Fig. 17.8 between surface and first indent

surrounded by significant amounts of white retained austenite is visible, which is consistent with the data of Fig. 4.18 predicting a volume percent retained austenite of 25 to 35 in a 1.3% C steel. In order to avoid surface retained austenite and the more brittle plate martensite form, gas carburization is generally done with the carbon potential of the gas set at approximately 0.8%.

The strong interest in energy conservation in recent times has led to many innovations in gas carburizing, and several of these are briefly discussed.

The Boost Diffuse Technique. In this technique, the carbon potential is initially set near the saturation value for a short time, which raises the %C near the surface well above the desired 0.8; this is the boost phase. Then, the carbon potential is reduced to 0.8 causing the %C to drop back to the 0.8 level at the surface as carbon atoms diffuse into the bulk of the steel. The method can reduce carburization time at 900 °C (1650 °F) by approximately 20%.

High-Temperature Carburizing. Increasing the temperature of the carburization process dramatically reduces the time required to achieve a given case depth. For a number of reasons, the rates of the two steps of carbon transport just discussed both increase very quickly as the temperature is raised. However, grain growth is a problem, and the technique is limited to steels with appropriate alloy additions for retarding grain growth, such as 8620 (Chapter 8, "Control of Grain Size by Heat Treatment and Forging"). High-temperature carburizing is sometimes done using a boost-diffuse cycle, with the boost step done at approximately 1040 °C (1900 °F) and the diffuse step either at the high temperature or at normal carburizing temperatures of approximately 900 °C (1650 °F). See Ref 17.6 for more discussion.

Plasma and Vacuum Carburizing. Atmospheric pressure is 760 mm of mercury, or 14.7 lb/in.^{2}. If electrodes are sealed into either end of a glass tube, the pressure inside the tube is reduced to between 1 to 20 mm with a vacuum pump, and then approximately 1000 direct current volts are applied between the electrodes, the tube lights up like a neon light. Some of the gas molecules are being ionized, and their outer electrons are energized to higher levels. When they fall back to their lower-energy states, light is emitted, and the energized gas is called a plasma or a glow discharge. A problem with trying to carburize steels using only natural gas (methane, CH_4) is that the rate of step 1 of the carburizing process is much lower than that obtained with endo gas. However, the step 1 rate for pure methane can be increased dramatically by either raising the temperature from 900 to 1050 °C (1650 to 1920 °F) and/or by forming a plasma in the gas surrounding the workpiece.

With both vacuum and plasma carburizing, samples are first heated to a temperature of approximately 1030 to 1050 °C (1890 to 1920 °F) in a vacuum, and then a mixture of methane (or propane) and nitrogen (which acts only as a carrier gas) is introduced into the chamber. Both carbon transport steps of the carburizing process then occur for a relatively short time. This is the boost phase. In the case of plasma carburizing, a plasma surrounding the workpiece significantly increases the rate of carbon transport to the surface in step 1. Because of this effect, the time of the boost phase can be reduced from 50 min for straight vacuum carburizing to only 10 min for plasma carburizing, with equivalent results. The diffuse phase is then carried out at the same temperature or slightly lower temperatures by simply removing the gas with vacuum pumps and holding the sample at temperature. Example results for plasma carburizing versus conventional carburizing are presented in Fig. 17.11. It is seen that to achieve equivalent case depths of 1.4 mm took 360 min with the conventional gas carburizing technique compared to only 40 min with plasma carburizing. In addition to the reduced time for these techniques over the conventional process, there are several other advantages discussed in Ref 17.2 (p 348 and 352). Major disadvantages are the high capital cost of the vacuum equipment and the restriction to alloy steels, such as 8620, to avoid grain growth.

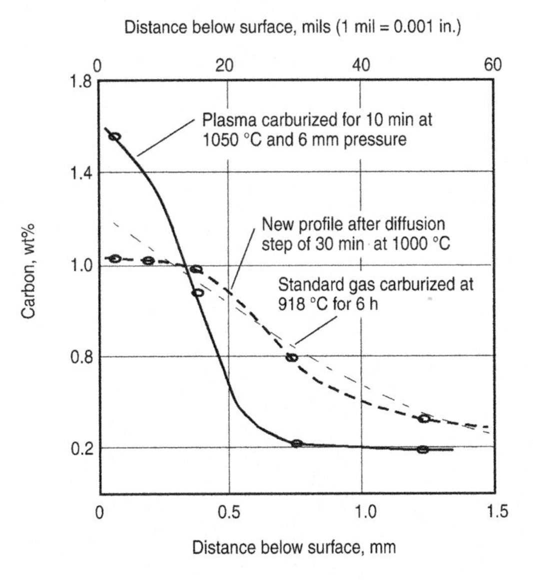

Fig. 17.11 Carbon profiles of a plasma-carburized 1020 steel after the boost and diffusion steps compared to the profile produced by standard gas carburization. Source: Ref 17.2, p 355

Other Gas Sources. As discussed in Chapter 7, "Diffusion—A Mechanism for Atom Migration within a Metal," the endo gas used for carburizing in the United States consists of approximately 40% H_2 + 20% CO + 40% N_2 by volume, and it is easily generated from natural gas. In countries that do not have ready supplies of natural gas, other techniques are used to generate similar gas mixtures of H_2, CO, and N_2. One technique generates the gas by feeding pure nitrogen and evaporated methanol (CH_3OH) into the furnace. Union Carbide calls their process UCAR (Ref 17.7), and the British Oxygen Company process is called Endomix (Ref 17.8). A second technique feeds liquid streams of methanol and acetone (CH_3COCH_3) into the furnace, a process called Carbomag or sometimes drip feed (Ref 17.9). These processes claim energy savings over endo gas of 10 to 20%.

Carbonitriding

This process is very similar to the standard carburizing process. One simply adds approximately 3 to 8% ammonia (NH_3) to the hydrocarbon gas. The process is carried out at lower temperatures (800 to 900 °C, or 1470 to 1650 °F) than plain carburizing, and the case depths are smaller, usually less than 0.75 mm (0.03 in., or 30 mils). Nitrogen atoms diffuse into the surface of the steel along with the carbon atoms so that there is an increase in both carbon and nitrogen content in the surface layers. The major advantage of the process over straight carburizing is that the hardenability of the steel increases with increasing nitrogen content. Figure 17.12 presents Jominy end quench data

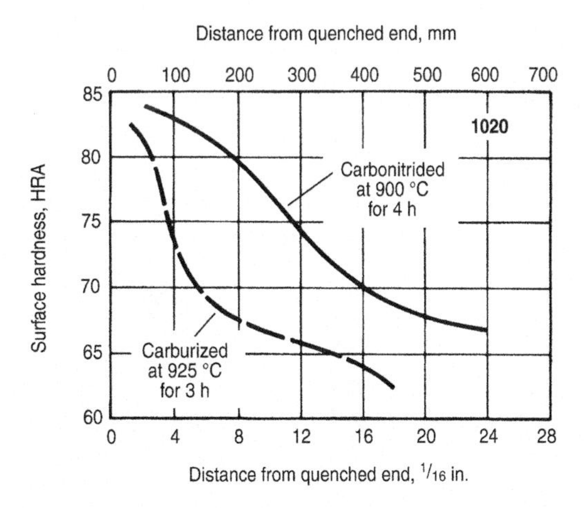

Fig. 17.12 Jominy data comparing a carbonitrided and a carburized 1020 steel. Source: Ref 17.1, p 58

showing the significantly increased hardenability produced by the addition of nitrogen. In order to obtain the increased surface hardness with carburization, it is, of course, necessary that the hardenability of the steel being carburized be adequate so that the case fully transforms to martensite on quenching. The increased hardenability with nitrocarburizing sometimes allows the use of cheaper plain carbon steels and/or oil quenching rather than water quenching. Perhaps the major disadvantage of nitrocarburizing is a reduction in impact strength, and hence, the process is not generally used on parts such as gear teeth. For further information, see Ref 17.1 (p 58) and 17.2 (p 376).

Nitrocarburizing (Ferritic Nitrocarburizing)

The iron-nitrogen phase diagram is shown in Fig. 17.13. By comparing this to the iron-carbon diagram of Fig. 4.4, several similarities are noticeable. In the iron-carbon diagram, the pearlite eutectoid occurs at 727 °C (1340 °F) and 0.77% C, whereas a eutectoid occurs in the iron-nitrogen diagram at 590 °C (1095 °F) and 2.3% N. In the iron-carbon system, the phases present below the eutectoid temperature are alpha iron + Fe_3C, while in the iron-nitrogen system, they are alpha iron + Fe_4N, which is usually labeled γ′. Both Fe_3C and Fe_4N are chemical compounds, but Fe_4N differs from Fe_3C in that its composition is not fixed. Whereas the composition of Fe_3C is limited to its chemical formula, as shown at 450 °C in Fig. 17.13, the composition of Fe_4N (γ′) ranges from 5.7 to 6.1% N. The phase on the far right is labeled epsilon (ε), which is also considered to be a compound, and it is quite hard and has a wide composition range. In the nitrocarbuizing process, nitrogen is diffused into the surface at a temperature of close to 570 °C. At this temperature, the iron is present in the alpha form (ferrite), so the process is often called ferritic nitrocarburizing.

The nitrogen can be supplied either from the gas phase or from cyanide salt baths; however, only the gas phase technique is discussed here. The nitrogen is introduced as ammonia (NH_3). The ammonia can be added to a hydrocarbon gas such as endo gas. Interestingly, even though the process is called nitrocarburizing, it is sometimes done by adding ammonia to pure nitrogen.

Nitrogen gas (N_2) is inert to iron at high temperatures. The N_2 molecule does not dissociate easily and allow nitrogen atoms to diffuse into the iron. However, when ammonia dissociates at high temperatures at the surface of iron, it forms nitrogen atoms that do enter into the iron matrix. The chemical reaction is written as:

$$NH_3 \rightarrow \tfrac{3}{2}H_2 + [N]$$

where the bracket around the N means a nitrogen atom dissolved into the iron matrix. As discussed previously in reference to Fig. 7.7 and 17.7, thermodynamic principles can be used to calculate the carbon potential in the iron-carbon system. Similarly, a nitrogen potential in the iron-nitrogen system can be calculated, and the results are shown in Fig. 17.14. The nitriding

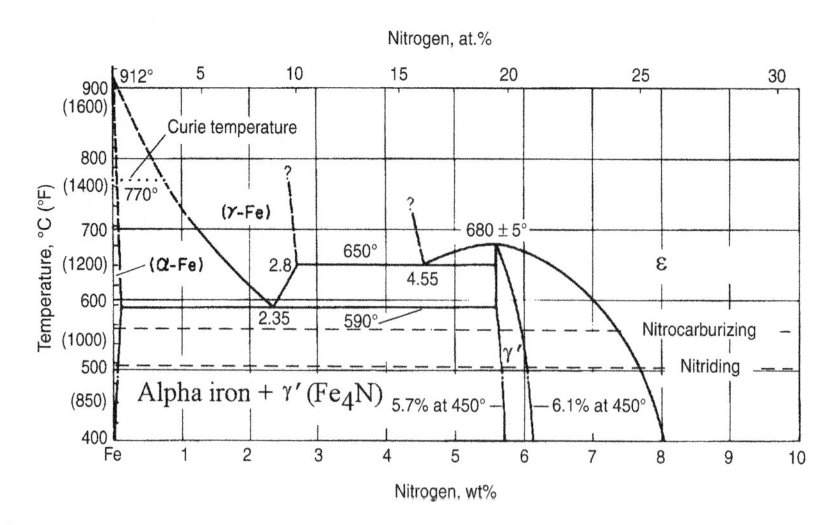

Fig. 17.13 Iron-nitrogen phase diagram. Source: Ref 17.10

potential, r, shown on the right axis, can be controlled by controlling the gas composition. The vertical line at 570 °C (1060 °F) can be used to predict what occurs at the surface of the steel during nitrocarburizing. As r increases from 0 to approximately 0.2, the surface nitrogen content will increase from 0 to the maximum that can dissolve in alpha iron, approximately 0.1% N. As r is increased from 0.2 to approximately 1, a layer of Fe_4N (γ') will form on the surface. When r increases above approximately 1, the surface will become coated with epsilon (ε) phase.

During ferritic nitrocarburizing, the nitriding potential is controlled to deliberately produce a layer of the ε phase on the surface, as shown in Fig. 17.15. This figure presents a typical case where a surface ε layer approximately 18 μm thick is obtained after 3 h at 570 °C (1060 °F). Below the surface layer, the pearlite-ferrite structure of the initial steel is visible. The ε phase is quite hard, providing excellent wear properties. Notice that the hardness is not due to martensite forming on quenching, as is the case in the other surface treatments that have been discussed. Nevertheless, it is common to quench nitrocarburized parts in oil, because this improves the fatigue resistance. It is interesting to note that experiments have shown that adding a hydrocarbon gas to the ammonia increases the carbon level in the ε layer but has little effect on hardness, wear, fatigue, or corrosion resistance. However, the hydrocarbon addition speeds up the rate of nitride layer formation, and it helps eliminate formation, of Fe_4N (γ') regions from within the ε surface layer. One of the major advantages of the process is the excellent dimensional stability. See Ref 17.1 (p 69) for an introductory discussion and Ref 17.2 (p 425) for a detailed discussion.

Nitriding

Like nitrocarburizing, the source of the nitrogen for nitriding can be either a gas or a cyanide salt bath. Also like nitrocarburizing, the process is done at temperatures below the eutectoid temperature, in this case, usually approximately 500 to 525 °C (930 to 980 °F). During the gas nitriding process, the nitrogen potential, r, of Fig. 17.14 is controlled to values just below that needed to avoid forming a layer of Fe_4N (γ') on the steel surface. Formation of a surface layer of Fe_4N (γ'), called a *white layer*, is not desirable and is a processing problem with nitriding. Several techniques have been developed to control it. Quenching is not employed with nitriding, so the high surface hardness does not result from martensite formation. Rather, it is necessary to use steels that contain alloying elements that will form nitride particles. Remember that the high hardness developed during tempering of tool steels, which is called secondary hardening (Fig. 10.10), is developed by the formation of nanometer-sized carbide particles formed with reactive alloying elements such as molybdenum, chromium, and vanadium. Nitrided steels achieve their high surface hardness by a similar

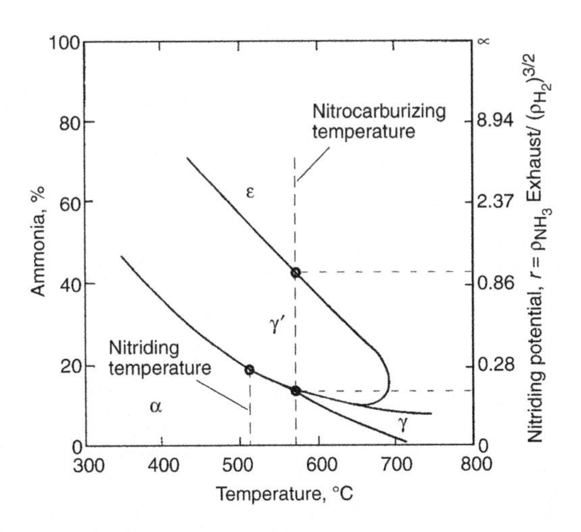

Fig. 17.14 Nitrogen potential versus temperature. Source: Ref 17.11, p 260

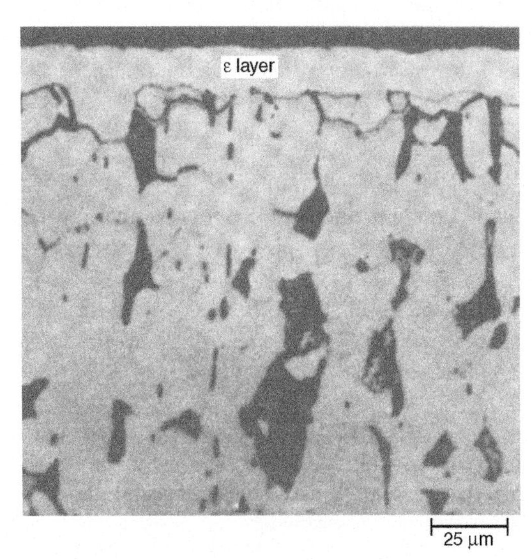

Fig. 17.15 Micrograph of a 1018 steel after nitrocarburizing at 570 °C (1060 °F) for 3 h and oil quenching. Source: Ref 17.2, p 425

effect where, in this case, the particles formed are nitrides. Hence, the process is only useful with steels containing reactive alloying elements.

The process provides excellent dimensional stability, because there are no quenching stresses and no volume changes due to formation of a new phase on quenching. Disadvantages of the process are that long times are required, cases are thin, and process control must be exact to minimize formation of Fe_4N (γ') layers on the surface. There are five groups of steels that lend themselves to nitriding: special aluminum-containing alloys called Nitralloy, high-chromium-molybdenum steels, the low-alloy steels 41*xx*, some tool steels such as H-11, and some stainless steels, such as 430, 446, and the precipitation-hardening stainless steels.

Both nitriding and nitrocarburizing can be done in a vacuum chamber using the technique described previously to generate a plasma. When applied to nitriding, the technique is often called ion nitriding as well as plasma nitriding. The operation can be done with nitrogen gas, because the plasma effectively ionizes the nitrogen, and ammonia is not required. The nitrogen potential is controlled by controlling the composition of the input gases, N_2, H_2, and CH_4. Because of the relatively low processing temperatures compared to carburizing, sometimes the samples are heated directly to the processing temperature by the heat of the plasma, thereby eliminating the need for a furnace within the vacuum system. As expected, the plasma reduces processing time and has several other advantages (Ref 17.2, p 420 and 431).

Table 17.1 presents a brief summary of the variation of the characteristics of the surface diffusion treatments discussed here. The qualitative values listed were collected from various references, and the boldface characteristics highlight optimal values. Although not mentioned in the previous discussion, the table illustrates that all of these techniques produce a compressive surface stress that is beneficial to fatigue life.

Summary of the Major Ideas in Chapter 17

1. A steel bar with a Rockwell C hardness in the 60s will have excellent wear properties. However, if the bar has this hardness all the way to its center, it will not be tough enough for many applications. Therefore, several techniques have been developed to selectively harden the surface of steel parts while maintaining softer and tougher cores. These techniques can be partitioned into those that use plating, such as chrome plating, and those that modify the surface layers either by selective surface heating or by diffusing atoms into the surface with or without subsequent heat treating. For both surface modification techniques, it is common to refer to a *case depth*, which is often defined as the depth below the surface where the hardness decreases to a value of HRC = 50.
2. The most common techniques for surface heat treatment are flame hardening and induction hardening. In both techniques, the surface is very rapidly heated much hotter than the interior, forming a rim of austenite near the surface that then transforms to martensite on quenching. The chapter focuses on the induction hardening process. This technique induces an electric current into the outer layer of the steel by using a very high-frequency current. Figures 17.1 to 17.3 present an example illustrating how the case depth is controlled and measured with induction surface hardening.

Table 17.1 Summary of characteristics of surface diffusion treatments

Characteristic	Carburizing	Carbonitriding	Nitriding	Nitrocarburizing
Case depth	Medium	Shallow	Shallow	Shallow
Compressive stress level	Medium-high	High	**Very high**	Medium-high
Scoring resistance	Medium-high	High	**Very high**	**Very high**
Impact loading characteristics	**Excellent**	Fair	Fair	Fair
Bending fatigue strength	Good	Good	Good	Fair/good
Resistance to seizure (antigalling)	Good	Good	**Excellent**	**Excellent**
Freedom from quench cracks	Excellent	Excellent	Not a problem	Excellent
Dimensional control (distortion)	Fair	Good	**Excellent**	**Excellent**
Cost of steel required	Low-medium	**Low**	Medium-high	Low-high
Processing time	Short	Short	Long	Short
Reduction case hardness on tempering	Yes	Somewhat	No	No
Improvement in corrosion resistance of nonstainless steels	No	No	Yes	Yes

3. The names of the several kinds of surface diffusion techniques follow the type of atom that is diffused into the steel surface: carburizing for carbon diffusion, nitriding for nitrogen diffusion, both carbonitriding and nitrocarburizing for carbon + nitrogen, and boronizing for boron diffusion. Of these techniques, carburizing is most widely used in industry. With this technique, the %C is increased locally at the surface, and, on quenching, a high hardness martensite forms at the surface, while the core remains soft and tough.
4. Carbonitriding is essentially carburizing with a small amount of nitrogen added. The added nitrogen improves hardenability, which sometimes allows cheaper plain carbon steels to be used, but there is some loss of impact resistance.
5. With nitrocarburizing and nitriding, the high surface hardness does not result from a martensite case, and consequently, quenching is not required. The processes are done at relatively low temperatures, approximately 570 °C (1060 °F) for nitrocarburizing and 525 °C (980 °F) for nitriding. With gas nitriding, the nitrogen is introduced using ammonia gas, and the high hardness results from formation of nanometer-sized nitride particles that form from the addition of reactive elements such as chromium, molybdenum, or vanadium to the steel. With nitrocarburizing, the nitrogen potential is increased, and a hard nitride phase, called the epsilon (ε) phase, is produced on the surface, as shown in Fig. 17.15. Both of these processes produce very hard surfaces with minimal dimensional changes. Table 17.1 presents a summary of the various characteristics of the surface diffusion processes discussed in the chapter.

REFERENCES

17.1. *Heat Treater's Guide: Practices and Procedures for Irons and Steels,* ASM International, 1995

17.2. *Heat Treating,* Vol 4, *ASM Handbook,* ASM International, 1991

17.3. H.E. Boyer, Ed., *Case Hardening of Steel,* ASM International, 1987

17.4. *Heat Treating, Cleaning and Finishing,* Vol 2, *Metals Handbook,* 8th ed., American Society for Metals, 1964

17.5. S.L. Semiatin and D.E. Stutz, *Induction Heat Treatment of Steel,* American Society for Metals, 1986

17.6. G.O. Ratliff and W.H. Samuelson, High Temperature Carburizing is Routine Process at Shore Metals, *Met. Prog.,* Vol 108, Sept 1975, p 75

17.7. R.W. Russell, Nitrogen Based Heat Treating at Platt Saco, *Met. Prog.,* Dec 1979, p 46

17.8. R.G. Bowes, B.J. Sheehy, and P.F. Stratton, A New Approach to N-Based Carburizing, *Heat Treat. Met.,* 1979.3, p 53

17.9. U. Wyss, Drip Feed Carburizing, *Met. Prog.,* Nov 1978, p 24

17.10. *Metallography, Structures and Phase Diagrams,* Vol 8, *Metals Handbook,* 8th, ed., American Society for Metals, 1973

17.11. *Source Book on Nitriding,* American Society for Metals, 1977

APPENDIX A

Temperature Measurement

ONE OF THE MOST IMPORTANT measurements that needs to be made in the heat treatment and forging of steel is the measurement of the temperature of the steel. These measurements are most often made using either thermocouples or infrared pyrometers. A very useful resource for both temperature-measuring equipment and information about temperature measurements is the Omega Company (Ref A1). They will supply free copies of *The Omega Temperature Handbook* that contains information on temperature measurement along with both thermocouple and infrared pyrometer products.

Thermocouples

Perhaps the most common thermocouple used to measure temperature during the heat treatment of steel is the Chromel-Alumel thermocouple. The device is very simple, and the general principles of its operation are discussed using this material as an example. A Chromel wire is welded to an Alumel wire, and the weld bead is placed at the location in a furnace where temperature is to be measured. Chromel is the trade name for an alloy made up of nickel plus chromium, and Alumel for an alloy that is mostly nickel. Electrical leads, often made from copper wire, are attached to the ends of the thermocouple wires protruding out of the furnace, and these leads connect to a digital voltmeter, as shown in Fig. A1. Alumel is attracted to a magnet and Chromel is not, so when hooking up to the voltmeter, the negative lead (Alumel) can be determined using a magnet.

Thermocouples take advantage of a physical phenomenon that occurs in nature, called the thermoelectric effect. The effect occurs when wires of any two dissimilar metals are welded together. It produces a small direct current voltage that can be detected with a high-impedance voltmeter (similar to a digital voltmeter) connected to the wire ends opposite the weld bead. For the configuration in Fig. A1, a voltage will be registered on the voltmeter if the temperature of the weld bead (called T_H and hot junction in Fig. A1) is different from the temperature of the copper junctions (called T_{cj} and cold junction in Fig. A1).

Figure A2 presents the thermoelectric output voltage for Chromel-Alumel. An equation for this curve can be found in *The Omega Temperature Handbook*. The book also contains long tables listing the voltages at given temperatures in both Centigrade and Fahrenheit. Figure A2 shows that as the temperature rises, the output voltage increases in almost a straight line. If the

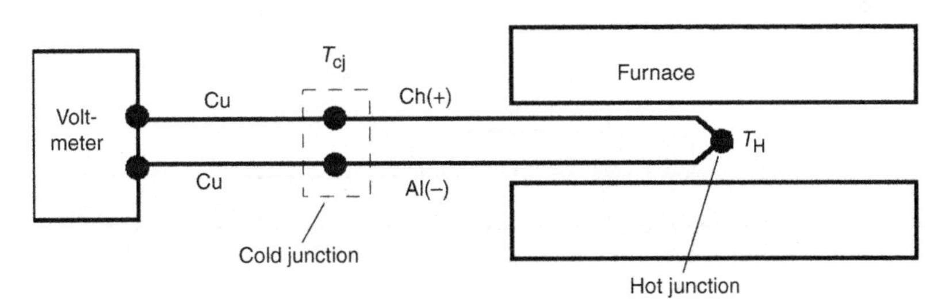

Fig. A1 Common setup for measuring furnace temperatures with a Chromel/Alumel thermocouple.

hot temperature, T_H, is 750 °C (1380 °F), the figure indicates that a voltage of V_H equal to approximately 32 mV will be produced. However, the hookup in Fig. A1 will not measure this voltage because of a complication that is important to understand. The voltage measured by the voltmeter in the circuit of Fig. A1 is reduced by the voltage called V_C in Fig. A2. Hence, the measured voltage is given as:

$$V = \text{measured voltage} = V_H - V_C \quad \text{(Eq A1)}$$

The voltage V_C is often called the cold junction voltage. It is the voltage given by the line in Fig. A2 at the temperature of the junction of the copper lead wires and the Chromel-Alumel wires, T_{cj}. If the cold junction is held at 0 °C (by immersion in an ice bath), then $V_C = 0$, and the measured voltage = V_H, the value given in the tables. However, the cold junction temperature is usually room temperature, T_{Rm}, but this temperature, T_{Rm}, will not be constant because room temperature can vary, and, if the junction is close to the furnace, it can rise significantly. Consequently, to produce reliable temperature measurements with a thermocouple, a cold junction compensation must be applied. Figure A3 shows a thermocouple hookup that uses an electronic cold junction compensator. The electronic compensator contains a battery and circuit components that measure the local temperature. The unit becomes the cold junction, and it automatically generates the required cold junction voltage, V_C, and adds it to the voltage given in Eq A1, so that now the measured voltage at the voltmeter is equal to V_H no matter how the cold junction temperature may change.

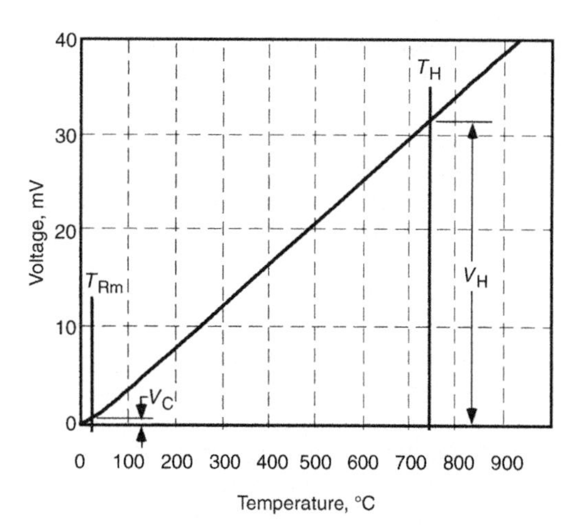

Fig. A2 Voltage output of Chromel-Alumel thermocouples

Several commercial devices are available from places such as Omega that display the temperature from a thermocouple directly on a digital instrument. Some of these devices require the use of a cold junction compensator in the circuit, as shown in Fig. A3, while others incorporate the compensator within themselves, so that the thermocouple wires can be hooked up directly to the device. If the device (or the cold junction compensator) is located far from the thermocouple, it will be necessary to use some type of hook-up wire to connect the two. If standard copper hook-up wire is used, small errors will be generated if the temperature at the thermocouple connection is not the same as at the device connection. This problem can be overcome by using the thermocouple wire as the hook-up wire. Alternatively, special so-called compensation wire can be purchased. It is generally cheaper than thermocouple wire and has a composition formulated to avoid problems produced by temperature variations between the hook-up connections.

There are various types of thermocouples that are commercially available, and Table A1 presents a list of several of these. It is common to refer to the various thermocouples with the type letter code shown in Table A1. For example, the Chromel-Alumel thermocouple is a type K thermocouple. One of the problems with using thermocouples in hot furnaces is that the metal thermocouple wires will oxidize. If the oxide films extend too far below the surface, the thermocouple will no longer operate

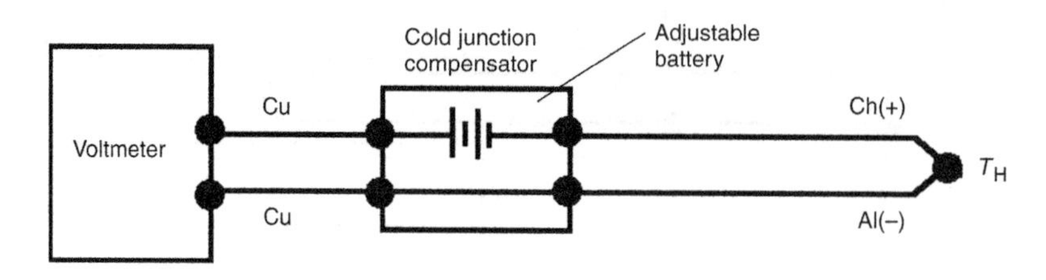

Fig. A3 Cold junction compensator automatically adds the required V_C voltage into the thermocouple circuit

correctly. Oxidation rates increase rapidly with rising temperature, and Table A1 lists the maximum recommended temperatures for use of the various thermocouple types in air. The temperature limits shown in Table A1 are for a normal lifetime. The lifetime may be extended by using very large-diameter thermocouple wire. Another problem with using thermocouples at high temperatures is that grain growth will occur and make the wires brittle. For very high temperatures, such as those used for forging and for melting steel, the best-choice thermocouple is one of the platinum varieties, which will not oxidize. The type B thermocouple has the advantage of reduced grain growth and therefore will resist breakage during use better than the other platinum thermocouples.

Thermocouples come in a variety of forms. They can be custom-made for particular applications from spools of wire. For high-temperature applications, it is necessary to thread the wires through some type of ceramic insulator so that the wires do not make electrical contact anywhere except at the weld bead. Figure A4 shows a homemade thermocouple where the wires are threaded through a two-hole ceramic insulator. For lower-temperature measurements, one can get by using fiberglass-insulated thermocouple wires that are available commercially.

Metal-sheathed thermocouples are also widely available commercially. As shown in Fig. A5, these thermocouples have an external metal sheath, and the wires are insulated from the sheath and each other by a finely powdered ceramic material that gives adequate electrical insulation up to the recommended maximum temperature of the thermocouple, often close to the melting temperature of the sheath metal. Various sheath metals are available that have excellent oxidation resistances at high temperatures, such as the nickel-base alloys Inconel and Hastelloy. These metal-sheathed thermocouples can be used to higher temperatures than unclad thermocouples. Omega sells a type K metal-sheathed thermocouple rated for a maximum temperature of 1205 °C (2201 °F). These thermocouples are supplied with an electrical connector on the end, as shown in Fig. A5, and various connector sizes and types are available. When using these thermocouples, the cold junction is at the connector. If it is possible that the connector may become warm, it is important that either it be connected directly to a cold junction compensator or a meter containing such a compensator. If it is necessary to use lead wire to connect to the compensated measuring device, one may simply use lengths of the thermocouple wire. Alternatively, spools of compensating wire mentioned previously are commercially available at cheaper prices for the various thermocouple types.

Table A1 Some common thermocouple materials

Type	Elements	Comments	Maximum temperature °C	Maximum temperature °F
K	Chromel/ Alumel	Most popular type	982	1800
J	Iron/ Constantan	High output	482	900
E	Chromel/ Constantan	High output	593	1100
T	Copper/ Constantan	High output but low temperature limit	260	500
R	Pt/Pt-13%Rh	High temperature, no oxidation	1450	2642
B	Pt-6%Rh/ Pt-30%Rh	Similar but less grain growth	1700	3092

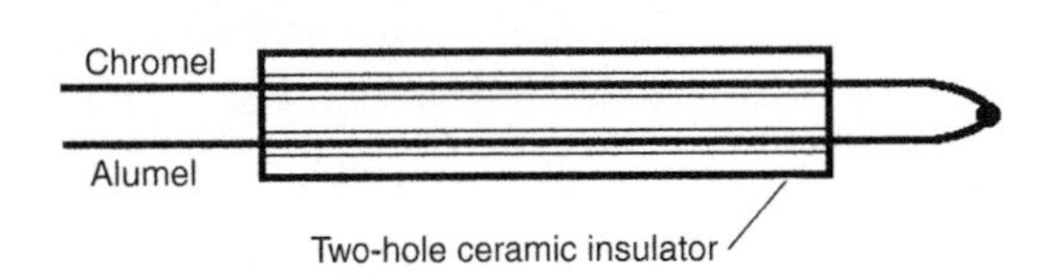

Fig. A4 Basic thermocouple

Radiation Pyrometers

Blacksmiths have been heating and quenching steels for many centuries, and they have been able to control the temperature of the hot steel by observing the color of the light (radiation) coming from it. Table A2 presents a typical approximation of the color of a hot body and its temperature. The color transitions on heating, from red to orange to yellow to white, are easily seen by observing the heatup of a furnace. It is necessary to keep the room dark to observe the first red color listed in the table. Dramatic advances in the understanding of the nature of light emitted from hot surfaces were made by physicists at the beginning of the 20th century.

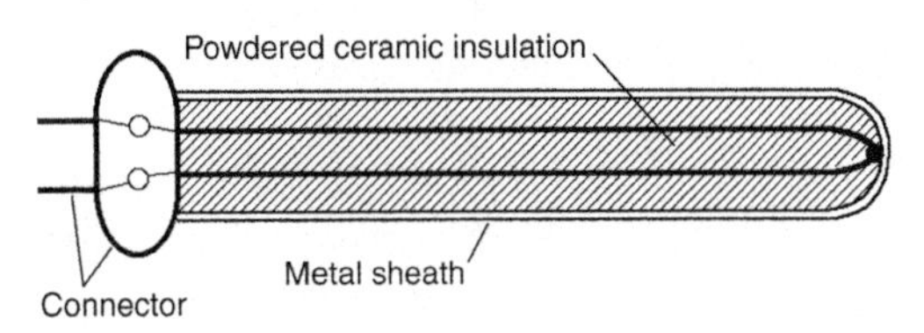

Fig. A5 Metal-sheathed thermocouple

Table A2 Approximate temperature colors

Color	Temperature °C	Temperature °F
Faint red	500	932
Blood or dark red	550–625	1022–1157
Dark cherry	635	1175
Full cherry red	700	1292
Bright cherry	800	1472
Dark orange	900	1652
Orange	950	1742
Full yellow or lemon	950–1000	1742–1832
Light yellow	1100	2012
White	1150 and up	2102 and up

Source: Ref A2

Devices called pyrometers use the emitted radiation to measure the temperature of a hot body more accurately than can be done by a well-trained blacksmith. The author's experience with infrared pyrometers while working with bladesmiths is that they can estimate temperatures fairly consistently, within 20 °C or so.

The light that is seen coming from a white-hot piece of steel is the same kind of light that is seen coming from the stars and the sun. It is a form of radiating energy that physics books call electromagnetic radiation. It consists of combined alternating electric and magnetic fields that give rise to the name electromagnetic. Light rays are thought of as waves propagating in a particular direction (at the speed of light, or 186,000 mile/s) and having electrical field intensities that vary, as shown in Fig. A6. It is customary to call the wavelength, the distance between peaks or valleys, by the Greek letter lambda, λ. The energy of the light wave becomes higher as λ becomes smaller. If the wavelength of the light is restricted to a narrow band of wavelengths in a range that is visible to the human eye, the brain distinguishes, or sees, specific colors.* Table A3 lists the bands of wavelengths of the common colors. However, as illustrated in Fig. A7, if the light contains wavelengths of several color bands, it appears white to the human eye. Also, human eyes can only see light having wavelengths over a relatively narrow range, generally referred to as the optical range. Light with wavelengths slightly larger than this range is called infrared light, while light with wavelengths shorter is called ultraviolet light.

Heated objects radiate energy, and it is possible to measure the temperature of an object

*For a clear description of how our brains assign the colors, see Ref A3.

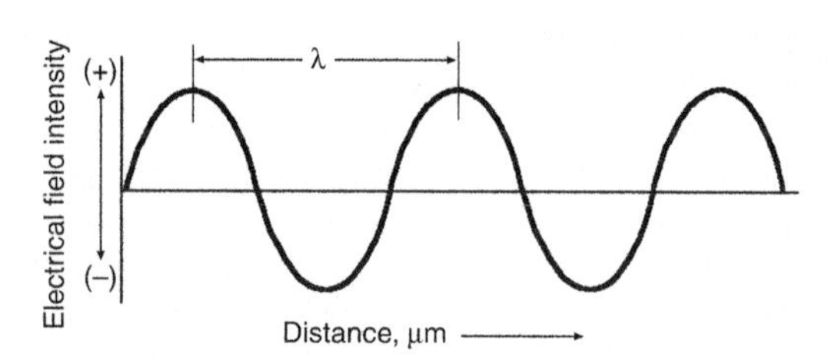

Fig. A6 Variation of electrical field intensity of a light wave of wavelength λ

Table A3 Wavelengths of color bands

Color	Wavelength band(a), microns
Red	0.65–0.77
Orange	0.59–0.65
Yellow	0.55–0.59
Green	0.48–0.55
Blue	0.43–0.48
Violet	0.36–0.43

(a) 25 microns = 1 mil = 0.001 in.

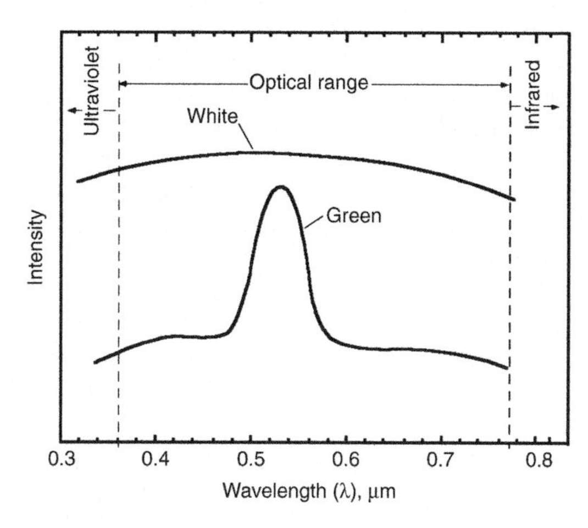

Fig. A7 Intensity versus wavelength of a white light and a green light

by measuring an intensity of the radiation being emitted from it. If the device that measures the temperature examines radiation in the optical range, it is called an optical pyrometer. Infrared pyrometers became available with the development of solid-state detectors in the 1960 to 1970s that efficiently measure radiation in the infrared range (shown to the right in Fig. A7). These devices are very effective at measuring the temperature of hot metals and furnaces, and they have now largely replaced optical pyrometers, because they are much more convenient to use.

To understand how radiation pyrometers work, one may start with the definition of a cavity radiator, which is shown in Fig. A8. The opening in the ideal cavity radiator is so small that all light coming into it is absorbed in its walls, and it is customary to call it a perfect black body. The light emerging from the cavity radiator is called black-body radiation, and its intensity can be measured experimentally. One obtains a value of power per area of the opening, or watts/m^2 or (joules/s)/m^2, where "/m^2" means per square meter. Physics books call this quantity the radiancy, R, while engineering heat-transfer books call it the heat flux, Q. The wavelengths, λ, of the emerging radiation vary over a wide range. The light can be spread into its wavelength components using spectroscopy techniques, making it is possible to measure R versus λ.

In a first approach, one would normally just plot R versus λ to observe the dependence. However, by plotting the ratio of R/λ versus λ, a fascinating result is observed: *the resulting curve is found to fit data from all cavity radiators*, independent of the material from which they are made. The ratio is called the spectral radiancy, R_λ, and the universal plots have the form shown in Fig. A9. The units of spectral radiancy are sometimes given as (watts/m^2)/micron, where the micron refers to the wavelength of the radiation and the m^2 to the area of the radiator in square meters. In Fig. A9, however, the units have been changed to (watts/in.^{2})/micron, so that if the spectral radiancy is multiplied by the wavelength of interest, one obtains the watts per square inch (power per area) emitted at that wavelength.

Figure A9 shows why the color of hot steel changes in the sequence red-orange-yellow-white as the temperature rises. The intensity of the emitted light shifts to lower wavelengths (higher energies) as the temperature rises. The highest intensity in the optical range always occurs (at least up to 3000 °C) at the red end of the range. Hence, as the metal heats, the first light intense enough for the eye to see is in the red band. Further heating increases the intensity of the smaller wavelengths visible to the eye, so that the colors change progressively through the color bands of Table A3 until there are enough of the shorter wavelengths visible to mix the wavelengths to the degree that the brain sees white light. So, colors beyond the yellow band of Table A3 are never seen, because, as objects become hot enough for green wavelengths to be sufficiently intense for detection by the eye, the mix of red-orange-yellow wavelengths also present causes the brain to see the light as white.

The radiation being emitted from the surface of metals and alloys is not black-body radiation. The spectral radiancy versus wavelength for the light emitted from hot steel surfaces will have a functional shape similar to that for black-body radiation, but it will be depressed, as shown for unoxidized steel at 2000 °C (3630 °F) in Fig. A10. It is common to characterize the reduced radiation intensity simply as a fraction of the black-body radiation, using a ratio called the emissivity. If the fractional drop in spectral radiancy is measured at a specific wavelength, then the emissivity is called the spectral emissivity:

$$\text{Spectral emissivity} = e_\lambda = \frac{(R_\lambda)_{\text{surface}}}{(R_\lambda)_{\text{cavity}}} \quad \text{(Eq A2)}$$

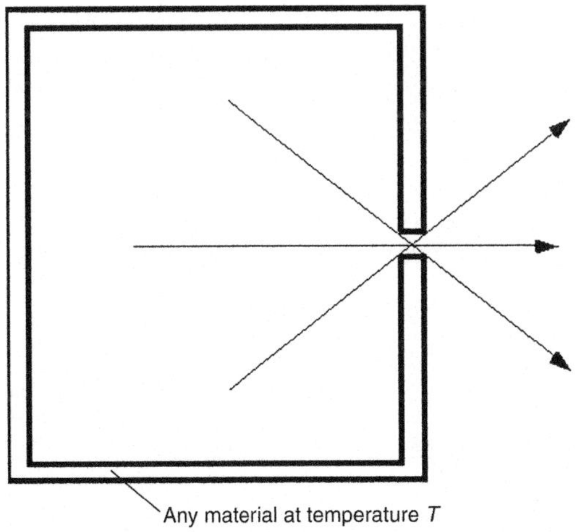

Any material at temperature T

Fig. A8 Cavity radiator

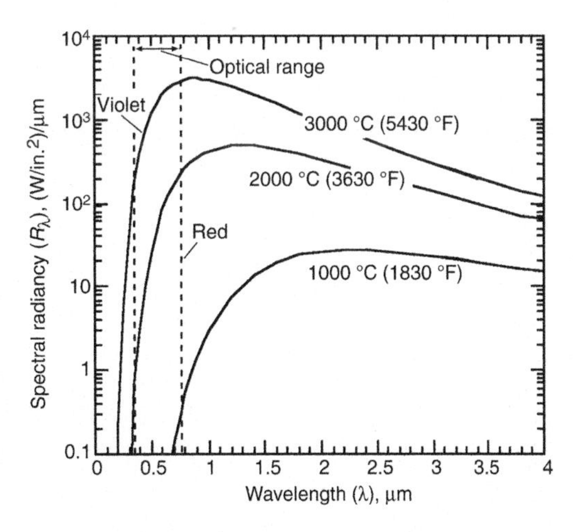

Fig. A9 Spectral radiancy versus wavelength for a cavity radiator (black-body radiation) at three temperatures

where the wavelength, λ, must be specified. For example, the spectral emissivity at a wavelength of 0.65 μm for the steel in Fig. A10 is $e_\lambda = 43/122 = 0.35$. At $\lambda = 2.0$ μm, e_λ falls to $66/331 = 0.20$. This result shows a general trend for the spectral emissivity of metals: it falls progressively further below 1.0 as λ increases.

For metals, the values of the emissivities are very sensitive to the presence of surface films, such as oxides, on the metal. Table A4 lists values of the spectral emissivities for steel and copper in both the oxidized and unoxidized conditions, and it is clear that oxidation has a large effect on emissivity. Heat treated and forged steels almost always are oxidized to some extent, unless special laboratory conditions are used. These data clearly illustrate the importance of making reliable emissivity corrections when using pyrometers to measure steel temperatures during heat treating and forging operations. The temperature indicated by the infrared pyrometer is an apparent temperature that will be lower than the true value, depending on how far the emissivity falls below 1.0.

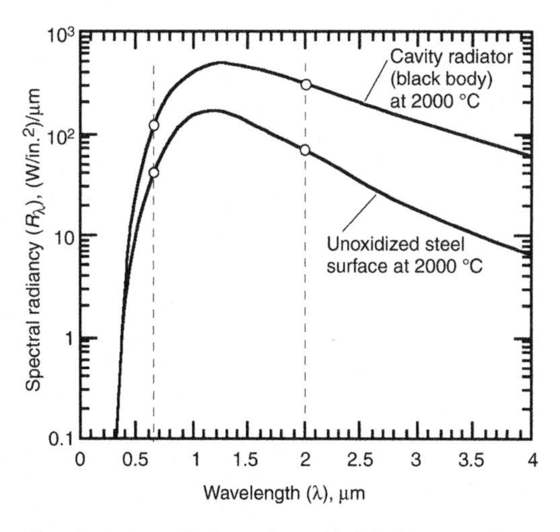

Fig. A10 Spectral radiancy from a black body (cavity radiator) versus that from the surface of steel at 2000 °C (3630 °F)

Table A4 Emissivity values

Material	e_λ ($\lambda = 0.65$ μm)
Steel (unoxidized)	0.35
Steel (oxidized)	≈0.7
Copper (unoxidized)	0.10
Copper (oxidized)	≈0.7

Source: Ref A2, p 490

There are several approaches to making emissivity corrections, three of which are discussed here. In the first method, a hole is drilled into the surface of the sample with an aspect ratio (length/diameter ratio) greater than approximately 5. The hole diameter must be large enough to allow for the pyrometer optics to focus inside. Radiation coming from such a hole gives a good approximation of black-body radiation, and it can be assumed that the emissivity is one. A second method is to simply attach a thermocouple to the sample under study and experimentally measure a correction factor for the pyrometer under the conditions of the measurement. This is the most accurate method, assuming that the thermocouple bead is attached securely enough to the steel that it is at the same temperature as the steel surface being imaged in the pyrometer. A third method is to use a two-color pyrometer. The two-color pyrometer measures the spectral radiancy at two specific wavelengths, λ_1 and λ_2, and then determines the ratio, $R_{\lambda_1}/R_{\lambda_2}$. This ratio is less dependent on changes in emissivity than the values of R_λ themselves. Therefore, two-color pyrometers require a much smaller emissivity correction than single-color pyrometers. However, they are more expensive.

Infrared pyrometers are easy to use and are becoming available in a great variety with many useful features, such as laser beam pointers that locate the position where the unit is focused. Information on the variety of instruments available and technical use of these instruments is available from the commercial company Omega (Ref A1). Table A5 presents information on infrared pyrometers for typical temperature ranges that were available in the late 1990s. A general rule of thumb is to use instruments with the lowest λ range available for metal samples. As shown previously, this reduces the emissivity correction, which leads to improved accuracy. Both the single- and two-color pyrometers have

Table A5 Typical infrared pyrometer ranges available

Spectral range (microns where works)	Temperature range of operation: Minimum °C	Minimum °F	Maximum °C	Maximum °F
0.76–1.06	600	1110	3000	5430
1.0–1.6	250	480	2000	3630
8–14	50	58	1000	1830

a dial-in knob that allows one to increase the apparent temperature to the corrected temperature corresponding to the emissivity set on the dial. So, if the emissivity is constant and the required correction has been measured, the apparent temperature on the digital readout can be automatically adjusted to give the correct values for emissivities less than 1.0.

REFERENCES

A1. Omega Engineering Inc., Stamford, CT, www.omega.com (accessed May 28, 2007)

A2. G.L. Kehl, *Metallographic Laboratory Practice*, McGraw-Hill, New York, 1949

A3. S. Zeki, Chapter 23, *A Vision of the Brain,* Blackwell Publishing Ltd., 1993

APPENDIX B

Stainless Steels for Knifemakers

CHAPTER 13, "STAINLESS STEELS," DISCUSSED two specific steels that are used for making knives of stainless steel, Sandvik 12C27 and Uddeholm AEB-L. There are several other stainless steels often used by knifemakers, and Table B1 presents most of the popular types. These steels often contain the addition of molybdenum. In addition to enhancing passivity, as mentioned in Chapter 13, molybdenum also improves toughness in the tempered condition.

In the discussion of AISI 440C in Chapter 13, the 1100 °C Fe-Cr-C ternary phase diagram (Fig. 13.10) was used to illustrate the expected composition of the austenite prior to quenching. To simplify the presentation, the addition of the 0.75% Mo in this steel was ignored by plotting the composition on the 1100 °C Fe-Cr-C ternary phase diagram. Because molybdenum is a strong carbide-forming element, it would be expected that the 0.75% Mo addition would make small changes to the pure Fe-Cr-C phase diagram. The diagram in Fig. 13.10 was produced by combining experimental measurements with theoretical thermodynamic calculations by a group of Swedish researchers headed by M. Hillert (Ref 13.10). This work led to the development of a sophisticated computer software program called ThermoCalc. At the request of the author, Dr. A. Kajinic at Crucible Research in Pittsburgh, PA, has used this program to calculate several Fe-Cr-C isothermal diagrams with a constant amount of molybdenum added. Figure B1 presents the isothermal Fe-0.8Mo-Cr-C diagram. This diagram matches the molybdenum level in 425M and provides a good approximation to the 0.75% Mo of the AISI 440 series of stainless steels.

Figure B1 presents the isothermal sections at both 1100 °C (2010 °F) (the solid lines) and 1000 °C (1830 °F) (the dashed lines). Comparing this diagram to that of the non-molybdenum diagrams of Fig. 13.12, it is seen that the carbon saturation line at 1000 °C is shifted just slightly to lower %C values. Notice that a larger change has occurred in the position of the $\gamma + K_1 + K_2$ region (shaded regions) that lies between the $\gamma + K_1$ and $\gamma + K_2$ regions. This is illustrated well by looking at the position of the overall composition of 440C alloy located at the solid dots on the two diagrams, Fig. 13.12 versus Fig. B1. (The open circle labeled 440C gives the austenite composition at 1100 °C.) The molybdenum addition shifts the $\gamma + K_1 + K_2$ region down and to the right on these diagrams, which

Table B1 Stainless steels often used for knives

Steel	Manufacturer	%C	%Cr	%Si	%Mn	%Mo	Other
440A	AISI steel	0.7	17	1	1	0.75	...
440B	AISI steel	0.85	17	1	1	0.75	...
440C	AISI steel	1.1	17	1	1	0.75	...
12C27	Sandvik	0.6	13.5	0.4	0.4	...	...
AEB-L	Uddeholm	0.65	12.8	0.4	0.65	...	...
DD400	Minebea	0.61	12.9	0.32	0.67	...	...
425M	Crucible	0.54	14.2	0.8	0.5	0.8	...
154CM	Crucible	1.05	14	0.3	0.5	4	...
ATS55	Hitachi	1	14	0.4	0.5	0.6	0.4Co, 0.2Cu
ATS34	Hitachi	1.05	14	0.35	0.4	4	...
AUS6	...	0.6	13.8	1	1	...	0.13V, 0.49Ni
AUS8	...	0.73	13.81	0.5	0.5	0.2	0.13V, 0.49Ni
AUS10	...	1.03	13.8	1	0.5	0.2	0.13V, 0.49Ni

favors formation of the K_1 carbide over the K_2 carbide for a given alloy composition. This result may be expected to reduce wear resistance (because K_2 is the harder carbide, see Table 13.10) if the non-molybdenum steels did contain mostly K_2 carbides, as predicted by Fig. 13.10. However, experiments (Ref 13.11) indicate non-molybdenum steels have mainly K_1 carbides, so this effect is unlikely. However, the molybdenum addition should produce some improvement in the corrosion resistance at the same level of %Cr. Calculations (Ref B1) predict that the austenite formed at 1100 °C (1830 °F) in 440C should contain approximately 0.73% Mo in addition to 12.2% Cr.

The alloy 325M is shown in Fig. B1, and it is seen to fall essentially on the tie line for AISI 440B (dashed line connecting closed and open circles labeled 440B). However, its overall composition lies much closer to the carbon saturation line, so that one would expect this alloy to be similar to 440B, except the carbides should be present in a smaller volume fraction and the fraction of primary carbides should be negligible. The ThermoCalc predictions for the 1100 °C (2010 °F) austenite of 325M are 13.5% Cr, 0.475% C, and 0.77% Mo (Ref B1). Thus, one would expect this alloy to be not quite as hard as the AEB-L alloy discussed at length in Chapter 13, because of the slightly lower %C, but a little better in corrosion resistance due to the molybdenum in the austenite.

The alloys ATS34 from Hitachi and 154CM from Crucible are essentially the same alloy with the relatively high molybdenum addition of 4%. The ThermoCalc prediction of the 1100 °C (2010 °F) isotherm of the Fe-C-Cr-4Mo system (Ref B1) is presented in Fig. B2. At 1100 °C (2010 °F), the isothermal section of Fig. B2 predicts that the alloy will consist of austenite plus the K_1 carbide. Additional calculations find that the composition of the austenite will be 10.6% Cr, 3.4% Mo, and 0.58% C. Because the overall composition of 154CM lies off of the carbon saturation line by approximately the same amount as 440C, it is expected that the volume fraction of carbides will be similar. The similar overall %C and total %Cr + %Mo in these two steels will probably result in the same problem of formation of large primary carbides during the solidification process.

Research on 154CM at Crucible (Ref B1) found that the %Cr in the austenite produced at 1065 °C (1950 °F) is approximately 10%. This result agrees well with the predicted ThermoCalc value of 10.6% at 1100 °C (2010 °F), because a chromium content of slightly less than 10.6 would be expected at the lower austenitizing

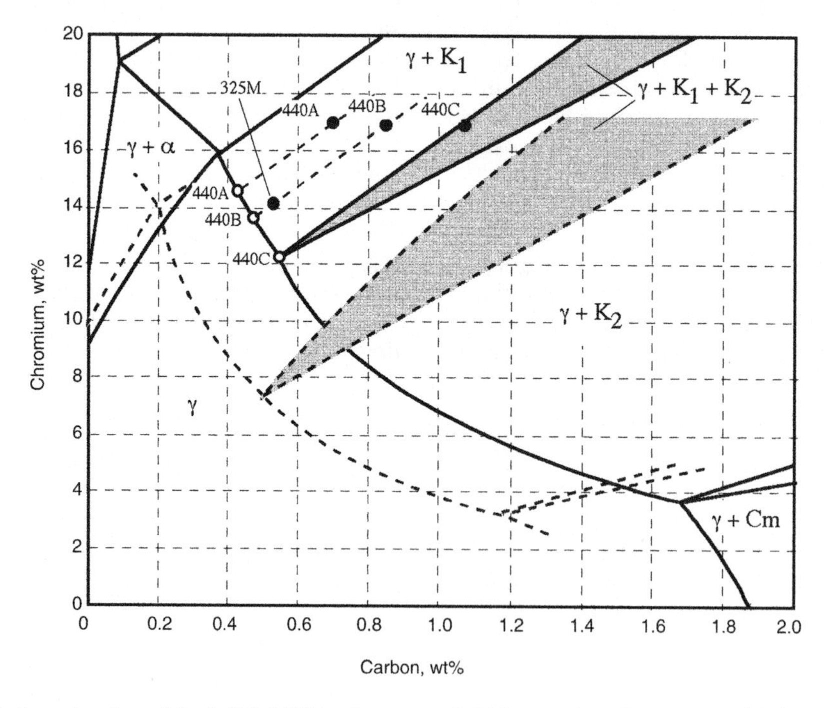

Fig. B1 Isothermal sections of the Fe-C-Cr-0.8%Mo alloy system. Solid lines at 1100 °C (2010 °F) and dashed lines at 1000 °C (1830 °F). ThermoCalc diagram provided by A. Kajinic. Source: Ref B1

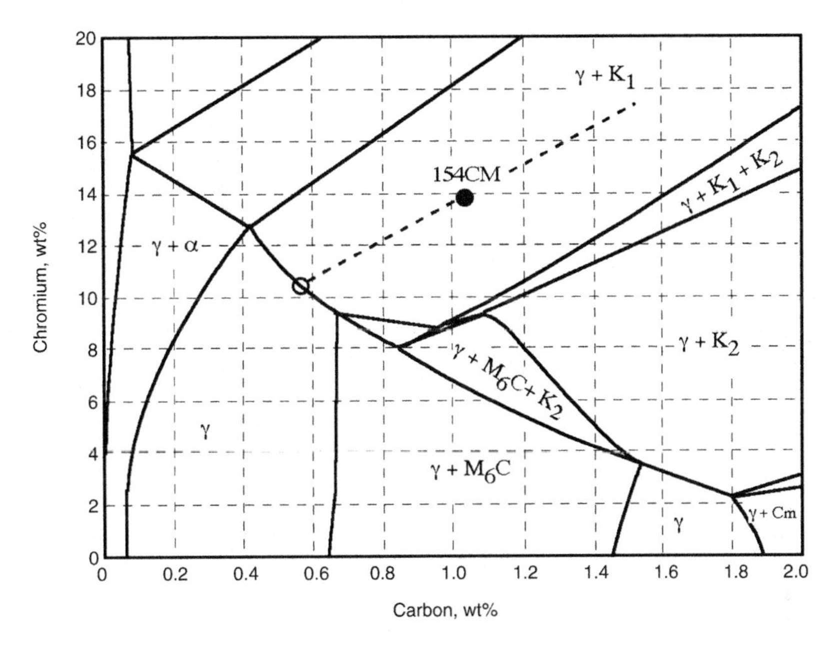

Fig. B2 Isothermal section of the Fe-C-Cr-4.0%Mo alloy system at 1100 °C (2010 °F). Source: Ref B1

Table B2 Additional stainless steels available from Sandvik

Steel	Overall		In austenite at 1000 °C (1830 °F)		Relative volume fraction carbides (is multiple of amount in AEB-L)
	%C	%Cr	%C	%Cr	
Uddeholm AEB-L	0.65	12.8	0.59	12.3	1.0 (standard value)
Sandvik 12C27M	0.52	14.5	0.52	14.5	No carbides at 1000 °C (1830 °F)
Sandvik 12C27	0.60	13.5	0.56	13.2	0.7
Sandvik 13C26	0.65	13	0.58	12.5	1.1
Sandvik 19C27	0.95	13.5	0.60	11.3	5.6

temperature of 1065 °C (1950 °F). The fact that the chromium level lies below the 12% Cr value generally desired for good passivity may indicate that this steel would have poor corrosion resistance. However, the 3.4% Mo present in the austenite should compensate for this reduction. Experiments at Crucible (Ref B1) have confirmed this to be pretty much the case.

From this discussion, it appears that the two steels discussed in Chapter 13, Uddeholm AEB-L and Sandvik 12C27, along with the similar steels of Table B1 (DD400 and AUS6) provide the best combination of properties desired in a knife blade:

- An as-quenched hardness in the 63 to 64 Rc range, which should provide high wear resistance
- An adequate level of chromium in the austenite formed prior to quenching to provide good corrosion resistance, a little above the minimum 12% Cr
- The presence of fine arrays of the $K_1 + K_2$ chromium carbides to enhance wear resistance plus the absence of the larger primary chrome carbides that promote pullout at sharpened edges

Sandvik manufactures a series of stainless steels having compositions close to the value of the 12C27 that was considered in Chapter 13. Table B2 presents a comparison of these steels to that of the Uddeholm AEB-L that was studied in Chapter 13. The overall compositions shown in the table were plotted on Fig. 13.10, and the predicted values of %C and %Cr in austenite at 1100 °C (2010 °F) are shown in the fourth and fifth columns of the table. The volume fraction carbide in the 1000 °C (1830 °F) austenite can be determined by measurement of the distance of the overall composition from the carbon saturation line. The carbide fraction in the Uddeholm AEB-L, which can be estimated from Fig. 13.14, was taken as a standard, and the final column of the table presents the factor indicating the carbide fraction relative to this standard. For example, the high carbon content in Sandvik 19C27 produces 5.6 times more carbides at 1000 °C (1830 °F) than are found in AEB-L. This steel will produce the highest hardness in the Sandvik series, but the carbides may be larger than desired on the cutting edge due to formation of primary carbides resulting from the increased carbon level. Also, the corrosion resistance will be the poorest due to a %Cr of only 11.3. As shown in Chapter 13, the as-quenched hardness, percent retained austenite, and volume fraction carbides in AEB-L are very sensitive to heat treat temperature, time, and quench rate. Because the compositions of the Sandvik 12C27 and 13C26 are so similar to AEB-L, it seems likely that the properties of these three steels may be more sensitive to the austenitization heat treatment than to choice of composition, unless precise heat treat conditions are used. The 12C27M of the Sandvik series should have the best corrosion resistance due to the highest %Cr in the austenite, but the lowered %C will produce the lowest as-quenched hardness.

As seen in Table B1, the Minebea steel, DD400, has a very similar composition to the Uddeholm and Sandvik steels of Table 13.8. A paper by Rideout (Ref B2) compared this steel to 440C for use in bearings hardened to the HRC = 61 to 64 range. The DD400 bearings are reported to have longer bearing life and reduced noise and vibration levels. Micrographs are presented to show large primary carbides present in 440C and their absence in DD400. The improved bearing properties are attributed to the absence of the primary carbides. As explained in Chapter 15, "Solidification," primary carbides are formed from the interdendritic liquid during solidification. Increased levels of %C will lead to primary carbide formation due to the combination of high chromium content and nonequilibrium freezing in these stainless steels. The absence of primary carbides in the DD400 steel and the AEB-L experiments described in Chapter 13 indicate that decreasing the %C from approximately the 1% level of 440C to approximately 0.6% in steels containing approximately 13% Cr is sufficient to reduce primary carbide formation to negligible levels during solidification.

REFERENCES

B1. A. Kajinic, Crucible Research, Pittsburgh, PA, personal communication, 2002

B2. J. Rideout, Bearing Steel Bests Type 440C, *Adv. Mater. Process.*, Dec 1992, p 39

Index

R

S

T

U

V

W

Y

The periodic table

		Hex. (Rm. Temp.) BCC (Hi. Temp.)		BCC		Mixed		
1 1.008 **H** Hydrogen -259								
3 6.939 **Li** Lithium 0.53 181 B	4 9.012 **Be** Beryllium 1.9 1289 H							
11 22.99 **Na** Sodium 0.97 97.8 B	12 24.31 **Mg** Magnesium 1.7 650 H							
19 39.10 **K** Potassium 0.86 63.7 B	29 40.08 **Ca** Calcium 1.6 842 F-B	21 44.96 **Sc** Scandium 3.1 1541 H-B	22 47.90 **Ti** Titanium 4.5 1670 H-B	23 50.94 **V** Vanadium 6.0 1910 B	24 52.00 **Cr** Chromium 7.1 1863 B	25 54.94 **Mn** Manganese 7.2 1246 cc-F-B	26 55.85 **Fe** Iron 7.8 1538 B-F-B	27 58.93 **Co** Cobalt 8.9 1495 H-F
37 85.47 **Rb** Rubidium 1.5 39.5 B	38 87.62 **Sr** Strontium 2.6 769 F-B	39 88.91 **Y** Yttrium 4.3 1522 H-B	40 91.22 **Zr** Zirconium 6.4 1855 H-B	41 92.91 **Nb** Niobium 8.4 2469 B	42 95.94 **Mo** Molybdenum 10.2 2623 B	43 (98) **Tc** Technetium 11.5 H	44 102.9 **Ru** Ruthenium 12.4 2334 H	45 102.9 **Rh** Rhodium 12.5 1963 F
55 132.9 **Cs** Cesium 1.9 28.4 B	56 137.3 **Ba** Barium 3.6 727 B	**La**	72 178.5 **Hf** Hafnium 13.3 2231 H-B	73 181.0 **Ta** Tantalum 16.7 3020 B	74 183.9 **W** Tungsten 19.3 3422 B	75 186.2 **Re** Rhenium 21.0 3186 H	76 190.2 **Os** Osmium 22.6 3033 H	77 192.8 **Ir** Iridium 22.6 2447 F
87 223 **Fr** Francium 27 B	88 226 **Ra** Radium 5.5 700 B	**Ac**	104 (261) **Rf** Rutherfordium	105 (262) **Ha** Hahnium	106 (263)	107 (262)		109 (266)

[Rare Earth Metals, mostly Hex at room temperature]

57 138.9 **La** Lanthanum 6.6 918 H-F-B	58 140.1 **Ce** Cerium 6.8 798 F-B	59 140.9 **Pr** Praseodymium 6.6 931 H-B	60 144.2 **Nd** Neodymium 7.0 1021 H-B	61 (147) **Pm** Promethium	62 150.4 **Sm** Samarium 7.5 1074 H-B
89 (227) **Ac** Actinium 10.1 F	90 232.0 **Th** Thorium 11.7 1755 F-B	91 231.0 **Pa** Protactinium 15.4 T	92 238.0 **U** Uranium 19.0 1135 O-T-B	93 (237) **Np** Neptunium 20.4 O-T-B	94 (244) **Pu** Plutonium 19.8 640 M-O-F-T-B

Key	
Atomic Number	Atomic Weight
Atomic Symbol	
Element Name	
Density (g/cm^3)	Melting Temp. (°C)
Crystal Structure* (Rm. Temp. to melting Temp.)	

*F = face centered cubic, B = body centered cubic, H = hexagonal, O = orthorhombic,

of the elements

FCC		Hex						
								2 4.003 **He** Helium -272
			5 10.81 **B** Boron 2.4 2092 R-T-R	6 12.01 **C** Carbon 2.2 4500 H	7 14.01 **N** Nitrogen -210	8 16.00 **O** Oxygen -219	9 19.00 **F** Fluorine -223	10 20.18 **Ne** Neon -249
			13 26.98 **Al** Aluminum 2.7 660 F	14 28.09 **Si** Silicon 2.3 1414 C	15 30.97 **P** Phosphorus 1.8 44.1	16 32.06 **S** Sulfur 2.1 115.2	17 37.45 **Cl** Clorine -101	18 39.95 **Ar** Argon -189
28 58.71 **Ni** Nickel 8.9 1455 F	29 63.54 **Cu** Copper 8.9 1085 F	30 65.37 **Zn** Zinc 7.1 420 H	31 69.72 **Ga** Gallium 5.91 29.8	32 69.72 **Ge** Germanium 5.4 962	33 74.92 **As** Arsenic 5.7 817	34 78.96 **Se** Selenium 4.6	35 79.91 **Br** Bromine 3.1 -7.3	36 83.80 **Kr** Krypton
46 106.4 **Pd** Palladium 12.0 1555 F	47 107.9 **Ag** Silver 10.5 962 F	48 112.4 **Cd** Cadmium 8.6 321 H	49 114.8 **In** Indium 7.3 157 T	50 118.7 **Sn** Tin 7.3 232 T	51 121.8 **Sb** Antimony 6.6 631 R	52 127.6 **Te** Tellurium 6.2 450	53 126.9 **I** Iodine 4.9	54 131.3 **Xe** Xenon
78 195.1 **Pt** Platinum 21.5 1769 F	79 197.0 **Au** Gold 19.3 1064 F	80 200.6 **Hg** Mercury 14.8 -38.9 H	81 204.4 **Tl** Thallium 11.9 304 H-B	82 207.2 **Pb** Lead 11.3 327 F	83 209.0 **Bi** Bismuth 9.8 271 R	84 (210) **Po** Polonium 9.5	85 (210) **At** Astatine	86 (222) **Rn** Raydon

63 152.0 **Eu** Europium 5.2 822 B	64 157.3 **Gd** Gadolinium 7.9 1313 H-B	65 158.9 **Tb** Terbium 8.3 1356 H-B	66 162.5 **Dy** Dysprosium 8.5 1412 H-B	67 164.9 **Ho** Holmium 8.8 1474 H	68 167.3 **Er** Erbium 9.0 1529 H	69 168.9 **Tm** Thulium 9.3 H	70 173.0 **Yb** Ytterbium 6.5 819 F-B	71 175.0 **Lu** Lutetium 9.8 1663 H
95 (243) **Am** Americium 11.9 H	96 (247) **Cm** Curium	97 (247) **Bk** Berkelium	98 (251) **Cf** Californium	99 (254) **Es** Einsteinium	100 (257) **Fm** Fermium	101 (256) **Md** Mendlevium	102 (259) **No** Nobelium	103 (260) **Lw** Lawrencium

T = tetragonal, R = rhombohedral, M = monoclinic, cc = complex cubic